Formation and Evolution of Galaxies and Large Structures in the Universe

NATO ASI Series

Advanced Science Institutes Series

A series presenting the results of activities sponsored by the NATO Science Committee, which aims at the dissemination of advanced scientific and technological knowledge, with a view to strengthening links between scientific communities.

The series is published by an international board of publishers in conjunction with the NATO Scientific Affairs Division

A	Life Sciences	Plenum Publishing Corporation
B	Physics	London and New York
C	Mathematical and Physical Sciences	D. Reidel Publishing Company Dordrecht, Boston and Lancaster
D	Behavioural and Social Sciences	Martinus Nijhoff Publishers
E	Engineering and Materials Sciences	The Hague, Boston and Lancaster
F	Computer and Systems Sciences	Springer-Verlag
G	Ecological Sciences	Berlin, Heidelberg, New York and Tokyo

[*NATO advanced study institute on ...*]

Formation and Evolution of Galaxies and Large Structures in the Universe

Third Moriond Astrophysics Meeting

edited by

Jean Audouze

Institut d'Astrophysique de Paris, France

and

Jean Tran Thanh Van

Laboratoire de Physique Théorique,
Université Paris Sud (Orsay), France

D. Reidel Publishing Company

Dordrecht / Boston / Lancaster

Published in cooperation with NATO Scientific Affairs Division

Proceedings of the NATO Advanced Study Institute on
Formation and Evolution of Galaxies and Large Structures in the Universe
La Plagne, France
March, 1983

Library of Congress Cataloging in Publication Data

NATO Advanced Study Institute on Formation and Evolution of Galaxies and Large
    Structures in the Universe (1983 : La Plagne, France)
    Formation and evolution of galaxies and large structures in the universe.

    (NATO ASI series. Series C, Mathematical and physical sciences ; v. 117)
    "Proceedings of the NATO Advanced Study Institute on Formation and
Evolution of Galaxies and Large Structures in the Universe, La Plagne, France,
March 1983"—T.p. verso.
    Includes indexes.
    1. Galaxies—Congresses. I. Audouze, Jean. II. Tran Thanh Van, J.
III. Title. IV. Title: Large structures in the universe. V. Series: NATO ASI
series. Series C, Mathematical and physical sciences ; vol. 117.
QB857.N38   1983   523.1'12   83–21115
ISBN 90–277–1685–4

Published by D. Reidel Publishing Company
P.O. Box 17, 3300 AA Dordrecht, Holland

Sold and distributed in the U.S.A. and Canada
by Kluwer Academic Publishers,
190 Old Derby Street, Hingham, MA 02043, U.S.A.

In all other countries, sold and distributed
by Kluwer Academic Publishers Group,
P.O. Box 322, 3300 AH Dordrecht, Holland

D. Reidel Publishing Company is a member of the Kluwer Academic Publishers Group

CONTENTS

PREFACE

This book is the third volume belonging to the series of proceedings of the Moriond Astrophysics meetings.

It follows "Cosmology and particles" published in 1981 by les Editions Frontières and "The birth of the Universe" which appeared in 1982 thanks to the same publisher.

This workshop took place on march 1983 at La Plagne and as in 1981 and 1982 took advantage of the simultaneous presence of the particle physicists and the astrophysicists. As said in the title of the book the participants have presented their most recent views on the evolution of large structures and galaxies and the relevance of these questions on particle physics and cosmology. Among the many developments which are at the origin of the thirty five papers gathered here, three main themes have been touched on by several speakers during the workshop. They are (i) the influence of the mass of neutrinos and other "inos" predicted by the "Supersymmetry" theories, which are the more likely candidates for the hidden mass of the Universe, (ii) the dynamics of triaxial galaxies and (iii) the possible occurence of pregalactic stars with a debate on their hypothetical nucleosynthetic role. Many different views have been presented on these three topics (and many others) which prove the liveness of the physical cosmology.

For the editor it has been quite difficult to find a reasonable and logical order to arrange the content of the book. The contributions of the participants to this very lively and exciting workshop have been put in six different sections.

In the first section the reader will find four papers dealing with the very early Universe and the influence of the particle physics on it. After a summary on the influence of inflation and supersymmetry on cosmology written by K.A. Olive (CERN), R. Hakim and S. Collin (Meudon) draw our attention on the present arbitrary status concerning the quark-hadron transition occuring during the very early Universe. P. Salati

J. Audouze and J. Tran Thanh Van (eds.),
Formation and Evolution of Galaxies and Large Structures in the Universe, ix–xiii.
© *1984 by D. Reidel Publishing Company.*

(Annecy) shows that systems of two heavy leptons should be extremely heavy (M > 15 Gev) to fulfill the conditions put by the early Universe. To end up this section R. Coquereaux (CERN) provides an interpretation of the cosmological constant (appearing in the Einstein equations) in terms of fields theories and discusses again the physical meaning of this constant and its implications on large structures of the Universe.

The second section is devoted to large structures and especially the pancakes which appear when adiabatic perturbations generate and propagate at the end of the radiative era. R. Bond (Cambridge) and collaborators study the shocks, the cooling and the fragmentation suffered by the pancakes in neutrino dominated universe and show that the cooling still remains a problem in such models. A. Szalay (Budapest) et al consider the same picture of pancakes including massive neutrinos and estimate the temperature fluctuations which should appear in the black body relic radiation as a consequence of this picture. They can already conclude that very large pancakes can be ruled out from such estimates. This contribution is followed by a review presented by N. Mandolesi (Bologna) on the determinations of the distorsions of the cosmic background radiation. S. White (Berkeley) et al have performed N body simulations show that the neutrino scales exceed by very large factors the scale of galaxies for any choice of cosmological parameters (Ω and H the hubble constant especially). According to these authors massive neutrinos do not seem to solve the problem of the occurence and evolution of large structures. Then C. Hogan (Caltech) suggests that pregalactic sources of energy such as quasars or protogalaxies may have influenced the development of large scale galaxy distributions. These distributions might be due to this early activity rather than the direct effect of adiabatic or isothermal fluctuations. Finally after a short comment made by R. Schaeffer (Saclay) concerning the relevance of pancakes on the statistical correlation function, P. Flin (Krakow) shows how to use the brightness profiles of galaxy clusters, their ellipticities and position angles to eliminate (rather than unfortunately to support) models concerning the origin of galaxies.

The third section deals with massive halos and more generally with invisible matter. D.J. Hegyi (Ann Arbor) reviews the evidence in favour of the fact that particles involved in supersymmetry theories appear to be the most promising candidates to constitute the massive halos. J.R. Primack and G.R. Blumenthal (Santa Cruz) favour the cold neutrinos as constituents of the dark matter which is also, for them, obviously non baryonic. P.J.E. Peebles (Victoria and Princeton) puts the emphasis more on a hierarchy model than on a pancake model and suggests that if the Universe is dominated with dark matter it leads on a very natural way to the formation of galaxies and especially of globular clusters. D.N. Schramm and K. Freese (Chicago) review the cosmological constraints which can be put on the mass of neutrinos and other "inos" (coming from supersymmetry theories). C. Norman (Leiden) and J. Silk (Paris) propose a model of formation of dwarf galaxies triggered by the efficiency of massive neutrino accretion by baryonic cores : the preexisting baryonic core accreting dark halos (made of massive neutrinos) can collapse and form such galaxies. Finally P. Vader (Yale) propose that massive halos may inhibit star formation processes : this would mean than blues galaxies should have halos more massive than red galaxies in which star formation process has been more efficient.

Then in the fourth section the occurence of pregalactic (population III, IV etc) stars are discussed in several contributions : M. Rees (Cambridge) proposes the idea that non baryonic matter promote the condensation of small scale pregalactic objects which would lead eventually to galaxies and clusters of galaxies. J. Silk (Paris) makes the prediction that the first stars were massive (100 solar masses). Similarly H. Zinnecker and S. Drapatz (Garching) suggest that thermal instabilities are able to induce fragmentation at very early stages and lead to a first generation of massive population II stars. These papers are followed by the quite provocative argument presented by J. Audouze and J. Silk (Paris) according whom the light elements might be produced in a pregalactic era rather than during the Big Bang. They speculate about the possibility for massive hydrogen made stars to generate He cosmic rays impinging on pure H interstellar medium and producing Deuterium without overproducing too much lithium. The proposal of M. Rees (Cambridge) to produce D and Li in spallation and neutron capture processes occuring around pregalactic black holes in a gas where He has been produced during the Big Bang is certainly less "radical" than the previous one. Finally the early nucleosynthesis which is discussed in these two papers is also considered in the contribution of C. Gry (Verrières le Buisson) et al. In this paper it is argued that infall of processed material (i.e. having been mixed with the out product of stellar nucleosynthesis) occuring during the evolution of galaxies might alleviate the difficulties encountered by the Standard Big Bang nucleosynthesis.

The two last sections deal with the evolution of galaxies. Section V gathers the papers concerned by the structure and the dynamical evolution of galaxies. T. de Zeeuw (Leiden) provides the reader with inhomogeneous triaxial mass models describing stellar systems of that shape. A. Wilkinson (Manchester) reports

on the first attempt made to fit a N-body potential with an
Eddington potential in order to make a global fit of triaxial
N-body model. C. Norman (Leiden) provides a self consistent
treatment of the dynamical evolution of spheroidal systems with
a central black hole. P. Crane (ESO Garching) compares and
shows a good agreement between the galaxy distributions deduced
from Zwicky catalogue of galaxies to a galaxy distribution
function based on statistical thermodynamics of gravothermal
objects. The two next papers deal with the physics of the
galatic disks : R.G. Carlberg (Toronto) shows that the heating
by transient spiral waves provides a natural explanation to the
mean radial velocity dispersion and therefore the vertical scale
height of the galactic disks. E.G. Lacey (Cambridge) analyzes
the heating and therefore the increase of the thickness of
stellar galactic disks by the scattering of the stars induced by
massive objects (giant molecular clouds of the disk or massive
black holes in the halo).

Finally the two last papers of this section are concerned
by the rotation of galaxies (and also clusters). G. Efsthatiou
(Cambridge) and J. Barnes (Berkeley) review the recent
observations of the rotational properties of galaxies and
clusters and propose that these objects can be set in rotation
by tidal torques of neighbors. R. Wyse (Princeton) and B. Jones
(Meudon) study the correlation between the rotational properties
of the galaxies and their surface brightness (the high rotation
elipticals and also the brightest) in terms of dissipational
collapse.

Besides the contribution of Gry et al the four last papers
which constitute section VI deal with chemical evolution of
galaxies J.W. Truran (Urbana-Champaign) provides a guided tour
in the domain pertaining both to the nucleosynthesis and the
galactic evolution. C. Chiosi (Padova) and F. Matteucci
(Frascati) reported on two different works : in the first, they
built models of chemical evolution of dwarf irregular galaxies
based on the stochastic formation process of stars as suggested
by Gerola and Seiden (Yorktowns). In the second, they show that
the chemical yields of C, O and Fe should vary during the
galactic history in order to account for the observed abundances
population I and II stars. The last paper of these proceedings
is the report of observations of specific galactic nuclei
(intermediate Seyfert type) called liners performed by B.E.J.
Pagel (Herstmonceux) with the Anglo Australian Telescope.

I hope that the reader will enjoy these texts as much as
the participants enjoy the presentations of these contributions
and the very lively discussions on these fascinating topics. The
burden of the organization of this meeting and the edition of
this book has been shared with many individuals who are thanked

more specifically in the french "Avant-propos". I just want to
end up this preface by mentioning that this meeting has been
made possible thanks to the generosity of the Scientific Affair
Division of NATO. This book comes from the work, the talent and
the friendness of all the authors whom I want to thank most
warmly for their effort.

Jean Audouze

AVANT-PROPOS

Les rencontres de Moriond en Astrophysique sont organisées depuis 1981 par J. Tran Thanh Van et moi-même en France, il convenait donc que ce livre contienne quelques lignes de notre "belle langue". C'est principalement grâce à l'obstination et la clairvoyance de mon ami et aussi du comité d'organisation comprenant P. Crane, T. Gaisser and D. Hegyi que depuis trois ans les astrophysiciens rencontrent les physiciens des particules dans les Alpes françaises et discutent de sujets d'intérêt commun. Comme on pourra le constater dans ce livre, ces rencontres ont donné lieu à des échanges et des discussions très intenses sur des sujets qui sont à la pointe de l'astrophysique théorique et observationnelle et qui s'adressent aussi de façon directe aux physiciens des particules. Ces rencontres doivent répondre à l'un des voeux de la direction générale du Centre National de la Recherche Scientifique qui souhaite voir s'instaurer des rapprochements de plus en plus constructifs entre les astrophysiciens, les physiciens nucléaires et les physiciens des particules.

Cette troisième rencontre d'astrophysique de Moriond a été soutenue sur le plan financier par l'organisation du Traité de l'Atlantique Nord. Je tiens à remercier en particulier le Dr. M. di Lullo responsable du programme des ateliers OTAN pour son aide et sa compréhension. B. Jones, C. Norman et J. Silk m'ont grandement aidé grâce à leurs suggestions sur le programme scientifique et sur les collègues à inviter. S. Corbin, B. Leloup, B. Lemaire, M.C. Pantalacci, M. Steinberg, K. Tranh Than Van ont apporté une aide essentielle dans la solution des divers problèmes matériels qui n'ont pas manqué de se poser à l'occasion de la tenue de cette rencontre.

Je tiens à exprimer ma reconnaissance à tous mes collègues qui ont rédigé des articles aussi complets et intéressants sur des sujets particulièrement "chauds" en astrophysique et en physique des particules aujourd'hui. Elisabeth Vangioni - Flam, enfin, a pris en charge l'indexation de l'ouvrage. Je l'en remercie très vivement.

Jean Audouze

LIST OF PARTICIPANTS

Van ALBADA T.	Kaptein Astronomical Institute - Groningen
AUDOUZE J.	Institut d'Astrophysique - Paris
BERGERON J.	Institut d'Astrophysique - Paris
BOND R.	Institute of Astronomy - Cambridge
BOUCHET F.	Physique Théorique - Ecole Polytechnique - Palaiseau
BRUNS K.	Max Planck Institute - Munich
CARLBERG R.	University of Toronto
CHIOSI C.	Instituto d'Astronomia -Padova
COLLIN S.	Observatoire de Paris-Meudon
COQUEREAUX R.	CERN - Genève
CRANE P.	ESO - Garching
DAR M.	University of Pennsylvania -Philadelphia
EFSTHATIOU G.	Institute of Astronomy - Cambridge
FANTANGELO P.	Instituto di Astronomia - Frascati
FLIN P.	Observatorium Astronomiczne - Krakow
FREESE K.	University of Chicago
GAISSER T.	Bartol foundation - Newark
GRY C.	LPSP - Verrières le Buisson
HAKIM R.	Observatoire de Paris-Meudon
HANSEL D.	Physique Théorique - Ecole Polytechnique - Palaiseau
HEGYI D.	University of Michigan - Ann Arbor
HOGAN C.	California Institute of Technology - Pasadena
LACEY C.	Institute of Astronomy - Cambridge
MANDOLESI N.	Consiglio Nazionale delle ricerche - Bologne
MATTEUCI F.	Instituto de Astrofisica Spaziale - Frascati
MAZURE A.	Observatoire de Paris-Meudon
NORMAN C.	Sterrewacht - Leiden
OLIVE K.	CERN - Genève
PAGEL B.E.J.	Royal Greenwich Observatory - Hertsmencenx
PEEBLES P.J.E.	Princeton University
PRIMACK J.	University of California - Santa Cruz
REES M.	Institute of Astronomy - Cambridge
SALATI P.	LAPP - Annecy
SCHAEFFER R.	Service de Physique Théorique - CEN de Saclay
SCHRAMM D.N.	University of Chicago
SILK J.	Institut d'Astrophysique - Paris
SYGNET J.F.	Physique Théorique - Ecole Polytechnique - Palaiseau

SZALAY A.   Eotvos University - Budapest
TARRAB I.   Institut d'Astrophysique - Paris
TRURAN J.W.   University of Illinois - Urbana-Champaign
VADER P.   Yale University - New Haven
VITTORIO N.   Instituto di Astronomia - Roma
WHITE S.   University of California - Berkeley
WILKINSON A.   University of Manchester
WYSE R.   Princeton University Observatory
de ZEEUW T.   Sterrewacht - Leiden
ZINNECKER H.   Max Planck Institute - Garching

I

COSMOLOGY OF THE VERY EARLY UNIVERSE
AND FUNDAMENTAL INTERACTIONS

PRIMORDIAL INFLATION AND SUPER-COSMOLOGY

Keith A. Olive

CERN, Geneva, Switzerland

ABSTRACT

A complete locally supersymmetric model for the early universe is reviewed. It begins with primordial inflation just after the Planck time. The (non-trivial) breaking of SU(5) is discussed in detail, with specific emphasis on baryon **generation at** $T \sim O(10^7)$ GeV and monopole suppression (no longer accomplished by inflation). Gravitational effects are taken into account through N = 1 supergravity and play an essential role. What one is left with is a problem-free scenario containing all the benefits of Guth's original inflation as well as density perturbations of a desirable magnitude for the formation of galaxies, a large baryon to photon ratio, and a possibly observable flux of magnetic monopoles. By inserting only two scales, the Planck scale and the supersymmetry breaking scale, both the weak and GUT scales are produced.

I. GUTs, COSMOLOGY AND INFLATION

The advent of Grand Unified Theories (GUTs)[1] has certainly been a cornerstone for our understanding of the very early stages of the Universe. GUTs have, and continue, to resolve questions that were once thought only answerable by an initial condition. The most celebrated of these questions is the origin[2] of the small baryon asymmetry present in the Universe today. Pre-GUT models were only able to speculate that the net baryon to photon ratio today[3]

$$\eta = \frac{n_b}{n_\gamma} \approx (3-6) \times 10^{-10} \tag{1.1}$$

3

J. Audouze and J. Tran Thanh Van (eds.),
Formation and Evolution of Galaxies and Large Structures in the Universe, 3–28.

is due to an initial condition or that initially $\eta \sim O(1)$ and that sometime during the evolution of the Universe a large amount of entropy was produced. The GUT alternative is clearly preferred as it ascribes the origin of η to microphysical processes.

Along with its solution to the problem of baryon generation, GUTs predict the existence of magnetic monopoles (GUMs)[4]. GUMs will be produced whenever a simple group such as SU(5) breaks down to a gauge group containing an explicit U(1) factor [e.g. SU(3) × × SU(2) × U(1)]. The mass of these GUMs will be

$$M_{GUM} \sim M_G/\alpha_G \sim 10^{16} \, GeV \tag{1.2}$$

where M_G is the GUT scale $[O(10^{15})$ GeV$]$ and α_G is the GUT gauge coupling constant. The number density of GUMs produced during the GUT phase transition is, however, very large[5]. Roughly one can estimate the density of GUMs produced by assuming that about one GUM is produced per causally connected region or horizon volume at the time of the transition

$$n_{GUM} \sim (2t_c)^{-3} \tag{1.3}$$

where t_c is related to the critical temperature by

$$t_c \sim 0.3 \, M_p \, T_c^{-2} \, N(T_c)^{-1/2} \tag{1.4}$$

where $M_p \approx 1.2 \times 10^{19}$ GeV is the Planck mass and $N(T_c)$ is the number of light degrees of freedom at T_c. When compared to the number of photons we have

$$\frac{n_{GUM}}{n_\gamma} \sim \left(\frac{10 \, T_c}{M_p}\right)^3 \tag{1.5}$$

In order to remain consistent with the cosmological bound $n_{GUM}/n_\gamma \lesssim 10^{-24}$, eq. (1.5) seems to require[5,6] $T_c \lesssim 10^{10}$ GeV. Naively one expects $T_c \sim M_G$, thus overproducing GUMs.

The standard big bang model, also contained a number of problems thought to be unrelated to GUTs. These include the curvature[7], isotropy[7], rotation[8] etc. problems. As we will see, GUTs offer a possible explanation to them all. These problems can be most easily understood when one considers the equations governing the evolution of Friedmann-Robertson-Walker models. The metric for this class of models is

$$ds^2 = dt^2 - a^2(t) \left[\frac{dr^2}{1-kr^2} + r^2(d\theta^2 + \sin^2\theta \, d\varphi^2)\right] \tag{1.6}$$

where a(t) is the scale factor and k = 0,±1 is a measure of curvature. The Einstein equations yield

$$H^2 \equiv \left(\frac{\dot{a}}{a}\right)^2 = \frac{8\pi\rho}{3 M_p^2} - \frac{k}{a^2} + \frac{\Lambda}{3} + \cdots \qquad (1.7)$$

with possible contributions coming from anisotropies such as shear or rotation. In eq. (1.7), ρ is the total mass-energy density and Λ is the cosmological constant.

It is known that today, the universe is neither curvature-dominated nor vacuum-dominated. In particular, we know that the curvature term in (1.7) is less than the matter term so that we can write

$$\frac{k}{a^2} < \frac{8\pi\rho}{3 M_p^2} \qquad (1.8)$$

We also know that in an adiabatically expanding Universe a $\sim$ 1/T and we can form a dimensionless constant

$$\hat{k} \equiv \frac{k}{a^2 T^2} < \frac{8\pi \Omega \rho_c}{3 M_p^2 T_0^2} \qquad (1.9)$$

where $\Omega \equiv \rho/\rho_c$, ρ_c = 1.88 × 10^{-29} gcm^{-3} is the critical energy density and T_0 = 2.7-3 °K is the present temperature of the microwave background radiation. Using $\Omega < 2$ we see that

$$\hat{k} \lesssim 2 \times 10^{-58} \qquad (1.10)$$

This is the curvature problem[7].

The curvature problem is typical of a problem whose solution is explained by inflation[7]. Remember that this problem (1.10) arose when we assumed an adiabatically expanding Universe (i.e. a $\sim$ 1/T). If however, there was some epoch in which this was not true and that we can let a grow by a factor 10^{28} with little or no change in T, the problem is solved. A temporary De Sitter phase has just this property. If the Universe at some stage where dominated by a cosmological constant eq. (1.7) indicates that a $\sim$ e^{Ht} rather than the power law type expansions associated with matter or radiation dominated Universes.

Once again we see that GUTs can save the day. Inherrent in GUTs are phase transitions. In particular, these phase transitions are associated with a change in the vacuum energy density and if for some reason the Universe supercooled and remained in the false vacuum below T_c, the Universe would become vacuum-dominated. The vacuum energy density would act like a cosmological constant and the Universe would enter an approximate De Sitter phase[7,9]. This is the well known inflationary Universe scenario[7]. The basic idea

was that when the energy density in radiation $\rho \sim T^4$, fell below
the vacuum energy density V_0 of the symmetric phase [e.g. SU(5)],
the Universe would begin to expand exponentially. The phase
transition to the broken phase [e.g. SU(3) × SU(2) × U(1)] would
then occur through the formation of bubbles whose formation rate
is given by[10]

$$P \sim D^2 e^{-B}$$
                                                                (1.11)

where D is a characteristic mass scale and B is the action for
tunnelling. When $P \sim H^4$ the transition is over. The requirement
however, that the exponential expansion last long enough (28 orders
of magnitude) requires B to be very large. In fact, it has been
shown[11] that in order to have enough inflation B is so large that
the transition never finishes. This results in a highly irregular
Universe containing a few bubbles of the broken phase imbedded in
a still exponentially expanding Universe in the symmetric phase.
     The solution of this difficulty is called the new inflationary
Universe[12,13]. In this revised scenario, the scalar field which
drives the phase transition [the 24 for SU(5)] tunnels through the
barrier responsible for supercooling *before* the inflation begins.
Suppose the scalar potential were very flat so that the field took
a long time to reach its vacuum expectation value (v.e.v.)
$v = \langle 0|\phi|0 \rangle$ at the global minimum after tunnelling. Then inside a
single bubble the vacuum energy density would still be approximately
constant and the bubble would expand exponentially. If a single
bubble expanded by 28 orders of magnitude, we need no longer be
concerned with "finishing" the phase transition. Eventually the
field will reach the global minimum and reheat the Universe by
converting the vacuum energy density to radiation.
     Indeed, such flat potentials do exist. They are the Coleman-
Weinberg[14] potentials which are derived by considering first order
radiative corrections to the tree potential. Their properties with
respect to cosmology have been discussed in detail[15] and here I
shall only concentrate on their inflationary capabilities[12,13].
In general the Coleman-Weinberg potential may be written as

$$V(\varphi) = A \varphi^4 \left(\ln \frac{\varphi^2}{v^2} - \frac{1}{2} \right) + D\varphi^2$$
                                                                (1.12)

where

$$A = \frac{1}{64 \pi^2 v^4} \left[\sum_B g_B m_B^4 - \sum_F g_F m_F^4 \right]$$
                                                                (1.13)

and $g_{B(F)}$ is the number of boson (fermion) helicity states of mass
$m_{B(F)}$. The expression (1.13) for A takes into account all possible
first order corrections due to fermions (with a relative minus sign)

as well as bosons. The effective mass^2 is given by

$$D = \frac{1}{2}\left(m_0^2 + cT^2 + bR - 3\lambda \langle \varphi^2 \rangle \right)$$ (1.14)

where m_0 is a bare mass, cT^2 is a gauge group dependent finite temperature correction, bR represents the coupling of ϕ to scalar curvature $R = R^\mu_\mu$ ($b = \frac{1}{6}$ for conformal couplings), $-\lambda/4$ is the ϕ^4 self-coupling and $\langle \phi^2 \rangle$ is the quantum expectation value of ϕ^2 in curved space[16,17].

In order to establish whether or not sufficient inflation can occur in an SU(5) Coleman-Weinberg model, I will follow the proposal of Hawking and Moss[18], that the tunnelling event occurs in a spacially homogeneous manner over a horizon volume. The tunnelling action in this case is given by[18]

$$B = \frac{3 M_p^4}{8}\left[\frac{1}{V_0} - \frac{1}{V_1} \right] \approx \frac{3 M_p^4}{8}\left[\frac{V_1 - V_0}{V_0^2} \right]$$ (1.15)

for $V_1 - V_0 \ll V_0$, where V_0 is the value of the potential at the origin (or symmetric minimum)

$$V_0 = \frac{1}{2} A v^4$$ (1.16)

and V_1 is the value of the potential at the local maximum (separating the two minima at $\phi = 0$ and $\phi = v$)[19]

$$V_1 \approx V_0 + D^2/4A \; \ell n \left(\frac{4 v^2 A}{D} \right)$$ (1.17)

so that[20]

$$B = 3 M_p^4 D^2/8 A^3 v^8 \, \ell n \left(\frac{4 v^2 A}{D} \right)$$ (1.18)

In order for the Hawking and Moss[18] picture to be correctly realized[21], the action B must be $\gg$ 1. (Otherwise the transition might take place before the De Sitter expansion begins.) Although this may not be a sufficient condition[22] (i.e. other small actions may exist) it is necessary for inflation. The second requirement for inflation is that the timescale to roll over from the local maximum to the global minimum is long enough to guarantee sufficient inflation. The roll over timescale is

$$\tau = 3H/2D$$ (1.19)

and to have enough inflation we must have:

$$\exp(H\tau) > 10^{28}$$
$$3H^2/2D > 65 \tag{1.20}$$

If we combine the two requirements, $B \gg 1$ and $H\tau > 65$, we see that[20]

$$\left(\frac{8A^3}{3}\right)^{1/2} \frac{v^4}{M_p^2} \ln^{1/2}\left(\frac{4v^2A}{D}\right) < D < \frac{4\pi Av^4}{130 M_p^2} \tag{1.21}$$

where we have used

$$H^2 = \frac{8\pi V_0}{3 M_p^2} = \frac{4\pi Av^4}{3 M_p^2} \tag{1.22}$$

In standard SU(5), A is determined. The dominant contributions to A come from the gauge bosons X and Y, so that with $g_B = 36$ and $m_{X,Y} = (25/8) g^2 v^2$ we have

$$A = \frac{5625}{1024 \pi^2} g^4 \approx 5 \times 10^{-2} \tag{1.23}$$

for $g^2/4\pi = 1/41$. In the range of interest the ln term in (1.21) is about 5 and in order for the double inequality to be consistent we must have

$$A \lesssim \frac{1}{4}\left(\frac{\pi}{130}\right)^2 \approx 10^{-4} \tag{1.24}$$

hence the first major stumbling block for inflation in SU(5). There are, however, several other reasons why standard SU(5) will not provide an inflationary scenario. The upper limit on the effective mass^2 D in (1.21) implies a mass for the scalar field $O(10^8)$ GeV. This requires an enormous amount of finetuning as scalar masses will generally pick up corrections $O(10^{15})$ GeV. Linde[17] has also pointed out that unless λ in (1.14) is $\lesssim 5 \times 10^{-3}$ scalar field fluctuations will drive the transition without sufficient inflation (the ϕ^4 coupling here is an order of magnitude larger).

The most devastating blow to the inflationary scenario in standard SU(5) comes from the production of density fluctuations[23-26]. GUTs do very well in producing[27] the desired

Harrison-Zel'dovich[28] scale independent spectrum of fluctuations. Without going into too much detail, it can be shown[23-26] that the magnitude of density fluctuations in SU(5) is:

$$\frac{\delta\rho}{\rho} \simeq \left(\frac{8A}{3\pi^3}\right)^{1/2} \ell n^{3/2}(Hk^{-1}) \sim O(1-10) \qquad (1.25)$$

for $Hk^{-1} \sim 10^{21}$ where k is the wave number of the fluctuation. Of course, what we want is a value $\delta\rho/\rho \sim O(10^{-4})$ in order to remain compatible with limits on the isotropy of the background radiation and still remain strong enough to eventually produce galaxies.

In the remaining sections of this contribution, I will try to argue that although the above looks very discouraging, the inflationary scenario is by no means dead. I will begin by discussing the motivations for considering supersymmetry and supergravity in section 2. In section 3, I will discuss a specific model for inflation in the context of supergravity. Finally, I will follow through the post-inflationary period through the breaking of SU(5) in section 4. Much of what will be discussed in the following is a result of a very fruitful collaboration with J. Ellis, D.V. Nanopoulos, M. Srednicki and K. Tamvakis.

II. SUPERSYMMETRY, SUPERGRAVITY AND COSMOLOGY

Supersymmetry[29] is a symmetry which directly relates bosons and fermions. Supersymmetric theories[30] are classified according to the number of charges N, present in the theory. For gauge theories with maximum spin = 1 fields one is restricted to $N \leq 4$. Local supersymmetry or supergravity[31] is consistent with gauge theories for $N \leq 8$. In the simplest supersymmetric theory (N = 1), the particle content is divided into two types of representations. The first is a gauge supermultiplet, which associates to each gauge boson a gauge fermion. That is, in addition to the gauge bosons such as the photon, gluons, $W^{\pm}$ and Z, there will be photinos, gluinos, winos and zinos. The self-interactions of these particles are determined by the gauge coupling constant. The second type of representation is the chiral supermultiplet which includes the quark and lepton fields and their associated scalar partners and the Higgs fields along with their fermionic partners. The chiral superfields will have both gauge interactions as well as self-interactions which are determined by a superpotential f. For a renormalizable theory, the superpotential can be at most cubic in the fields.

In a globally supersymmetric theory[29], the effective potential describing the self-interactions of the chiral superfields is derived from the superpotential f as follows

$$V(\varphi_i) = \sum_i |f_{\varphi_i}|^2 \qquad (2.1)$$

where

$$f_{\varphi_i} = \partial f / \partial \varphi_i \tag{2.2}$$

for all fields ϕ_i. Global supersymmetry is unbroken if

$$f_{\varphi_i} = 0 \quad \forall \ \varphi_i \tag{2.3}$$

(in the present discussion, I will neglect all gauge interactions; similar conditions for these exist as well).

In a locally supersymmetric theory, the effective potential[32] is somewhat more complicated

$$V(\varphi_i) = \left[\sum_i |f_{\varphi_i}|^2 - \frac{3}{M^2} |f|^2 \right] exp\left[\sum_i |\varphi_i|^2 / M^2 \right] \tag{2.4}$$

where the generalized derivative f_{ϕ_i} is now

$$f_{\varphi_i} = \partial f / \partial \varphi_i + \varphi^* f / M^2 \tag{2.5}$$

where $8\pi M^2 = M_p^2 = G_N^{-1}$ is the inverse of the gravitational constant. Note that regardless of the superpotential f, the exponential in (2.4) renders $V(\phi_i)$ non-renormalizable. Thus, because of the non-renormalizability of supergravity, there is no longer any use to the restriction to, at most, cubic terms in f. Indeed, we shall make much use of this extra freedom. Local supersymmetry is also unbroken for $f_{\phi_i} = 0$.

Aside from its asthetic value, supersymmetry is of course very useful for curing maladies in particle physics and as we shall see, in cosmology. Non supersymmetric gauge theories seem to require scalar fields for spontaneous symmetry breaking. Apart from this, there is no other reason (theoretical or experimental) for their existence. Supersymmetry has them already built in. But the immense popularity of supersymmetry is because of the possibility it offers of solving the gauge heirarchy problem[33]. Stated simply, the gauge heirarchy problem is twofold. On the one hand, one would like to understand why there is the apparant discrepancy in the mass scales

$$M_W \ll M_G \ll M_P \tag{2.6}$$

Secondly, in ordinary GUTs, when one calculates radiative corrections to scalar masses (which are responsible for generating the scales M_W and M_G) they tend to always push the light scalars (with

masses M_W) up to the heavy ones (with masses M_G). In order to preserve the heirarchy, an enormous amount of finetuning must take
place.

Supersymmetry has the possibility of solving[33] the second
aspect of the heirarchy problem. Because of the existence of
so-called non-renormalization theorems[34], which guarantee cancellations (to all orders in perturbation theory) when calculating
radiative corrections in an exactly supersymmetric theory. Thus,
once a particular value of a parameter is set at the tree level,
it stays there. In a broken supersymmetric theory there will be
corrections proportional to the mass splitting between bosons and
fermions

$$m_B^2 - m_F^2 = M_S^2 \varepsilon \qquad (2.7)$$

where M_S is the scale of supersymmetry breaking and ε is some
coupling constant. These cancellations have also been employed in
attempts to cure the strong CP problem[35].

In analogy with spontaneous symmetry breaking of gauge theories,
global supersymmetry is broken with the appearance of a goldstino.
In local supersymmetry, the goldstino is eaten by the gravitino,
the superpartner of the spin 2 graviton, through the superhiggs
effect[36,37]. The mass of the gravitino is determined in much the
same way as that of gauge boson in ordinary gauge theories

$$m_B = g \langle \varphi \rangle \qquad (2.8)$$

where g is a gauge coupling constant and $\langle \phi \rangle$ is the v.e.v. of the
scalar field breaking the gauge symmetry. The corresponding order
parameter for supersymmetry is M_S^2 and the gravitino mass is given
by[37]

$$m_{3/2} \sim G_N^{1/2} M_S^2 \sim M_S^2 / M_P \qquad (2.9)$$

In particular ε in (2.7) should be $O(m_{3/2}/M_P)$ so that mass splittings
and mass corrections should be[38]

$$\delta m \sim M_S^2 / M_P \lesssim M_W \qquad (2.10)$$

in order to preserve the hierarchy. In fact in this way, broken
supersymmetry generates[39,40] the weak interaction scale. In
section 4, we will see how the GUT scale is also generated by M_S.

Let us now see how supersymmetry affects the cosmological
inflationary scenario[20,41]. If one recalls, one of the major
difficulties with the new inflationary model stemmed from the dis-
crepency between (1.23) and (1.24). Radiative corrections in a
supersymmetric theory, however, are greatly altered. In exact
supersymmetry, the no-renormalization theorems[34] tell us that first
order radiative corrections of the Coleman-Weinberg type are exactly
zero. Indeed by glancing at the expression (1.13) for the ϕ^4

self-coupling A, we realize that for each boson of mass m_B and total helicity g_B, there is an assoicated fermion of mass $m_F = m_B$ and total helicity $g_F = g_B$ (for each N = 1 chiral superfield there are two scalars per fermion and for gauge superfields, one gauge boson per gauge fermion). Hence with exact supersymmetry

$$A \equiv 0 \tag{2.11}$$

We know, however, that supersymmetry must be broken in nature and the mass degeneracy between bosons and fermions be lifted. The splitting between masses is just (2.7) which we will leave arbitrary for the moment. [It may even be different for different supermultiplets[39,40].] The superheavy gauge boson contribution to (1.13) then becomes[20]

$$A = \frac{g_{B(F)}}{64 \pi^2 v^4} \left[m_B^4 - m_F^4 \right] \approx \frac{g_{B(F)}}{32 \pi^2 v^4} \left[m_B^2 - m_F^2 \right] m_B^2$$

$$\approx \frac{g_{B(F)}}{32 \pi^2 v^4} m_B^2 M_s^2 \varepsilon \tag{2.12}$$

For the SU(5) gauge bosons X and Y, we have[20]

$$A = \frac{75}{32 \pi^2} g^2 \frac{M_s^2}{v^2} \varepsilon \tag{2.13}$$

The twin constraint (1.21) which insures sufficient inflation translates to

$$\frac{1}{2} g^3 \frac{M_s^3 \varepsilon^{3/2} v}{M_p^2} \lesssim D \lesssim 2 \times 10^{-2} g^2 \frac{M_s^2 \varepsilon v^2}{M_p^2} \tag{2.14}$$

or to remain consistent

$$g M_s \varepsilon^{1/2} \lesssim 5 \times 10^{-2} v \sim 10^{14} \, GeV \tag{2.15}$$

which is trivially satisfied in almost any supersymmetric model.

The remaining problems discussed in the previous section are also easily seen to disappear. The technical problem of finetuning the mass terms in (1.14) to low values is cured by realizing that the deadly corrections can now be kept small $O(M_s^2 \varepsilon)$. Even the constraint coming from the overproduction of density perturbations disappears. Equation (1.25) becomes

$$\frac{\delta\rho}{\rho} \sim \frac{5}{2\pi^{5/2}} \cdot \frac{g\,M_s\,\varepsilon^{1/2}}{v} \cdot \ln^{3/2}\!\left(H k^{-1}\right) \tag{2.16}$$

which for $v \sim 10^{15}$ GeV and $g \sim 0.7$ will yield $\delta\rho/\rho \sim 10^{-4}$ for $M_S\,\varepsilon^{\frac{1}{2}} \sim 2 \times 10^{10}$ GeV. Hence a glimmer appears that inflation and supersymmetry will go well together.

The major effects, until now, of supersymmetry were that it i) produced a larger barrier to keep the action B large and insure sufficient supercooling and ii) by lowering the vacuum energy density of the symmetric state, it produced a flatter potential allowing for a longer roll-over time scale. The above, however, was not a supersymmetric model for inflation, but rather an indication of which way things will go. In the following sections, I shall look closely at real supersymmetric models and their potential for supplying a complete and problem-free cosmological scenario.

III. PRIMORDIAL INFLATION

As we have seen, in order to preserve[38] the gauge hierarchy, any realistic supersymmetric model will need relatively small mass splittings $O(M_W)$. This would imply a supersymmetry breaking scale

$$M_s^2 \sim M_W M_P \sim \left(10^{10}\ GeV\right)^2 \tag{3.1}$$

At energy scales at or above the GUT scale, the effects of supersymmetry breaking will be negligible. It seems reasonable, therefore to consider inflation in the context of (nearly) exact supersymmetry. First order corrections to the tree potential may then be done away with entirely. Let us furthermore, for reasons which will become clear shortly, consider the field which drives inflation to be a chiral superfield gauge singlet ϕ (the inflaton) which receives a v.e.v. $\langle 0|\phi|0\rangle = \mu$ with

$$M_G < \mu < M_P \tag{3.2}$$

Before specifying a particular superpotential for the inflaton, suppose that the derived effective zero temperature potential can be put in the polynomial form[41]

$$V(\varphi) = \alpha\varphi^4 - \beta\varphi^3 + \gamma\varphi^2 + \delta \tag{3.3}$$

Starting with a general potential of this form (which can hopefully be derived in a supersymmetric theory), we can derive what conditions must be imposed on the parameters α, β, γ, and δ in order to achieve inflation. To begin with, we assume that all subsequent

phase transitions will produce a change in the vacuum energy density
which is negligible when compared with the one discussed here.
This will be true, of course, if the inflation scale is much larger
than the GUT scale for example. Thus at the global minimum of the
potential (at $\langle 0|\phi|0 \rangle = \mu$) we will require that the cosmological
constant vanish. The effective cosmological constant (after the
transition) is just the value of the potential at the minimum

$$\Lambda = \frac{8\pi}{3} V(\mu) / M_P^2 \tag{3.4}$$

thus we require

$$V(\mu) = 0 \tag{3.5}$$

We must also insure that at $\phi = \mu$ the potential has a minimum i.e.,

$$\frac{\partial V}{\partial \varphi}(\mu) = 0 \quad \text{and} \quad \frac{\partial^2 V}{\partial \varphi^2}(\mu) > 0 \tag{3.6}$$

The vacuum energy density at the origin or "symmetric" minimum is
just $V(0) = \delta$. Finally, the potential must possess a local maximum
near the origin. Its position is solved to be at

$$\varphi_1 \approx 2\gamma / 3\beta \tag{3.7}$$

The value of the potential at the local maximum is just

$$V_1(\varphi_1) \approx \frac{4}{27} \frac{\gamma^3}{\beta^2} + \delta \tag{3.8}$$

Up to this point, I have not discussed anything but the zero
temperature potential. In general, however, one must consider
finite temperature corrections to the potential of the form

$$\delta V_T = \frac{T^2}{8} \partial^2 V / \partial \varphi \partial \varphi^* \tag{3.9}$$

(In general the field ϕ will be complex; it is considered real here
only for simplicity). At high temperatures prior to the onset of
inflation, these are of course important and must be considered.
For all temperatures such that $T^2 \ll \gamma/\alpha$, the corrections take on
a simple form,

$$\delta V_T = c T^2 \varphi^2 \tag{3.10}$$

$$c = 3\alpha - \frac{9}{4} \beta^2 / \gamma \tag{3.11}$$

Furthermore, during inflation, these may be neglected if at the Hawking temperature

$$T_H^2 = \frac{H^2}{4\pi^2} = \frac{2}{3\pi} \frac{\delta}{M_p^2} \tag{3.12}$$

which is the lowest temperature attainable during the De Sitter phase[18,42,43], we have $cT_H^2 \ll \gamma$.

Let us now determine under what conditions we have inflation. The first requirement[21] was that the action be large. In this case, we have

$$B = \frac{M_p^4 \gamma^3}{18 \beta^2 \delta^2} \gg 1 \tag{3.13}$$

while the condition for a long roll-over time scale is

$$\frac{\gamma}{\delta} < \frac{4\pi}{65 M_p^2} \tag{3.14}$$

To get an idea of what these conditions (3.13) and (3.14) mean, it is useful to scale the parameters β, γ, δ by dimensionless quantities

$$\hat{\beta} = \beta/\mu \; ; \; \hat{\gamma} = \gamma/\mu^2 \; ; \; \hat{\delta} = \delta/\mu^4 \tag{3.15}$$

Equation (3.13) then becomes

$$B = \frac{M_p^4 \, \hat{\gamma}^3}{18 \mu^4 \, \hat{\beta}^2 \hat{\delta}^2} \gg 1 \tag{3.16}$$

and eq. (3.14) becomes

$$\frac{\hat{\gamma}}{\hat{\delta}} < \frac{4\pi}{65} \left(\frac{\mu^2}{M_p^2} \right) \tag{3.17}$$

where μ remains the v.e.v. of the field ϕ at the global minimum.

We can clearly see now the effect of the inflationary scale. If we choose $\mu \sim 10^{15}$ GeV, we must then fine tune $\hat{\gamma}/\hat{\delta} \leq 10^{-9}$ and furthermore $\hat{\beta}^2/\hat{\delta} \leq 10^{-14}$! Thus if we suppose at best that $\alpha \sim \hat{\beta} \sim \hat{\delta}$, [remember eq. (3.6)], all of these must be $0(10^{-14})$ and $\hat{\gamma} \sim 10^{-25}$. Clearly, rather absurd values for coupling constants. Notice, however, what happens when $\mu = M_p$. Equation (3.17) tells us that $\hat{\gamma}/\hat{\delta} \leq 10^{-1}$ and $\hat{\beta}^2/\hat{\delta} \leq 10^{-6}$ which is more typical of known Yukawa couplings for example. These observations are what we feel to be the motivations for primordial inflation[41]. As the v.e.v. of the field which drives inflation becomes larger than the GUT scale and moves towards the Planck scale fine tunings disappear while choices for couplings become more natural.

Before moving on to a specific model, it is important to note what happens to the magnitude of density fluctuations[44]. The basic property that these perturbations are essentially scale-independent remains intact. What we find, however, is that it is not longer the ϕ^4 coupling which governs the magnitude, but rather the cubic coupling β. We find that for the potential of the form (3.3)

$$\frac{\delta\rho}{\rho} \approx \frac{1}{(2\pi^3)^{1/2}} \frac{\beta}{H} \ln^2(Hk^{-1})$$

(3.18)

In terms of $\hat{\beta}$, we must require $\hat{\beta}^2 \sim 10^{-12} (\mu/M_p)^2 \hat{\delta}$ to have $\delta\rho/\rho \sim 10^{-4}$. This again is a very stiff requirement especially for $\mu \ll M_p$. As we will see shortly, supergravity will ease this constraint as well as previous ones.

As primordial inflation implies $\mu \sim M_p$, we must ask ourselves if first order gravitational (FOG) effects will start to become important. Indeed, Planck time physics remains very uncertain. One possibility, however, is considering the effects of local super-symmetry or supergravity. The hope, of course, is that eventually all quantum gravitational effects will be understood in the context of, for example, an N = 8 supergravity theory. In some, as yet unknown, way at scales at or below the Planck scale, one may be left, however, with a residual N = 1 supergravity theory. We will make, therefore, a large assumption; all FOG (and higher) effects are taken into account when one works in the context of supergra-vity. In particular, gravitational interactions will be included by considering non-renormalizable terms in the superpotential[45] and at scales below M_p we are left with a workable effective theory.

To treat inflation in simple supergravity[46], let us start with the most general superpotential for a single scalar field, the inflaton ϕ,

$$f_I = m^3 \left(\sum_n \frac{\lambda_n}{n+1} \left(\frac{\phi}{M}\right)^{n+1} + \lambda' \right)$$

(3.19)

where m is an (as yet) unspecified mass parameter, $M = M_p/\sqrt{8\pi}$ and the λ_i are all dimensionless coupling constants. From eqs. (2.4) and (2.5), we can derive the effective potential and put it in the form (3.3). There will, of course, be terms higher than fourth order, but at scales $\phi \ll M$ we can neglect (?) these. In order for the superpotential (3.19) to provide a satisfactory inflationary picture, we must insure that at the origin we have a true minimum and that there is a second minimum at $\langle 0|\phi|0\rangle = \mu$ which we will take to be equal to M. A supersymmetric minimum at M requires[46]

$$\sum_n \frac{\lambda_n}{n+1} = \sum_n \lambda_n = 0 ; \quad \sum_n n \lambda_n \neq 0 \qquad (3.20)$$

A minimum at the origin is found if $\lambda' = 2\lambda_1$.

Without passing through all the details, a similar analysis to the general one given above yields the necessary constraints on the parameters

$$\lambda_0 > 260 \ \lambda_2 \qquad (3.21a)$$

$$-\lambda_3 m^3 << 4 \times 10^{-3} \qquad (\lambda_3 < 0) \qquad (3.21b)$$

$$\lambda_0 m^3 << 10^{-3} \qquad (3.21c)$$

for the choice $\lambda_1 = -\lambda_3$; and this easily achieved for m = 10^{-2} M, $\lambda_0 \approx 1$, $\lambda_1 = -\lambda_3 \approx 10^{-1}$ and $\lambda_2 \approx 10^{-3}$. This is of course only a single choice for parameters of which many are possible and still give density perturbations $\delta\rho/\rho \sim 10^{-4}$. Hence, we see that in the context of supergravity one can have an acceptable inflationary scenario without any severe adjustments to parameters.

At this point, we must remember however, that although primordial inflation can solve the standard cosmological problems such as curvature, isotropy etc., it says nothing about the monopole problem. Previously inflation was occurring during the SU(5) phase transition. In this case, because the number of GUMs per bubble is O(1) and we live in a single bubble, there is only about one GUM in our visible Universe. Primordial inflation is not associated with SU(5) and care must be taken with respect to the production of GUMs. In fact, the whole problem of breaking SU(5) and baryon generation is not trivial in the context of supersymmetry. In the following section, I will look at this problem as well as the evolution of the Universe subsequent to primordial inflation.

IV. SUPER-COSMOLOGY

Globally supersymmetric theories, all have one thing in common; the lowest vacuum state is the supersymmetric one with zero vacuum energy[47]. At zero temperature, supersymmetric GUTs may have several degenerate minima[48], e.g. SU(5), SU(3) × SU(2) × × U(1), SU(4) × U(1) etc. all with zero vacuum energy density. Unlike gauge symmetries, supersymmetry is broken at finite temperature[49] due to the different spin statistics of bosons and fermions. As the Universe cools, it may eventually choose a supersymmetric minima[50], but which one?

In a non-supersymmetric GUT, at very high temperatures, generally only a symmetric minimum [e.g. SU(5)] exists. As the temperature

drops, broken minima [e.g. SU(3) × SU(2) × U(1) or SU(4) × U(1)]
begin to appear. At a critical temperature, T_c, the value of the
potential at both the symmetric minimum and broken minimum are
equal. (There may be more than one critical temperature.) At
lower temperatures, the broken minima drop far below the symmetric
state and a phase transition occurs. What one tends to find[48] in
supersymmetric GUTs is that the symmetric state remains the global
minimum as the temperature drops. In a zero temperature, exactly
supersymmetric GUT, this state of affairs would (naïvely) remain
until T = 0, hence the Universe would never be able to get out of
the SU(5) phase. One might hope that because supersymmetry must
(somehow) get broken, things might change. But in order to preserve the gauge heirarchy, these effects should only be felt below
M_W, thus making baryon generation practically impossible.

Something else does occur, however. As the Universe cools,
at some point SU(5) interactions become strong. Indeed, when one
calculates the renormalization group equations for the SU(5) fine
structure constant, $\alpha = g^2/4\pi$, one can find[48] that $\alpha \sim 0(1)$ at
a temperature $T \sim \Lambda_{SU(5)} \sim 10^9\text{-}10^{10}$ GeV. To see what happens as
T approaches $\Lambda_{SU(5)}$, one has to look at the form of the finite
temperature effective potential

$$V_T(\Sigma) = V(\Sigma) + c_1 \Sigma^2 T^2 - c_2 T^4 \qquad (4.1)$$

where Σ is the 24 breaking SU(5), c_1 can be derived from the superpotential yielding $V(\Sigma)$ and $c_2 > 0$ is a constant depending on the
particle content[49]. At high temperatures, $T > \Lambda_{SU(5)}$, SU(5) has
many more light particle degrees of freedom than either SU(3) ×
× SU(2) × U(1) or SU(4) × U(1) and $c_2(5) > c_2(3,2,1)$, $c_2(4,1)$.
When T approaches $\Lambda_{SU(5)}$, the theory becomes "confining" and particle degrees of freedom being to disappear, thus lowering $c_2(5)$
relative to the other states. At this point the phase transitive
becomes possible. One may also arrange the Higgs sector in such
a way as to make $c_2(3,2,1) > c_2(4,1)$ so that SU(3) × SU(2) × U(1)
is actually preferred over SU(4) × U(1)[51]. [This choice in fact
is also preferred from the point of view of generating a baryon
asymmetry[52].]

Although a transition is now in principle possible because
the vacuum energy density of the broken state is lower than the
symmetric one, a large barrier separating the two phases may
prohibit the transition[50,53]. For example, if one starts with
the most general (renormalizable) superpotential for the adjoint Σ
alone[48]

$$f(\Sigma) = \frac{1}{2} M_\Sigma \, Tr \, (\Sigma^2) + \frac{1}{3} \lambda \, Tr \, (\Sigma^3) \qquad (4.2)$$

where $M_\Sigma \sim M_G \sim 10^{16}$ GeV, one finds that the bubble formation rate
of the broken phase is very small:

$$P \approx M^4 e^{-B} \tag{4.3}$$

where the action B in this case is[53]

$$B \approx .04 \left(M_\Sigma / T \right)^9 \tag{4.4}$$

P is largest when the transition first becomes possible, i.e. at $T \sim \Lambda_{SU(5)}$ so that

$$P \lesssim 10^{-10^{52}} \tag{4.5}$$

and hence no transition occurs.

In the context of global supersymmetry, this situation can be remedied by taking a very small value for both λ and M[51]. For small enough values, the barrier separating the broken and symmetric phases can be made arbitrarily small. In particular, we would like the barrier height to $\lesssim O(10^{10})$ GeV so that strong coupling phenomena at a temperature of the same order could drive the transition. To do this, however, one needs[41] $M_\Sigma \sim M_W$ and $\lambda \sim M_W/M_G$. Although such an assignment is technically possible, it seems very unnatural and contrary to the desire to keep all couplings $O(10^{-3}-1)$.

It is at this point that supergravity and the use of non-renormalizable terms can save us. Consider the superpotential[54]

$$f_S = \frac{a_1 X^4}{M} + \frac{a_2 X^2 Tr(\Sigma^3)}{M^2} + \mu^2(z + \Delta) \tag{4.6}$$

where $M = M_p/\sqrt{8\pi}$ and X and z are SU(5) singlets. The couplings a_1 and a_2 will be taken $O(1)$. The third term in (4.6) is needed to break local supersymmetry and will not be discussed here. The mass scale $\mu \sim M_S \sim 10^{10}$ GeV. Once again from eqs. (2.4) and (2.5), one can derive the effective potential for the fields X and Σ. Indeed, (4.6) was chosen so as to mimick the behaviour of (4.2). When one solves for the various minima of the effective potential one finds that in fact the $SU(3) \times SU(2) \times U(1)$ minimum is the lowest. Furthermore, it occurs when Σ takes the v.e.v.

$$\langle \Sigma \rangle \approx (\mu M)^{1/2} \tag{4.7}$$

This is in reality a prediction of the GUT scale

$$M_G^2 \approx (\mu M) \sim M_S M_p \sim (10^{15} GeV)^2 \tag{4.8}$$

At this minimum, the field X has an expectation value $\langle X \rangle = (\mu/M)^{3/4} M$ and gives a barrier height of $(\mu/M)^{5/4} M \sim 10^8$ GeV.

Thus we have been able to reproduce the scenario in global supersymmetry without small parameters by considering supergravity.

To complete the model of the evolution of the early Universe
in N = 1 supergravity, there must still be a mechanism for genera-
ting a baryon asymmetry[2] and a solution to the monopole problem[5].
The solution to the monopole problem is rather simple in this case.
Because the transition has been delayed down to $T \sim \Lambda_{SU(5)} \sim$
$\sim 10^9$-10^{10} GeV, this is in effect the critical temperature for the
SU(5) transition and the temperature at which monopoles are pro-
duced. Equation (1.5) would then say that

$$\frac{n_{GUM}}{n_\gamma} \sim \left(\frac{10 \, \Lambda_{SU(5)}}{M_P}\right)^3 \sim 10^{-24} - 10^{-27} \qquad (4.9)$$

which is consistent with the cosmological limit. In fact, coupled
together with the new inflationary scenario, primordial inflation
and supercosmology might be the only way to get an observable
number of monopoles. As monopoles will in general not be produced
thermally, the standard new inflationary scenario[12,13] would predict only one monopole per visible region.

To generate a baryon asymmetry is in general not quite so
easy. Detailed calculations[55,56] of the baryon asymmetry produced
in standard models, tend to show that a large asymmetry becomes more
and more difficult as the scale at which the asymmetry is produced
is lowered. In standard SU(5) one has[56]

$$\frac{k \, n_B}{s} \approx 2 \times 10^{-3} (\Delta B)[(3K)^{1.2} + 1]^{-1} \qquad (4.10)$$

where s/k is the specific entropy, ΔB is the net baryon number produced by a decay of a pair of Higgses H, $\bar{H}$ and K = 2.9 $\times$ 10^{17} α_H/M_H
is the ratio of the decay rate to the expansion rate at $T = M_H$.
The observed baryon to photon ratio today is related to kn_B/s by

$$\frac{n_B}{n_\gamma} \approx 7 (k \, n_B/s) \qquad (4.11)$$

In the standard picture, if the lightest Higgs boson were to have
a mass $O(10^{15})$ GeV with a Yukawa coupling $\alpha_H \sim 10^{-4}$ we would have
a suitable asymmetry $kn_B/s \sim 10^{-3}$ ΔB.

In supercosmology, however, to have any chance of generating
an asymmetry at 10^{10} GeV one needs Higgs bosons with masses of that
order that have baryon number violating interactions. In a non-
supersymmetric theory, this would result in a (too) low asymmetry
$kn_B/s \sim 4 \times 10^{-8}$ ΔB. In supersymmetric models, the situation at
first sight seems even worse[57]. The increased number of particle
degrees of freedom and larger couplings along with dimension 5
operators[30], drive things towards equilibrium and not away from
it. Naïvely, one might then expect[57] the baryon asymmetry to
become negligible in supersymmetric models with 10^{10} GeV Higgses.

This approach does not take into account a possibility which is available, i.e. to couple the H and $\bar{H}$ to an SU(5) singlet[51]. In this case, one can show that it is relatively simple to get a large asymmetry with low mass Higgses in supersymmetry.

To complete the scenario, consider the superpotential[54,58]

$$f_B = a_3 \frac{X \, Y \bar{H} H}{M} + a_4 M \bar{\theta} \theta + a_5 \frac{\bar{\theta} \Sigma^2 H}{M}$$

$$+ \; a_6 \frac{H \Sigma^2 \theta}{M} + a_7 \mu Y^2 + a_8 Y^3$$

$$+ \; Yukawa \; couplings \qquad (4.12)$$

where H and $\bar{H}$ are a 5 and $\bar{5}$ and Y is the additional SU(5) singlet which will be used to generate the baryon asymmetry. θ and $\bar{\theta}$ are a 50 and $\overline{50}$ used to separate the mass scales of the doublet and triplet in H and $\bar{H}$[59]. All couplings a_3-a_8 will be O(1). The complete superpotential for the "world" will then be

$$f = f_I + f_S + f_B \qquad (4.13)$$

Subsequent to inflation, as ϕ reaches its v.e.v., the Universe begins to fill up with the inflatons ϕ at an initial temperature

$$T_\phi \approx 10^{15} \, GeV \qquad (4.14)$$

Through a gravitational coupling which is generated when one considers the full superpotential (4.13), one can see that the inflatons can decay predominantly into the singlets Y through

$$\left(\frac{m_\phi}{M}\right) a_8 \, \phi \, Y^3 \qquad (4.15)$$

with a rate given by

$$\Gamma_\phi = m_\phi^3 / M^2 \qquad (4.16)$$

for

$$m_\phi = \frac{\partial^2 f}{\partial \phi^2} = \frac{m^3}{M^2} \sum_n n \lambda_n \approx 10^{13} \, GeV \qquad (4.17)$$

The decay $\phi \to 3Y$ takes place at a temperature around 10^{10} GeV.

Shortly afterwards, the SU(5) transition takes place at $T = \Lambda_{SU(5)} \sim 10^9$-$10^{10}$ GeV and the net baryon number of the Universe is $n_B/n_\gamma \approx 0$. Through the "a_3" coupling in (4.12), the singlets Y (which now dominate the Universe) decay only into an H, $\bar{H}$ pair. The decay rate for this process is[58]:

$$\Gamma_Y \sim a_3^2 \langle X \rangle^2 m_Y / M^2$$

$$\sim a_3^2 (\mu/M)^{5/2} M \qquad\qquad (4.18)$$

for $m_Y \sim \mu$ [from the "a_7" term in (4.12)]. This decay ($Y \to H, \bar{H}$) occurs at $T \approx 10^7$ GeV. Recall however, that prior to the decay $n_Y/n_\gamma \approx 1$ and so afterwards we must have $n_H/n_\gamma \approx 1$ as well. For $m_H \sim \mu \sim 10^{10}$ GeV at $T = 10^7$ GeV, this is very far from equilibrium. (The equilibrium value would go as $e^{-m_H/T} \ll 1$.) The resultant baryon asymmetry will be its maximum value $n_B/n_\gamma \approx 10^{-2}$ ΔB. It should be stressed that this value is not the result of fine tunings or small coupling constants, but rather a judicious coupling of a singlet[51] and the use of non-renormalizable interactions[54,58].

Finally, it is interesting to note, that without any extra effort, the gravitino problem[60] is also solved[58,61]. This problem arose[60] when one realizes that gravitinos which decay only gravitationally, do so very late, and if sufficiently abundant, produce too much entropy, possibly ruining the successes of big bang baryosynthesis and/or big bang nucleosynthesis. If, however, the initial abundance of gravitinos were reduced by inflation for example, the problem could be avoided[61]. In the above scenario, there is no gravitino problem for an inflaton mass $m_\phi < 10^{15}$ GeV[58]. Baryon generation required an inflaton mass $m_\phi \gtrsim 10^{13}$ GeV (in order to be able to produce Y's in abundance). We hope that it is not a coincidence that the parameters derived from demanding density perturbations $\delta\rho/\rho \sim 10^{-4}$ gave an inflaton mass $m_\phi \approx 10^{13}$ GeV!

V. SUMMARY AND DISCUSSION

What I have tried to show in the previous sections, is that supersymmetry, supergravity and cosmology all fit very nicely together. The big bang model itself has been able to explain the observed expansion of the Universe, the existence of the microwave background radiation and the origin of the light elements through nucleosynthesis. Before the inception of GUTs, it was not able to give any clue as to why the present density is *so* close to the critical density while at the same time being *so* old. It was also not able to explain the high degree of isotropy nor the small observed baryon asymmetry. It is by now no secret that GUTs have radically changed our understanding of the very early stages of the Universe. Baryon number violating interactions led to the solution of the baryon asymmetry, while postulating the existence of a past inflationary period can solve many of the remaining problems. Notwithstanding, the monopole problem and gravitino problem created by GUTs and supersymmetry can also be solved in the same framework.

What I have not discussed up until now is the value of the cosmological constant. Herein lies a remaining cosmological problem, perhaps even more severe than previous ones. Why is the cosmological constant so small? Put in dimensionless form, we know that in order that the expansion today is not dominated by Λ we must have

$$\frac{\Lambda}{M_p^2} \lesssim 10^{-121} \; ! \qquad (5.1)$$

Surely an astounding piece of fine tuning if there ever was one. The trouble is we cannot even say that this was an initial condition which stayed there. Each vacuum phase transition leaves a change in the value of Λ. N = 8 supergravity[62] generally begins with $\Lambda \sim -M_p^2$ (an anti-De Sitter space). This very large negative somehow must have become very positive just around the Planck time (if we want to keep the inflationary scenario). Whether or not this change was due to the breakdown of the higher supersymmetries by the formation of condensates is as yet unknown. In the present model, the inflationary epoch begins with $\Lambda/M_p^2 \sim 10^{-12}$ (a long way away from its present value). At the GUT phase transition Λ would have been down already to $\Lambda/M_p^2 \sim 10^{-40}$. The next change might have been at the SU(2) $\times$ U(1) transition which saw $\Lambda/M_p^2 \sim 10^{-68}$ followed perhaps by the quark hadron transition which left an effect, $\Lambda/M_p^2 \sim 10^{-80}$ until somehow it arrived at its present value (5.1). This can not be an accident. Is there a symmetry which guarantees a low value today? This question remains a still unanswered goal for cosmology.

Apart from the cosmological constant, (which is tuned here to be zero today), the model presented appears to be devoid of problems involving small parameters or fine tunings. There have, of course, been other attempts at supersymmetric inflationary[63] and GUT scenarios[64], most notably, those which are based on the reverse hierarchy scheme[65]. These scenarios have many features in common with this one (e.g. flat potentials) but seem to fail to produce an acceptable picture after inflation. General difficulties with this type of model will be discussed elsewhere[66]. The main features introduced in the present model are i) at high enough energies ($\gg M_S$), supersymmetry appears to be exact and we may do away with radiative corrections; and ii) the extensive use of non-renormalizable terms leading to effective interactions at scales lower than M_p. If supergravity is on the right track for unifying gravity with standard gauge interactions, then it should not seem surprising that a variety of small parameters previously not understood, stem from these non-renormalizable terms which scale with inverse powers of M_p. One can only be optimistic, that the so-far rapid success of applying supergravity to GUTs and cosmology is just beginning.

Acknowledgements

I would like to deeply thank J. Ellis, D.V. Nanopoulos, M. Srednicki and K. Tamvakis for a very enjoyable and productive collaboration. I would also like to thank G. Gelmini and M. Quiros for useful discussions.

REFERENCES

1) For reviews, see: J. Ellis, CERN preprint TH.3174 (1981), to be published in Proc. 1981 Les Houches Summer School;
   P. Langacker, Phys. Rep. 72C (1981) 185;
   D.V. Nanopoulos, in *Theory of Fundamental Interactions*, ed. G. Costa and R. Gatto, Proc. of the Int. School of Physics Enrico Fermi LXXXI (Soc. Italiana di Fisica, Bologna) 1982, p. 156.

2) A.D. Sakharov, Zh. Eksp. Teor. Fiz. Pisma Red. 5, (1967) 32.

3) K.A. Olive, D.N. Schramm, G. Steigman, M.S. Turner and J. Yang, Ap. J. 246 (1981) 557;
   J. Yang, M.S. Turner, G. Steigman, D.N. Schramm and K.A. Olive, Chicago EFI preprint (1983), *Primordial Nucleosynthesis: A critical comparison between theory and experiment.*

4) G.'t Hooft, Nucl. Phys. B79 (1974) 276;
   A.M. Polyakov, Zh. Eksp. Theor. Fiz. Pisma Red. 20 (1974) 194.

5) Ya.B. Zel'dovich and M.Y. Khlopov, Phys. Lett. 79B (1979) 239;
   J.P. Preskill, Phys. Rev. Lett. 43 (1979) 1365.

6) A.H. Guth and S.H. Tye, Phys. Rev. Lett. 44 (1980) 631.

7) A.H. Guth, Phys. Rev. D23 (1981) 347.

8) J. Ellis and K.A. Olive, CERN preprint TH.3399, to be published in Nature (1983).

9) D. Kazanas, Ap. J. 241 (1980) L59.

10) S. Coleman, Phys. Rev. D15 (1977) 2929;
    C. Callan and S. Coleman, Phys. Rev. D16 (1977) 1762.

11) A.H. Guth and E. Weinberg, Phys. Rev. D23 (1981) 826, Nucl. Phys. B212 (1983) 321.

12) A.D. Linde, Phys. Lett. 108B (1982) 389.

13) A. Albrecht and P.J. Steinhardt, Phys. Rev. Lett. 48 (1982) 1220.

14)   S. Coleman and E. Weinberg, Phys. Rev. D7 (1973) 1888.

15)   L. Abbott, Nucl. Phys. B185 (1981) 233;
      M. Sher, Phys. Rev. D24 (1981) 1699;
      P. Hut and F. Klinkhamer, Phys. Lett. 104B (1981) 439;
      A. Billoire and K. Tamvakis, Nucl. Phys. B200 [FS4] (1982) 329;
      K. Tamvakis and C.E. Vayonakis, Phys. Lett. 109B (1982) 283;
      G.P. Cook and K.T. Mahanthappa, Phys. Rev. D23 (1981) 1321 and
          D25 (1982) 1154;
      M.D. Pollock and M. Calvani, Phys. Lett. 117B (1982) 392.

16)   A. Vilenkin, Phys. Lett. 115B (1982) 91;
      A. Vilenkin and L.H. Ford, Phys. Rev. D26 (1982) 1231.

17)   A.D. Linde, Phys. Lett. 116B (1982) 335.

18)   S.W. Hawking and I.G. Moss, Phys. Lett. 110B (1982) 35;
      See also S.W. Hawking, Cambridge-DAMTP preprint *Euclidean*
      approach to the inflationary universe (1983).

19)   C.-G. Källman, private communication (1982).

20)   J. Ellis, D.V. Nanopoulos, K.A. Olive and K. Tamvakis, Phys.
      Lett. 118B (1982) 335.

21)   P.J. Steinhardt, Univ. of Pennsylvania preprint UPR-0192T
      (1982).

22)   S. Parke, Phys. Lett. 121B (1983) 313.

23)   S.W. Hawking, Phys. Lett. 115B (1982) 295;
      See also S.W. Hawking and I.G. Moss, Cambridge-DAMTP preprint
      Fluctuations in the inflationary universe (1982).

24)   A.H. Guth and S.-Y. Pi, Phys. Rev. Lett. 49 (1982) 1110.

25)   A.A. Starobinski, Phys. Lett. 117B (1982) 175.

26)   J.M. Bardeen, P.J. Steinhardt and M.S. Turner, Chicago EFI
      preprint 83-13 (1983).

27)   W.H. Press, Phys. Scr. 21 (1980) 702.

28)   E.R. Harrison, Phys. Rev. D1 (1970) 2726;
      Ya.B. Zel'dovich, Mon. Not. R. Astron. Soc. 160 (1972) 1P.

29)   Ya.A. Gol'fand and E.P. Likhtman, Pisma Zh. Eksp. Theor. Fiz.
          13 (1971) 323;
      D. Volkov and V.P. Akulov, Phys. Lett. 46B (1973) 109;
      J. Wess and B. Zumino, Nucl. Phys. B70 (1974) 39;

For a review see P. Fayet and S. Ferrara, Phys. Rep. 32C (1977) 249.

30) For a review, see D.V. Nanopoulos, *Grand Unification and physical supersymmetry*, Lectures given at the XIIIth GIFT International Seminar on Theoretical Physics, Girona, Spain 1983, ed. J.A. Grifols, A. Mendez and A. Ferrando (World Scientific Pub. Co., Singapore 1983).

31) For a review, see P. van Nieuwenhuizen, Phys. Rep. 68C (1981) 189.

32) E. Cremmer, B. Julia, J. Scherk, S. Ferrara, L. Girardello and P. van Nieuwenhuizen, Nucl. Phys. B147 (1979) 105;
E. Cremmer, S. Ferrara, L. Girardello, and A. Van Proeyen, Phys. Lett. 116B (1982) 231 and Nucl. Phys. B212 (1983) 413.

33) L. Maiani, *in* Proc. of the Summer School of Gif-sur-Yvette (1979) p. 3.
E. Witten, Nucl. Phys. B188 (1981) 513;
S. Dimopoulos and H. Georgi, Nucl. Phys. B193 (1981) 150;
N. Sakai, Zeit. für Phys. C11 (1982) 153.

34) J. Wess and B. Zumino, Phys. Lett. 49B (1974) 52;
J. Iliopoulos and B. Zumino, Nucl. Phys. B76 (1974) 310;
S. Ferrara, J. Iliopoulos and B. Zumino, Nucl. Phys. B77 (1974) 413;
M.T. Grisaru, W. Siegel and M. Rocek, Nucl. Phys. B159 (1979) 420.

35) J. Ellis, S. Ferrara and D.V. Nanopoulos, Phys. Lett. 114B (1982) 231.

36) D.V. Volkov and V.A. Soroka, JETP Lett. 18 (1973) 312.

37) S. Deser and B. Zumino, Phys. Rev. Lett. 38 (1977) 1433.

38) J. Ellis and D.V. Nanopoulos, Phys. Lett. 116B (1982) 133.

39) R. Barbieri, S. Ferrara and D.V. Nanopoulos, Z. Phys. C13 (1982) 267; Phys. Lett. 116B (1982) 16.

40) J. Ellis, L. Ibanez and G.G. Ross, Phys. Lett. 113B (1982) 983, CERN preprint TH.3382 (1982).
M. Dine and W. Fischler, Nucl. Phys. B204 (1982) 346.

41) J. Ellis, D.V. Nanopoulos, K.A. Olive and K. Tamvakis, CERN preprint TH.3404, Nucl. Phys. B in press (1983).

42) G. Gibbons and S.W. Hawking, Phys. Rev. D15 (1977) 2738.

43) R. Brandenberger and R. Kahn, Phys. Lett. 119B (1982) 75.

44) J. Ellis, D.V. Nanopoulos, K.A. Olive and K. Tamvakis, Phys.
     Lett. 120B (1983) 331.

45) J. Ellis, D.V. Nanopoulos and K. Tamvakis, Phys. Lett. 121B
     (1983) 123;
     J. Ellis, J. Hagelin, D.V. Nanopoulos and K. Tamvakis, SLAC
     preprint PUB-3042 (1983).

46) D.V. Nanopoulos, K.A. Olive, M. Srednicki and K. Tamvakis,
     Phys. Lett. 123B (1983) 41.

47) B. Zumino, Nucl. Phys. B89 (1975) 535.

48) M. Srednicki, Nucl. Phys. B202 (1982) 327;
     D.V. Nanopoulos and K. Tamvakis, Phys. Lett. 110B (1982) 449.

49) L. Girardello, M. Grisaru and P. Salomonson, Nucl. Phys. B178
     (1981) 331.

50) J. Ellis, C.H. Llewellyn Smith and G.G. Ross, Phys. Lett. 114B
     (1982) 227.

51) D.V. Nanopoulos, K.A. Olive and K. Tamvakis, Phys. Lett. 115B
     (1982) 15.

52) D.V. Nanopoulos and K. Tamvakis, Phys. Lett. 114B (1982) 235.

53) M. Srednicki, Nucl. Phys. B206 (1982) 132.

54) D.V. Nanopoulos, K.A. Olive, M. Srednicki and K. Tamvakis,
     Phys. Lett. 124B (1983) 171.

55) E.W. Kolb and S. Wolfram, Nucl. Phys. B172 (1980) 224, Phys.
     Lett. 91B (1980) 217.

56) J.N. Fry, K.A. Olive and M.S. Turner, Phys. Rev. D22 (1980)
     2953 and 2977; Phys. Rev. Lett. 45 (1980) 2074.

57) J.N. Fry and M.S. Turner, Chicago EFI preprint 83-10 (1983).

58) D.V. Nanopoulos, K.A. Olive and M. Srednicki, CERN preprint
     TH.3555 (1983).

59) C. Kounnas, D.V. Nanopoulos, M. Quiros and M. Srednicki, CERN
     preprint TH.3559 (1983).

60) S. Weinberg, Phys. Rev. Lett. 48 (1982) 1303.

61) J. Ellis, A.D. Linde and D.V. Nanopoulos, Phys. Lett. 118B
    (1982) 59.

62) B. De Wit and H. Nicolai, Nucl. Phys. B188 (1981) 98; Nucl.
    Phys. B208 (1982) 323.

63) A. Albrecht, S. Dimopoulos, W. Fischler, E. Kolb, S. Raby and
    P. Steinhardt, in preparation (1983);
    C.E. Vayonakis, Phys. Lett. 123B (1983) 396.

64) S. Dimopoulos and S. Raby, Los Alamos preprint LA-UR-82 (1982)

65) E. Witten, Phys. Lett. 105B (1981) 267.

66) J. Ellis, D.V. Nanopoulos, K.A. Olive, M. Srednicki and
    K. Tamvakis, in preparation (1983).

REMARKS ON THE QUARK-HADRON TRANSITION IN THE EARLY UNIVERSE

Rémi Hakim[*] and Suzy Collin[**]

Groupe d'Astrophysique Relativiste[*] and
Laboratoire d'Astrophysique[**]
Observatoire de Meudon
92195 Meudon Principal Cedex, France

1. INTRODUCTION
2. DYNAMICS OF THE QUARK/GLUON SYSTEM
3. STATISTICAL DESCRIPTION OF THE QUARK PLASMA
4. FLUCTUATIONS
5. RESULTS AND CONCLUSION

1. INTRODUCTION

These last years fascinating speculations on the consequences of
phase transitions, within the context of Grand Unified Theories,
in the early universe have been set forth and "predict" numerous
phenomena which might occur at temperatures as high as 10^{15}GeV...
a temperature far beyond present experimental possibilities.
Moreover, although Grand Unified Theories do represent a natural
generalization of the presently admitted theories (essentially,
the Weinberg-Salam model and Quantum Chromodynamics) they unify
theories whose experimental basis is not as rich as it should be
(think to the Higgs'particles, for instance), whose theoretical
content is far from being understood (e.g. the basic problem of
quark confinement is as yet unsolved) ; where calculations are
quite difficult to work out, owing to an intrinsic non-linearity,
and hence to be compared with high energy experiments ; and, finally, where approximations are not undercontrol.

However, at a less - slightly less - speculative level, one
may consider the state of matter at densities and/or temperatures
more in accordance with what is known at our laboratory scale :

29

J. Audouze and J. Tran Thanh Van (eds.),
Formation and Evolution of Galaxies and Large Structures in the Universe, 29–53.
© 1984 by D. Reidel Publishing Company.

in the early universe this would involve temperatures ranging from
$\sim$ .1 GeV to $\sim$ 100 GeV.

At these temperatures and/or densities the hadrons become
gradually close-packed (fig.1) and next transform into a quark
soup or, in more elegant words into a hot quark plasma.

This transition from hadrons to quarks (i.e. going back in
time) is generally assumed to be a phase transition and more specifically a first order phase transition (fig.1). Needless to say
that it could well be that there does not exist any such phase
transition whatsoever : think, for instance, to the analogy with
the ionization process of a neutral gas ; all thermodynamic quantities vary continuously from the atomic state ($\sim$ hadrons) to the
plasma state ($\sim$ quarks) and there is no phase transition at all.

It should be clear that, in order to get a definite and convincing answer to this question, the basic theoretical problems of
Quantum Chromodynamics must be solved. Therefore, we have to assume
the existence of such a first order phase transition.

The usual way to handle such a transition consists in calculating the equation of state of the quark plasma, the equation of
state of the hadron phase (with another set of theoretical assumptions) and finally link both curves via a Maxwell plateau (equality of pressures and chemical potentials in both phases ; see
fig.2).

The essential interest of phase transitions in the early universe - but not the only one - lies in the fact that they are associated with violent fluctuations which might subsequently be at the
origin of galaxy clusters or, more modestly, of galaxies, or even
more modestly, of stars beolonging to the so-called "population III"
or, at least, "seeds" for an eventual creation of galaxies (in the
latter case, the mechanism for such a creation remains to be found ;
see e.g. (1).

It follows that two main problems are to be considered. The
first one deals both with the critical temperature at which the
transition occurs (or, equivalently, at what time it occurs) and
also with is duration. It is clear, indeed, that one cannot create
objects more massive than the mass contained inside the horizon :
hence, the size of the horizon and the energy density at the time
of the transition play a basic role. Furthermore, the duration of
the transition determines in part the possibilty for the fluctuations to grow sufficiently. The second problem deals with the
spectrum of the fluctuations, a quantity generally put by hand at
the onset of galaxy creation models.

To these two main problems we should also add the question of

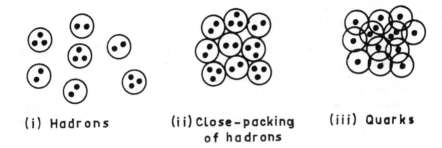

(i) Hadrons   (ii) Close-packing   (iii) Quarks
                    of hadrons

Figure 1 : Various densities of the hadron fluid.

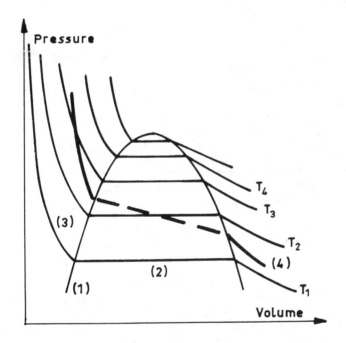

Figure 2 : Typical phase diagram for a first order phase transition;
curve (1) provides the transition region while curves
(2) represent the usual Maxwell plateau; curve (3) corresponds to the quark phase; curve (4) represents a
typical thermodynamical path of the early universe. The
intersection of curves (1) and (4) is the critical temperature we are looking for.

the <u>nucleation</u> of droplets of hadrons (possibly by "impurities" ;
see e.g. (in another context) (27) in the quark plasma during the
phase transition and also their dynamics.

Surprisingly enough the number of articles devoted to quarks
in the early universe is quite limited, and particularly those
which are connected with the quark/hadron phase transition. They
deal either with a zero-temperature universe (3), (4)[1] or with
the standard model (5) and, in the latter case the quarks may be
partially confined (6), (7), (8), (9) and (10) or fully confined
(11), (12), (13), (14) and (15).

In this talk, all these questions will be considered and discussed in a phenomenological way, the only reasonable one - in our
opinion - when taking into account the present status of the theory.
More specifically, we shall be concerned essentially with Olive's
approach.

2. DYNAMICS OF QUARK SYSTEMS

Quantum chromodynamics is generally considered as being <u>the</u> correct
theory of strong interactions and, as a matter of fact, <u>it</u> is quite
consistent with present day experiments. However, owing to its intrinsic non-linearity calculations are very difficult to perform.
Also, its vacuum structure is not yet understood and, more specifically, the confinement mechanisms of quarks and gluons are still
unknown. Thence the thermodynamics of the quark/hadron plasma cannot be obtained directly from Quantum Chromodynamics as long as its
theoretical problems remain unsolved.

Of course, preliminary calculations performed within perturbative expansions[2] have been carried out either at T=0°K (17), (18),
(19), or at finite temperatures (20) as to the <u>quark phase</u>. At T=0°K
the calculations have been pushed to the fourth-order in the coupling constant while at T ≠ 0°K only the third order has been obtained. Unfortunately, these last results are not particularly conclusive the more so since the various terms of the perturbative expansion are of the same order of magnitude, a circumstance that
casts some doubts on the convergence of the expansion. Moreover, in
the absence of an admitted confinement mechanism, the hadron phase
cannot be properly described within this perturbative frame-work.

Let us add that non-perturbative effects have tentatively been
considered (21) but the matter is still in a controversial state.

Consequently, if one wants deriving some results in view of
astrophysiqcal situations - neutron stars, quark stars, quark era
in the early universe - one must resort to some tractable <u>phenomeno-</u>
<u>logical</u> model. Such a model should contain both theoretical ingre-

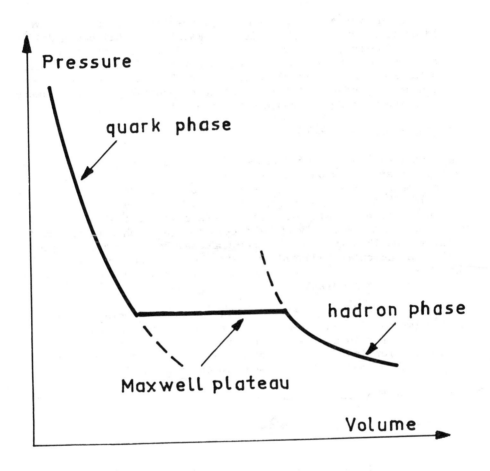

Figure 3 : The quark equation of state and the hadron one, calculated with different theories, are linked by a Maxwell plateau (equality of pressures and of chemical potentials).

dients from Quantum Chromodynamics and also some experimental input
such as, for instance, a description of confinement.

Among the various phenomenological possibilities at hand, we
should mention various bag models and, mainly, the MIT bag model
(22). Nevertheless, they are generally as difficult to handle as
the original quantum chromodynamics. However, and in spite of some
attempts (23) it is extremely difficult to evaluate the density and
temperature dependence of the bag because, precisely, of the absence
of a convincing model of quark confinement !

2.1. The quark-quark potential

The dynamical models considered - at a phenomenological level - in
this paper are based on the use of an ad hoc potential V(r). Indeed,
the hadron spectroscopy is quite reasonably well described with a
potential (except the pion mass, of course) inserted in a Schrödinger
- like equation. Usually, the parameters of the potential are fitted
so as obtain the ψ-family and the result is next checked on the
Y-family and then applied to other hadrons. In this way, the phenomenological potential adopted is tested between roughly .1 fm and 1fm.
Numerous potentials can be found in the litterature and some of them
are listed in Appendix A.

These phenomenological potentials generally contain essentially two terms : one of them is supposed to take account of quark
confinement (it is an ever increasing function of the radial coordinate r ; for instance a power law) while the second one may correspond to the one- gluon exchange in the static limit.

One of the most popular choice is

$$U(r_{12}) = - \vec{\lambda}_1 \cdot \vec{\lambda}_2 \ \{ - \frac{a}{r_{12}} + br_{12}\} \tag{1}$$

where a and b are positive constants to be fitted with the use of
the charmonium spectrum and where the eight matrices $\vec{\lambda}_i$ are the
well-known Gell-Mann matrices (the index i, in $\vec{\lambda}_i$ or in r_{12}, refers
to the i-th quark). It is sometimes called the "QCD potential" since its
Coulombian part corresponds to the one gluon exchange (in this case,
a is the QCD fine structure constant) while the linearly rising term
- the confining potential - seems to be suggested by lattice calculations (or simulations).

It should be noticed, however, that the confining part of this
potential (or of other ones) is a little bit pathological. First,
the $\vec{\lambda}_1 \cdot \vec{\lambda}_2$ matrix has opposite signs when the quarks 1 and 2 are
either in a singlet or an octet state, resulting in the absence of
a zero of energy in this last case (i.e. when the coefficient of r
is negative). Perhaps more important is the fact that, in the singlet

state case, arbitrarily high energies can be reached resulting in pair creations and hence in a softening of the potential, not so simple to handle. A supplementary drawback of the confining term is that it leads to unobserved van der Waals - like forces between hadrons (24). For instance, a linearly rising potential would lead to long range forces in $1/r$ in nuclear matter.

Consequently, we should be cautious while using such potentials and specially at low densities.

An improvement of the potential (1), whose Fourier transform is

$$\tilde{U}(k) = - \vec{\lambda}_1 \cdot \vec{\lambda}_2 \{ \frac{a}{k^2} + \frac{bc}{k^4} \} \qquad (2)$$

(c being a numerical constant), has been performed by Richardson (25) (see also (26)) who replaced the QCD fine structure constant a by the effective one obtained from the renormalization group, i.e. by

$$a \to \alpha_{eff}(k^2) \sim \frac{12\pi}{33-2n_f} \frac{1}{\ell n(k^2/\Lambda^2)} \qquad (3)$$

where n_f is the number of quark flavours that come into play at momentum k and where Λ is a scale parameter to be determined by experiment (and this is not an easy task...). This procedure has the advantage of embodying the important property of asymptotic freedom, an important theoretical and phenomenological ingredient. Finally, a simple interpolation between the two regimes (low k and large k) leads to

$$\tilde{U}(k) = - \frac{4}{3} \cdot \frac{12\pi}{27} \cdot \frac{1}{k^2} \cdot \frac{1}{\ell n(1+k^2/\Lambda^2)} \quad (n_f = 3) \qquad (4)$$

Such a potential has been used by Schöberl (27) to obtain the hadron spectroscopy.

Once a potential is chosen and its free parameters fitted with high energy data, one assumes that it is still as good for quark-quark interactions (apart from unessential factors $\vec{\lambda}_1 \cdot \vec{\lambda}_2$) whatever their flavor (the c and b quarks used to fit the parameters are heavy).

2.2. Quark masses

The next dynamical question to be discussed is the one of the quark masses. Which one should be introduced in the Schrödinger equation? Is it the constituent mass (most authors)? the current mass ((11), (12), (15), etc...)? the running mass (8)? In fact there is no

clear answer and the problem is quite delicate and complex. This
is the reason why, in Appendix A, the potentials given are generally accompanied with the masses used in the fitting of the parameters.

In our opinion, in order to be consistent with e.g. the QCD
potential (4) - where results arising from the group renormalization equations have been inserted - the running mass should be
used, perhaps in such a way as to reduce to the current mass at
high momenta and to the constituent mass at low momenta.

In fact, the renormalization group equations (see e.g. (28))
provide such a running mass ; unfortunately, they provide asymptotic
forms only (exactly as for the effective coupling constant) and,
moreover, they have two solutions. One of them is

$$m(p) \sim \frac{1}{\ell n^{\gamma}(p^2/\Lambda^2)} \tag{5}$$

while the other one, reads

$$m(p) \sim \frac{4m_o^3}{p^2} \ell n^{\gamma}(p^2/\Lambda^2) \tag{6}$$

(with $m_o \sim 300$ MeV) where

$$\gamma = \frac{12}{33-2n_f} \tag{7}$$

Theoretical arguments based on a possible spontaneous breakdown of
chiral symmetry led H. Pagels et al. (29) to favour the second
solution (6) which they parametrize as

$$m(p) = \frac{m(o)\Lambda^2}{\Lambda^2+p^2} \tag{8}$$

The fact that the renormalization group equations provide asymptotic
forms only renders extrapolations to low momenta somewhat doubtful,
owing to the fact that this corresponds to the non-perturbative
region where our present knowledge fails. However, near $p \sim o$ (or
a few hundred MeV) and at a phenomenological level, $m(p)$ is practically constant, at least as a kind of average (see fig.3). This
property is valid for the range of energies where the predicted
spectroscopy of hadrons is correctly described. For instance, for
the Richardson's potential (4), the constituent mass appears to be
a valid approximation for a few hundreds MeV.

To these considerations let us also add that non-perturbative
effects, such as instantons, lead to other forms for $m(p)$; for

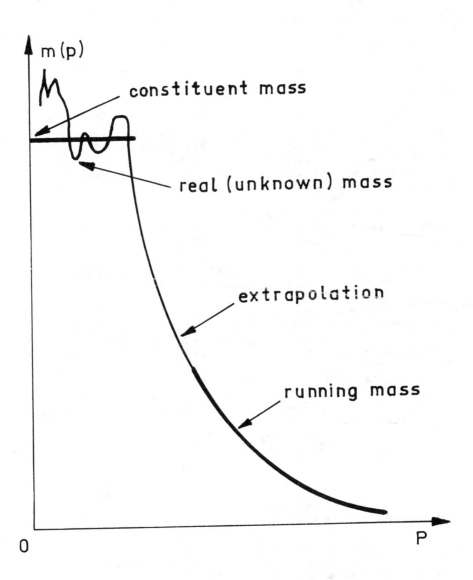

Figure 4 : Various quark masses to be (possibly) used in quark
phase calculations.

instance, the use of instantons (21) provides

$$m(p) \sim 1/p^{12}. \tag{9}$$

2.3. Discussion and possible improvements

We first notice that most of the potentials used in the litterature coincide more or less between . 1fm and 1 fm, i.e. where they
are actually tested ; more specifically, in the region where the
potential is approximately linear. This might perhaps be an indication that at low densities such a phenomenology of dense matter
could possibly contain some truth ...

However, at higher densities - of the order of $\sim (1/.2\text{fm})^3$ -
where the various potentials deviate from each other, different
results can be obtained.

On the other hand, this approach suffers from a number of draw-
backs which may be more or less cured.

(1) At high densities or equivalently, at short distances or at
high momenta, relativistic effects should be taken into account and
thus a Schrödinger equation approach is not adequate : a static
approximation has to be corrected. This can be done either by
taking account of relativistic corrections (30) or writing a two-
body Dirac equation for semi-relativistic quarks (31). Among many
other possibilities, it is also possible to write a Bethe-Salpeter
equation (in the ladder approximation) where the gluon propagator
is replaced by

$$= \frac{\alpha}{k^2} + \frac{bc}{k^4} \tag{10}$$

(where $k^2 \equiv \omega^2 - \vec{k}^2$, in equation (10)) and where the vertex is replaced by an "effective" one, i.e. where α is in fact $\alpha_{eff}(k^2)$ as
given by Eq. (3) above.

(2) As important as the preceding point is a possible dependence of various physical quantities, α_{eff}, $V(r)$, $m(k)$, etc... on
density and temperature.

For instance, it has been shown (32) that

$$\alpha_{eff}(k,T) \sim \frac{\alpha(0,0)}{1 + \frac{33}{48\pi^2} \alpha \ \ell n(k^2/\Lambda^2) + \frac{m^2[T]}{k^2}} \tag{11}$$

where $m^2[T] \sim T^2$; similarly, the same authors have calculated in
a non-perturbative way (as a self consistent solution of the Dyson-

Schwinger equations) the static limit of the gluon propagator : a simple Fourier transform of the latter quantity gives the potential and its density/temperature dependence.

Another attempt by Zhao Wanyun (33), within the context of perturbation theory, provides

$$V(r,T) = \alpha^{5/2} \cdot \frac{3}{32\pi^3} \cdot \frac{1}{T^2 r^2} \exp. [-2m[T].r] \qquad (12)$$

However, these attempts and several others are not yet completely conclusive.

(3) Another important problem linked with the passage from the non-relativistic case to the relativistic one, is the following. When one considers the Coulomb - like part of the potential (1), for instance, it should be considered as the fourth-component of the gluon field (it corresponds to the static limit of the one-gluon exchange term of a perturbative expansion).

On the other hand, it is well known that either a Lorentz - scalar potential or a four-vector potential lead to the same kind of non relativistic limit. Consequently, if the potential is to be considered as the fourth component of the gluon field A^μ, it must enter the relativistic equation under consideration through the gauge invariant combination

$$p^\mu \rightarrow p^\mu + ig \vec{\lambda} . \vec{A}^\mu \qquad (13)$$

while if it is a scalar, it must appear in the mass term as

$$m \rightarrow m + gV, \qquad (14)$$

assuming as usual a Yukawa - like coupling with the quark field.

To these two possibilities, one should add a third one, which is intermediate : the confining potential might be a Lorentz-scalar to be inserted in the quark mass as indicated in Eq. (14) while the "interacting" part - the gluonic one - should be considered as the fourth component of a four-vector.

These three cases may be summarized in the following three expressions for the relativistic energy :

$$(i) \quad \xi (\vec{p}) = \{\vec{p} + m^2(\vec{p})\}^{1/2} + V(r) \qquad (15)$$

$$(V \equiv A^\circ)$$

$$(ii) \quad \xi (\vec{p}) = \{\vec{p}^2 + [m(\vec{p}) + V(r)]^2\}^{1/2} \qquad (16)$$

(V = Lorentz scalar)

(iii) $\xi(\vec{p}) = \{\vec{p}^2 + m(\vec{p}) + V_{conf.}(r)]^2\}^{1/2} + V_{gluon}(r)$ (17)

(intermediate case).

As to this last point - considered by Boal et al. (15) - it
should be noticed that a scalar field is introduced ab initio in
the SLAC bag model (34) or in a simplified version (35) and that its
treatment is certainly simplier than the use of the hamiltonians (16)
or (17). However, despite the pseudo-Thomas-Fermi calculation of
Boal et al. (15), it leads to partial confinement only (36), (37) ;
see also (38) : quarks appear to be heavy at low densities and light
at higher ones. In any case, if a scalar field is introduced so as
to mimic confinement, it should be considered as the result of a
complex process (interactions with scalar gluonic modes ?) steming
from Quantum Chromodynamics.

3. STATISTICAL DESCRIPTION OF THE QUARK PLASMA

If the dynamical approach considered above were completely correct
and trustworthy, we should eventually find the thermodynamical
parameters that characterize the (possible) phase transition from
hadrons to quarks. This would demand a proper treatment of two-
body and three-body correlations. Usually two-body correlations can
be handled although with technical difficulties while three-body
correlations can be treated - in a non quantum framework - only in
quite particular cases. Needless to say that this is an almost impossible task in this context. However, one might perhaps think of
these three-body correlations as already included, at least by part,
in the phenomenological confining potential ? Furthermore, while
three-body forces appear to play an extremely small role in hadron
spectroscopy, at the temperatures where the transition is supposed
to take place ($\sim$ 280 MeV) only few baryons ($\sim$ 3% according to
Olive's calculations (11) and (12) are produced.

    Therefore we are led - as other - to deal with the simplest
tractable statistical approximations ; essentially mean field approximations such as Thomas-Fermi's, Hartree's or Hartree-Fock (see
e.g. (39)) which are quite easy to deal with. Fortunately, these
approximations are generally valid at high densities (in the Quantum
Chromodynamic case, the validity of the Hartree approximation at
high densities has been shown by E. Alvarez (40), leading thereby to
a more or less correct description of the quark phase. As the density
is lowered, this kind of approximation is less and less valid especially in the physically interesting region of the quark-hadron
phase transition (assumed to exist, of course).

It follows that the hadron phase presumably cannot be described from these quark dynamics and statistical approximations and another model must be used. This is at least as difficult as for the quark phase (!) so that we shall use, for the sake of comparison, the very simple model considered by Olive (11), (12). (Thomas-Fermi approximation for hadrons (mainly π-mesons) interacting via the exchange of various particles (ω-mesons as to the baryons)).

In this paper, where mean field approximations are used, a quark (or an antiquark) is moving freely in the average potential of the other quarks and antiquarks. As a consequence, one can get rid off all the λ-matrices algebra since

$$< U(r) > \equiv < \vec{\lambda}_1 \cdot \vec{\lambda}_2 \, V(r) > \tag{18}$$

simply reduces to the average $q\text{-}\bar{q}$ potential, owing to the fact that the system as a whole is in a color singlet state.

3.1. Statistical description of the quarks

Let us first consider the statistical state of the quarks, the gluons being dealt with later on. There are only five known flavors (u, d, s, c, b) and the corresponding quarks differ only by their masses (at the temperatures considered, only the u, d, s quarks play a role since $m_u \sim m_d \sim 350$ MeV and $m_s \sim 550$ MeV, while $m_c \sim 1.5$ GeV and $m_b \sim 4.5$ GeV).

Let us consider the species i ; its average occupation number is

$$< n_i(p) > = \frac{1}{\exp.[\beta(\xi_i(p)-\mu)] +1} \, , \tag{19}$$

where $\beta \equiv (k_B T)^{-1}$; where μ is the chemical potential, which we take to be zero in the primordial universe, due to the fact that, at high temperatures, the quarks coming from the $B \neq 0$ matter constitute only a small "impurity" compared with the ones produced by the background blackbody radiation. In Eq. (19) $\xi_i(p)$ is the excitation spectrum of the quasi-quarks within the medium.

(1) In the Thomas-Fermi approximation, first considered by Olive in this context, it reads

$$\xi_i(p) = \{\vec{p}^2 + m^2(\vec{p})\}^{1/2} + V(d), \; i = 1,2,\ldots,5 \tag{20}$$

where d is the average interquark distance, given by

$$\frac{4\pi}{3} d^3 = n_Q^{-1} \tag{21}$$

where n_Q is the total quark density, i.e. including both all species, all colors and also antiquarks :

$$n_Q = \sum_{i=1}^{i=5} n_i \tag{22}$$

with

$$n_i = g_Q \ d^3p \ < n_i(p) > , \tag{23}$$

where g_Q is the degeneracy factor

$$g_Q = 2 \ (\text{particle/antiparticle}) \times 2 \ (\text{spin}) \times 3 \ (\text{color}) \tag{24}$$
$$= 12.$$

Here two points are worth mentioning.

First and unlike what was done by (11) and (12), d should not necessarily be given by Eq. (21). For instance, at extremely high temperatures, the quark density is simply proportional to the photon density (there are only free particles in the medium and each of them contribute with the number of degrees of freedom they carry), i.e. by

$$n_Q \sim [g_Q/16] \ . \ n_\gamma , \tag{25}$$

where the photon density is given by

$$n_\gamma = a \ T^3 \tag{26}$$

(a being the Stefan constant), so that we could take d as being

$$d = \left(\frac{12}{g_Q a\pi} \right)^{1/3} \ \frac{1}{T} \ . \tag{27}$$

In fact, Eqs. (21) and (27) represent two extreme cases : the low temperature case and the high temperature one, respectively. In fact, the correct screening factor d should be calculated from a study of the proper oscillation modes of the quark plasma. This is evaluated below.

Next, in Eq. (22) or (25) only quark densities come into play: this is due to the fact that a given quark is sensible only to color irrespective of the species considered in the interaction. However, this has been criticized by Boal et al. (15) who argued that gluon contributions (and hence gluon density) should be included in the evaluation of d. Their argument essentially amounts to saying that since quarks feel colored states, gluons should necessarily be

taken into account. In fact, this argument rests also on their statistical treatment of gluons. Although it is perfectly right to
argue that way, we shall take the opposite view for consistency
reasons which we explain below.

Finally, Eqs. (19), (20), (22) and (23) lead to the following
self-consistent equation

$$n_Q = \sum_{i=1}^{i=5} \frac{g_Q}{2\pi^2} \quad dp \, \frac{p^2}{\exp.[\beta \xi_i(p)] + 1} \, , \tag{28}$$

where a priori $\xi_i(p)$ depends on n_Q through d. Its solution provides
the total quark density as a function of β which, once inserted into
$< n_i(p) >$, allows the determination of the various thermodynamical
quantities, such as the energy density

$$\rho = \sum_{i=1}^{i=5} \rho_i \tag{29}$$

with

$$\rho_i = \frac{g_Q}{2\pi^2} \int dp \, \frac{p^2.[p^2 + m_i^2(p)]^{1/2}}{\exp.[\beta \xi_i(p)] + 1} + \frac{1}{2} n_i \, V(d) \tag{30}$$

where the factor 1/2 occuring in the last term is due to the need
of avoiding a double counting of the interaction (note that in the
mean field approximations considered in this paper, it appears in
a natural manner). Similarly, the pressure is obtained as

$$P = \sum_{i=1}^{i=5} P_i \tag{31}$$

where

$$P_i = \frac{1}{3} < \vec{v} . \vec{p} > \tag{32}$$

the Hamiltonian used shows that

$$\vec{v} = \frac{\partial H}{\partial \vec{p}} \tag{33}$$

$$\vec{v} = \frac{\partial}{\partial \vec{p}} [\vec{p}^2 + m_i^2(\vec{p})]^{1/2} \, , \tag{34}$$

which, in the case where $m_i(p) = $ const., reduces as usual to

$$\vec{v} = \vec{p}/[\vec{p}^2 + m_i^2]^{1/2} \tag{35}$$

or

$$P_i = \frac{g_Q}{6\pi^2} \int dp \; \frac{p^3 (\partial/\partial p)[p^2 + m_i^2(p)]^{1/2}}{\exp. \, [\beta \, \xi_i(p)] + 1} \tag{36}$$

Finally the whole thermodynamics of the <u>quarks</u> is completly determined in this approximation and it remains to evaluate the contributions of the gluons.

(2) In the true Hartree approximation (see e.g. (39)), the average potential is given by

$$< V(r) > = \int d^3\vec{r}' \; n_Q(\vec{r}') \cdot V(\vec{r}-\vec{r}') \quad , \tag{37}$$

which reduces to

$$< V(r) > = n_Q \int d^3 r \; V(r) \tag{38}$$

in the case of a homogeneous thermal equilibrium, as considered here. In this case, the quasi-quark excitation spectrum is simply

$$\xi_i(p) = [p^2 + m_i^2(p)]^{1/2} + 4\pi \, n_Q \int dr \; r^2 \; V(r) \tag{39}$$

In fact, the integral in this last equation does not extend to infinity but to the <u>screening length</u> d to be taken either from Eq. (21) as Olive or from Eq. (27) as Wagnoner <u>et al</u>. (7) or rather in an intermediate way as is done below.

When one takes Eq. (21) for d and in the high density limit $n_Q \gg 1$, the last term of Eq. (39) can be written as

$$4\pi \, n_Q \int dr \; r^2 \; V(r) \sim 4\pi \, n_Q \, d^3 \; V(d)$$

$$\sim 4\pi \, n_Q \, \frac{3}{4\pi n_Q} \; V(d)$$

$$\sim 3 \; V \; [(3/4\pi n_Q)^{1/3}]. \tag{40}$$

Up to the factor 3, this is Olive's assumption (actually, if one takes his assumption seriously, his potential energy is <u>underestimated</u> by a factor 12, corresponding to the number of quarks felt by a given quark (i.e. 12 is roughly the average number of spheres tangent to a given one in an assembly of equal radius close-packed spheres)).

The thermodynamical quantities are calculated in a similar way as previously except for the energy density which reads (for the i-th species) (see (39)).

$$
\rho_i = \frac{g_Q}{2\pi^2} \int dp\; \vec{p}^2 \; <n_i(p)> \{\xi_i(p) - \frac{1}{2} n_Q \int d^3r\; V(r)\} \quad (41)
$$

$$
= \frac{g_Q}{2\pi^2} \int dp\; \vec{p}^2 \; <n_i(p)> [\vec{p}^2 + m_i^2(p)]^{1/2}
$$

$$
+ 2\pi\; n_Q \cdot n_i \int_0^d dr\; r^2\; V(r) \quad (42)
$$

so that

$$
\rho = \sum_{i=1}^{i=5} \frac{g_Q}{2\pi^2} \int dp\; \vec{p}^2 \; <n_i(p)> [\vec{p}^2 + m_i^2(\vec{p})]^{1/2}
$$

$$
+ 2\pi\; n_Q^2 \int_{t_c}^d dr\; r^2\; V(r) \; . \quad (43)
$$

These last equations show that in the Hartree approximation there is no need of any particular technique to deal with the double counting of the average potential energy.

(3) The next approximation which we consider in this paper is the Hartree-Fock one, because of the particular importance of exchange correlations at relatively low densities and/or temperatures. Also it has been proposed to deal with nuclear forces as the result of exchange forces between quarks (41).

In this case, the excitation spectrum is given by (39) the following integral equation

$$
\xi_i(p) = \xi_i^{Hartree}(p) - \frac{1}{(2\pi)^3} \int d^3p'\; V(\vec{p}'-\vec{p}) \cdot <n_i(\vec{p}')> , \quad (44)
$$

where $\tilde{V}$ is the Fourier transform of the potential under consideration. This self-consistent equation can be solved by iteration and the result is used in the calculation of the pressure and of the energy density.

3.2. Statistical treatment of the gluons

The statistical analysis given above for the quarks can be repeated mutatis mutandis for the gluons. In particular, in our mean field approximations, their occupation number $< n_g(\vec{k}) >$ is given by

$$
< n_g(\vec{k}) > = -\frac{1}{\exp. [\beta\omega(\vec{k})] - 1} \quad (45)
$$

where $\omega(\vec{k})$ is their excitation spectrum. From Eq. (45) the contributions to pressure, energy density and gluon density are found via standard formulae (one should also remember that there are eight kinds of gluons).

The essential problem is thus the one of the derivation of the excitation spectrum $\omega(\vec{k})$. Olive (11) and (12), and also Boal et al. (15), chase

$$\omega(\vec{k}) = |\vec{k}| + V(d) \quad , \tag{46}$$

which embodies the facts that (i) the gluons are massless and (ii) they are interacting via the potential V (the same as the quark-quark potential and with the same Thomas-Fermi approximation, in Olive's article ; note that V = bd, the confining potential. in his paper) which therefore contributes to the gluon energy $\omega(\vec{k})$ by the factor V(d).

This approach can be questioned on several points. First, when considered inside matter, the gluons become massive, although one could distinguish (see e.g. (32)) between an "electric" and a "magnetic" mass (in fact, they are of the same order of magnitude). Next, the gluons do not appear as such within the medium but only as modes (or quasi-gluons or plasmons) propagating in the quark plasma.

We are thus faced with two problems : (i) how to confine the gluons and (ii) how to calculate the modes (i.e. the dispersion relation) of the quark plasma ?

In fact, as to the first problem, there is no particular need to confine the gluons since they are just modes propagating in the quark plasma : if the quarks are confined (below a critical temperature) there is no longer any possible modes ! Hence, no quarks implies no gluons.

The second problem is much more difficult to deal with essentially because of the lack of control of the approximations used. For instance, the Hartree approximation and the use of color singlet states as physical states lead to an excitation spectrum (37) similar to the one of a quantum elctrodynamical plasma (42) and (43). Other non-perturbative approachs provide other results (32). In the spirit of the approach used by Olive or by Boal et al. (15) as well as in this article, we should actually use a spectrum derived from our dynamical model ; this would be consistent with what was done previously and does not present any technical difficulties.

However, a common characteristic of most of the spectra obtained is that they are of the form

$$\omega^2(\vec{k}) \sim \omega_p^2 + \vec{k}^2 \; , \tag{47}$$

where the plasma frequency ω_p^2 may differ slightly from model to model and also for "electric" and for "magnetic" modes. Eq. (47) may sometimes be a good approximation in the two limiting cases of long and short wavelengths. For the plasma frequency ωp, instead of the relativistic quantum plasma frequency (42) and (43).

$$\omega_{pi}^2 = \frac{4\pi\alpha}{m} \int \frac{d^3p}{\xi_i(p)} < n_i(p) > \; , \tag{48}$$

a natural generalization including both the results of the renormalization group equations and the excitation spectrum adopted for the quarks, can be used ; it reads

$$\omega_{pi}^2 = 4\pi \int \frac{d^3p}{\xi_i(p)} \frac{\alpha(p)}{m_i(p)} < n_i(p) > \; . \tag{49}$$

Note also that there exists as many gluonic modes as quark flavors and, usually, with a third order degeneracy.

Finally, we are in position to come back to the screening length d which should be derived directly from the quasi-gluons excitation spectrum. In fact, K. Kajantie et al. (32) found in their non perturbative QCD approach

$$\left\{ \begin{array}{l} d_{elect.} \sim (1/\alpha T^2)^{1/2} \\[2ex] d_{mag.} \sim \left\{ \dfrac{8\pi}{3T^2} \cdot \left(\dfrac{1}{\alpha^2}\right)^{3/2} \right\}^{1/2} \; , \end{array} \right. \tag{50}$$

which both behave as T^{-1}, showing thereby that Eq. (27) is not a so bad approximation. A more phenomenological approach for cases intermediate between the cold and the hot case is, more conventionally,

$$d \sim \frac{\omega_p}{V_{th}} \tag{51}$$

where V_{th} is a relativistic thermal velocity (see e.g. J.L. Synge (44) or R. Hakim et al. (45)).

4. FLUCTUATIONS

The calculation of fluctuations for various physical quantities of importance in cosmology, such as density or energy density, can be

performed in a particularly simple way within the framework of our
mean field approximations. In particular, their spectrum can be
obtained quite easily and also the typical coherence length associated with them.

Let us focuse our attention on the case of baryon number
fluctuations or, equivalently, on the fluctuations of the quark
density. Their spectrum $<\delta n^2>_{\omega,k}$ can be obtained from the spectrum
of the quasi-quarks, $<\delta n^2>^0_{\omega,k}$, propagating within the plasma and
from its "dielectric" constant ε $(\omega,\vec{k})$ (see e.g. (46) and (39) ;
the relativistic case has been considered by H. Sivak (47)) through

$$<\delta n^2>_{\omega,k} = \frac{<\delta n^2>^0_{\omega,k}}{|\varepsilon(\omega,k)|} \tag{52}$$

while the k-spectrum is given by

$$<\delta n^2>_k = \int d\omega \quad <\delta n^2>_{\omega,k}$$

$$= \int d\omega \quad \frac{<\delta n^2>^0_{\omega,k}}{|\varepsilon(\omega,k)|} \tag{53}$$

It should be noticed that this last equation actually contains two
terms : one of them represents the coherent fluctuations, i.e. of
the plasma waves, corresponding to the implicit pole terms in the
denominator $|\varepsilon(\omega,k)|$ while the other one is nothing but the thermal
fluctuations. The quasi-quark spectrum $<\delta n^2>^0_{\omega,k}$ is essentially a
quasi-free fermian spectrum and has been calculated by H. Sivak (47)
in the relativistic case as

$$<\delta n^2>^0_{\omega,k} = \frac{1}{e^{\beta\omega}-1} \frac{1}{(2\pi)^2} \sum_{a,\ell=\pm1} \int \frac{d^3p}{E_p \cdot E_p^a} \quad x$$

$$x(\delta_{\ell 1}-<n(p)>) \cdot \delta[E_p+\ell E_p^a+a\omega] \quad x$$

$$x[\frac{\omega^2-\vec{k}^2}{2} + 2E_p^2 + 2 \ a\omega E_p] \tag{54}$$

In this equation, one has set

$$\begin{cases} E_p \equiv E(\vec{p}) \\ E_p^a \equiv E(\vec{p}+a\vec{k}) . \end{cases} \tag{55}$$

Moreover, the $\delta_{\ell 1}$ - term represents a vacuum term which has to be dropped in our phenomenological approach. It can also be remarked that the term corresponding to $\ell=1$ being a high frequency term can be dropped in our cosmological context since we are essentially interested in more or less static modes and/or low wave lengths.

As to the calculation of the "dielectric" constant $\varepsilon(\omega,k)$ there is no particular problem in this approach, and only relativistic generalization (36), (49), of the usual calculation (46), (48), (39) must be considered.

As a final remark we may notice that, if we admit T.D. Lee's (50) suggestions as to the confinement mechanism of a color dielectric constant smaller than one in the hadron phase, then an enhancement of the density fluctuations should result (as shown an Eq.52)) in the transition region. This might be considered as an indication that the potential actually used leads to a confining phase transition, eventhough higher correlations were not dealt with.

5. CONCLUSION

The above discussion gives idea of numerous sources of uncertainties occuring in the dynamics and the statistical treatment of the quark-gluon plasma : uncertainties on the "correct" quark-quark potential (if any) ; uncertainties on the quark masses ; difficulties of a treatment of three-body correlations, etc. It should also be noticed that the hadron phase is as uncertain as the quark phase !

Consequently - still assuming the existence of a first order quark-hadron transition - it is not surprising that various calculations, resting on different assumptions, give rise to critical temperatures ranging from $\sim$ 150 MeV to 600 MeV ! On figure 5 we have plotted the pressure of the quark-gluon plasma versus the temperature and, for the sake of comparison the ideal gas case has been drawn. The "error bars" indicate ranges of pressure within which pass curves corresponding to different assumptions. Incidentally, some of our curves stop at a given temperature (of the order of 180 MeV to 250 MeV) where the self-consistent equation for n has no solution : they are indicated by dots and a hached region in the figure. These temperatures should not be interpreted as the critical temperatures we are looking for but rather as the indication of the breakdown of our statistical approximations.

Finally, even this phenomenological approach does not give credible answers to our question (as far as cosmology is concerned): this will be our (pessimistic) conclusion.

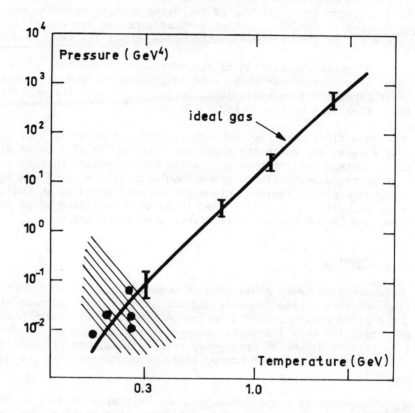

<u>Figure 5</u> : The pressure-temperature diagram of the quark phase
of the early universe; the ideal gas case has been
plotted for the sake of comparison. The "error bars"
indicate limits within which pass various curves corresponding to several assumptions discussed in the text.
The dots indicate some points where the self-consistent
equation for the density has no longer any solution,
for various assumptions.

ACKNOWLEDGEMENTS

We are quite indebted to Drs. A. Le Yaouank, L. Oliver, O. Pène
and J.C. Raynal for their patient explanations on quarks. We are
also indebted to Dr. H. Sivak for communicating his results prior
to publication.

APPPENDIX : Some phenomenological potentials

Ther are, at the present moment, dozens of more or less satisfactory phenomenological quark-antiquark potentials and we mention
only a few among them.

(1) Pure confinement

$$v(r) = br$$

(2) "QCD" potential

$$v(r) = -\frac{a}{r} + br$$

(3) Improved "QCD" potential (25)

$$v(r) = \frac{8\pi}{27} \Lambda \left[\Lambda r - f\frac{(\lambda r)}{\lambda r} \right]$$

where,

$$f(t) = 1-4 \int dq \frac{\sin qt}{q} \left\{ \frac{1}{\ln(1+q^2)} - \frac{1}{q^2} \right\}$$

with (27) Λ = 430 MeV, $m_u = m_d$ = 410 MeV, m_s = 625 MeV.

(4) "Gluon condensate" (51)

$$v(r) = -\frac{4\alpha}{3r} + \frac{8}{5} \left\{ \left[\left(\frac{3\alpha}{2r}\right) + \frac{5\pi^2 M_0^2 r^2}{72} \right]^{1/2} - \frac{3\alpha}{2r} \right\}$$

with α = .4 and M_0 = 330 MeV

(5) Martin's potential (52)

$$v(r) = - 8.064 \text{ GeV} + 6.870 \ r^{.1} \ (\text{GeV})^{1.1}.$$

FOOTNOTES

1) Although this article deals with the transition to a pion-con-
densed state, most (if not all) the results apply to the quark-

hadron phase transition as well.

2) Let us note, however, some Hartree-Fock calculations (16).

REFERENCES

(1) Jones, B.J.T., Jones, J.E., 1983, ed. The origin and Evolution
     of Galaxies, (D. Reidel, Dordrecht, Holland).

(2) Hosotany, Y., 1983, Phys. Rev. D27, p. 789.

(3) Lascher, G., 1979, Phys. Rev. Letts. 42, p. 1646

(4) Hogan, C.J., 1982, Astrophys. J. 252, p. 418.

(5) Weinberg, S., 1972, "Gravitation and Cosmology" (Wiley ; New
     York).

(6) Fowler, G.N., Raha, S., Weiner, R.M., 1980, Preprint.

(7) Wagoner, R.V., Steigman, G., 1979, Phys. Rev. D20, p. 825.

(8) Dicus, D.A., Pati, J.C., Teplitz, V.L., 1980, Phys. Rev. D21
     p. 922.

(9) Alvarez, E., Hakim, R., 1979, Phys. Rev. D19, p. 1696.

(10) Alvarez, E., Hakim, R., 1979, Astron. Astrophys. 80, p. 71.

(11) Olive, K., 1981, Nuclear Phys. B 190, p. 483.

(12) Olive, K., 1982, Nuclear Phys. B198, p. 461.

(13) Källmann, C.G., 1982, Phys. Letts. B112, p. 213.

(14) Källmann, C.G., Montonen, C., 1982, Phys. Letts. B115, p. 473.

(15) Boal, D., Schächter, J., Woloshyn, R., 1982, Phys. Rev. D26,
     p. 3245.

(16) Alvarez, E., 1981, Phys. Rev. D23, p. 1715.

(17) Kislinger, M.B., Morley, P.D., 1979, Phys. Reports 51, p. 63.

(18) Freedman, B.A., Mc Lerran, L.D., 1977, Phys. Rev. D16, p. 1130.

(19) Baluni, V., 1978, Phys. Rev. D17, p. 2092.

(20) Kapusta, J.I., 1979, Nuclear Phys. B148, p. 461.

(21) Shuryak, E.V., 1980, Phys. Reports 61, p. 71.

(22) Chodos, A., Jaffe, R.L., Johnson, K., Thorn, C.B., Weisskopf,
     V.F., 1974, Phys. Rev. D9, p. 3771.

(23) Müller, B., Rafelski, J., 1980, CERN preprint TH. 2928.

(24) Gavela, M.B., Le Yaouank, A., Oliver, L., Raynal, J.C.,
     Sood, S., 1979, Phys. Letts. B82, 431.

(25) Richardson, J.R., 1979, Phys. Letts. 82B, 272.

(26) Rafelski, J., Viollier, R.D., 1979, CERN Report TH. 2673.

(27) Schöberl, F., 1982, CERN Report TH. 3287.

(28) Politzer, H.D., 1974, Phys. Reports 14C, 129.

(29) Pagels, H., Stokar, S., 1979, Phys. Rev. D20, 2947.

(30) Miller, K.J., Olsson, M.G., 1982, Preprint MAD/PH/57.

(31) Childers, R.W., 1982, Phys. Rev. D26, 2902.

(32) Kajantie, K., Kapusta, J.I., 1982, CERN Report TH. 3284.

(33) Wanyun, Zha, 1982, ISAS Preprint 63/82/ E.P.

(34) Bardeen, W.A., Chanowitz, M.S., Drell, S.D., Weinstein, M.,
     Yan, T.M., 1975, Phys. Rev. D11, 1094.

(35) Rafelski, J., 1976, Phys. Rev. D14, 2358.

(36) Hakim, R., 1978, Riv. Nuovo Cim. 1, N°6.

(37) Hakim, R., 1981, Plasma Physics Techniques for QED and QCD

Systems in Statistical Mechanics of Quarks and Hadrons, ed.
H. Satz, North Holland.

(38) Diaz-Alonso, J., Hakim, R., 1978, Phys. Letts. 66A, 466.

(39) Kadanoff, L.P., Baum, G., 1962, Quantum Statistical Mechanics, Benjamin, New York.

(40) Alvarez, E., 1982, Phys. Letts. 110B, 315.

(41) Barry, G.W., 1977, Phys. Rev. D16, 2886.

(42) Hakim, R., Heyvaerts, J., 1978, Phys. Rev. A18, 1250.

(43) Hakim, R., Heyvaerts, J., 1980, J. Phys. A13, 2001.

(44) Synge, J.L., 1957, The Relativistic Gas, Amsterdam, N. Holland.

(45) Hakim, R., Mangeney, A., 1971, Phys. Fluids 14, 2751.

(46) Sitenko, A.G., 1967, Electromagnetic Fluctuations in Plasma, Academic Press, New York.

(47) Sivak, 1982, submitted to Ann. Phys.

(48) Ichimaru, S., 1973, Basic Principles of Plasma Physics, Benjamin, Reading Mass.

(49) Carruthers, P., 1983, Rev. Mod. Phys. 55, 245

(50) Lee, T.D., 1981, Particle Physics and Introduction to Field Theory, Harwood Acad. Publ., New York.

(51) Flory, C.A., 1982, SLAC-PUB 2889.

(52) Martin, A., 1980, Phys. Letts. 93B, 338.

(53) Klinkhamer, F.R., 1982, "Quark Liberation at High Temperature", in "The Birth of the Universe", 17-th Moriond Meeting ; J. Audouze and J. Tran Than Van Eds.

(54) Schramm, D.N., Crawford, M., Planetary Mass Black Holes and the Quark-Hadron Transition in "The Birth of the Universe ; 17-th Moriond Meeting, 1982, ed. J. Audouze and J. Tran Than Van (Editions Frontières).

(55) Schramm, D.N., Crawford, M., 1982, Nature 298, 538.

NOTE ADDED :

Although they are not directly related to the quark-hadron transition in the early universe, we would like to mention two interesting papers connected with the subject. In the first one (53) emphasis is put on results from Monte Carlo simulations of gauge theories on a lattice which, indeed, do exhibit first order phase transition interpreted as a quark-hadron transition. In the second article (54) (55) simulations with classical quarks interacting via a simplified Richardson (25) potential are used in a study of primeval fluctuations.

MASS BOUNDS ON A DILEPTONIC SYSTEM FROM COSMOLOGY

Pierre SALATI

L.A.P.P., BP 909, 74019 Annecy-le-Vieux, France

ABSTRACT:

We consider a system of two heavy (mass bigger than a few GeV) neutral leptons coupling only to the Z° neutral vector boson. The interaction may mix the two particles. We study the behaviour of that dilepton during its departure from chemical equilibrium at the very beginning of the Universe, assuming the standard hot Big-Bang model is correct. The present cosmic mass density of that dilepton is calculated. In order for this density not to exceed the upper limit of $5 \ 10^{-31} \text{g.cm}^{-3} \times \left(\dfrac{\text{Hubble Constant}}{50 \text{ km/s.Mpc}}\right)^{2}$, the dilepton mass is constrained. In particular, in the case of the quasi-antidiagonal coupling of the dilepton to the Z°, the dilepton mass would have to be greater than a lower bound of the order of 15 GeV and the gap between its two components would have to be less than an upper bound of the order of 10% of the dilepton mass.

This work has been done in Annecy-le-Vieux (France) at the Laboratoire d'Annecy-le-Vieux de Physique des Particules by P. Binétruy, G. Girardi et P. Salati.

J. Audouze and J. Tran Thanh Van (eds.),
Formation and Evolution of Galaxies and Large Structures in the Universe, 55–75.

1.   INTRODUCTION

Modern cosmology and the standard Big Bang Model provide us
a good mean to test high energy physics. At very early times,
the Universe was so hot that matter was composed of its primordial
constituents: particles like photons, electrons, quarks ...
Among them, leptons and their spectroscopy continue to be an
intriguing puzzle in particle physics. Some models predict heavy
neutrinos associated with the known light ones. More recently
supersymmetry gives birth to a new "zoology", each known particle
unfolding with its supersymmetric partner. In this paper we will
focuse our attention on neutral heavy (mass larger than a few GeV)
leptons. They only interact via weak interactions (gravitational
ones are neglected). They could be heavy neutrinos or supersymmetric partners of gauge bosons or (and) higgs bosons. As they
have not yet been detected in high energy accelerators and as the
theory is unable to predict their mass, we use the standard Big
Bang Model to derive mass bounds imposed by cosmological data like
the mean mass density of the Universe.

We denote by S such a particle. As it is neutral, its
electric charge is zero. It does not interact via electromagnetic
and strong interactions. It only couples to the Z° vector boson,
responsible for weak interactions. We just start with one lepton
S and we will recall the analysis done by Lee and Weinberg in 1977.
The aim of this paper is to deal with a more complicated situation
with two leptons S_1 and S_2, and we will compare our results
concerning a dileptonic system S_1-S_2 with those obtained in 1977
concerning only one lepton S. The coupling of S to the Z° is
described by the following diagram:

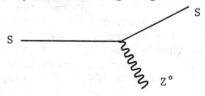

$$\bar{\psi} \cdot \frac{e}{\sqrt{2}\ \mathrm{Sin}2\theta_W}\ \gamma^\mu(1-\gamma_5)\ \cdot\ \psi\ Z^\circ_\mu$$

The heavy neutral lepton S is left (nothing is changed if it is
right). θ_W is the weak angle. We have used $\sin^2\theta_W = 0.229$.
We notice that S is stable.

When the temperature of the Universe T is higger than the
mass M of the lepton S, the population of S behaves like a
relativistic gaz of spin 1/2 particles. Basically, a detailed
study of the behaviour of the S-population when the temperature

T dropps under the mass M shows that under the double action of
an important freezing and dilution, the S-lepton is fossilized.
Moreover we know the present value of its mass density and we can
infer a limit upon its mass. 1 MeV is roughly the temperature at
which the rate of interaction of S with other particles becomes
less than the expansion rate of the Universe. So, as long as
$T > O(1 \text{ MeV})$ S is kept thermalized by collisions with ν, e,...
and is in thermal equilibrium. Moreover, as soon as $T < M$, the
S lepton begins to disappear. We have assumed that it only couples
to the Z°. So S annihilates only on a $\bar{S}$, its anti-particle, in
processes such as

$$S\bar{S} \rightarrow \ell\bar{\ell} \tag{1}$$

where ℓ and $\bar{\ell}$ are light fermions as ν, e, μ,... The annihilation
is mediated via a Z° gauge boson and the Feynman diagram responsible
for it is shown here:

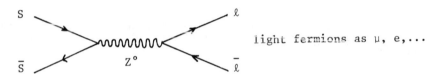

light fermions as μ, e,...

One has to keep in mind that we deal with a chemical reaction.
A $S\bar{S}$ pair can give light fermions. But the reverse process does
exist in the same time: a $\ell\bar{\ell}$ pair of large sufficient energy can
annihilate to give a $S\bar{S}$ pair. So the number density n of the S-
lepton relaxes toward its chemical equilibrium value n_0. n_0 is
got if the chemical reaction

$$S\bar{S} \leftrightarrow \text{light-fermions } \ell\bar{\ell} \tag{2}$$

is very rapid. Actually, $n=n_0$ when the Temperature T is large.
When T falls down M, the rate of this reaction is getting smaller
and smaller as $n=n_0$ decreases due to annihilation. Suddenly, when
$T \simeq O(M/20) =$ freezing Temperature T_f there are so few S leptons
that the reaction is stopped. This is a good example of quenching.
The number density n is frozen and, apart the expansion of the
Universe, n does not evolve. The S lepton population goes out of
chemical equilibrium. Therefore, as regards the behaviour of S-
leptons, the basic scenario is the following when the Temperature
decreases:

. T > Mass of lepton S. The S population is in thermal equilibrium
and in chemical equilibrium. $S\bar{S}$ pairs annihilate but they are
continuously recreated by reverse processes. The chemical reaction
$S\bar{S} \leftrightarrow \ell\bar{\ell}$ is so rapid that the number density n of S has relaxed

completely toward the chemical equilibrium value n_0. As T is
larger than M, the S behaves like a light ultra-relativistic gaz.

. $T_{freezing} \simeq O(M/20) < T <$ Mass of lepton S. The chemical
reaction $S\bar{S}-\ell\bar{\ell}$ is always very rapid. So S is in thermal and
chemical equilibrium and n is always equal to n_0 due to a very
fast relaxation. Mass effects appear so $n_0=n$ falls down very
rapidly.

. $T = T_{freezing}$. They are so few S leptons that the relaxation
is not sufficiently rapid to force n to be equal to n_0. n becomes
larger than n_0. There is a departure from the chemical equilibrium
for S.

. $O(1 \text{ MeV}) < T < T_{freezing}$. S leptons are no more in chemical
equilibrium. After some non negligible decrease, n × (Radius of
Universe)^3 has reached its present value. However, there is still
a thermal equilibrium.

. $T < O(1 \text{ MeV})$. There is no equilibrium at all. The S leptons
are fossilized. As there are stable we know there present mass
density.

To make this quantitative, the rate equation for n writes:

$$\frac{dn}{dt} = -\frac{3\dot{R}}{R} n - <\sigma v>n^2 + <\sigma v>n_0^2 \qquad (3)$$

n is the number density of S-leptons at time t, R is the cosmic
scale factor.

. The first term refers to the dilution, consequence of the
expansion of the Universe.

$$\frac{\dot{R}}{R} = \text{Hubble Constant at time } t = (8\pi\rho G/3)^{1/2} \qquad (4)$$

G is the gravitational constant. ρ is the energy density of
Universe at time t. We have

$$\rho = N_F \frac{\pi^2}{15} (kT)^4 \qquad (5)$$

with N_F an effective number of degrees of freedom, counting 1/2
and 7/16 respectively, for each boson or fermion species and spin
state. N_F depends on the physical content of the Universe at
time t.

. The second term refers to the annihilation rate of S in the
chemical reaction $S\bar{S} \leftrightarrow \ell\bar{\ell}$. $<\sigma v>$ is the average value of the $S\bar{S}$
annihilation cross-section times the relative velocity. For non-
relativistic velocities (when especially T < Mass of the S lepton),
the cross-section σ for the exothermic chemical reaction (1)

behaves like $1/v$, so $<\sigma v>$ is a temperature independent constant:

$$<\sigma v> = \frac{G_F^2 M^2}{2\pi} N_A \qquad (6)$$

G_F is the Fermi-coupling $1.15\ 10^{-5} GeV^{-2}$ and N_A is a dimensionless factor which depends both on the number of annihilation channels open and on the coupling of the light fermions to the Z°. If the coupling between a light fermion ℓ and the Z° is given by:

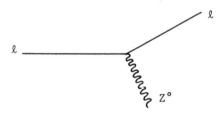

$$\bar{\psi}\ \cdot\ \frac{e}{2Sin2\theta_W}\ \gamma^\mu(a+b\gamma_5)\psi$$

therefore

$$N_A = \sum_{channels} \frac{a^2 + b^2}{2} \qquad (7)$$

As an example, for a mass of S of 5 GeV we have N_A = 14.6.

.   Finally the last term takes into account the reverse process $\ell\bar{\ell} \rightarrow S\bar{S}$, responsible for the relaxation of n toward n°.

Lee and Weinberg found that the present mass density of heavy neutrinos and antineutrinos can be very well represented by

$$\rho_S = (4.2\ 10^{-28} g/cm^3)\ \times\ (m_s(GeV))^{-1.85}\ (N_A/\sqrt{N_F})^{-0.95} \qquad (8)$$

We observe that the bigger the mass, the smaller the mass density of the fossilized heavy S leptons. Thus a lower bound is inferred from the requirement that the S mass density cannot exceed the upper limit of $2\ 10^{-29} g/cm^3$. They found that the lepton mass would have to be greater than a lower bound of the order of 2 GeV. More recently, Gunn et al. estimated that, due to their big mass, the heavy S leptons would collapse around galaxies and that their contribution to the total mass density of the Universe would be approximately 10% only. From data analysed by Peebles, it appears that the Universe is quasi-flat. The new upper limit on ρ_S is now:

$$(\Omega_s=10\%)\times(\rho_{total} \simeq \rho_{critical} = 5\ 10^{-30} g\ cm^{-3})\times(\frac{H}{50km/s.Mpc})^2 \qquad (9)$$

So the bound on the S mass is rather in the range 7-15 GeV than
2 GeV.

As many theories involve two heavy neutral leptons instead of
one, we reanalyse the cosmological implication of their mass
density. In the next section we analyse the different processes
by which the two leptons interact with themselves and with light
matter. We write the equation of evolution for their number
density. In section 3 we analyse the way they decouple from
chemical equilibrium for different kinds of coupling S-Z°-S.
Finally in section 4 we discuss one model, using our cosmological
bounds and we conclude.

2.   THE DILEPTONIC SYSTEM

We consider now a dilepton S_1 and S_2. Those are two neutral
heavy leptons with mass M_1 for S_1 and M_2 for S_2. The coupling
between the S and the Z° is given by the matrix Coup:

$$\frac{e}{\sqrt{2}\ Sin2\theta_W}\ (\overline{S_1 S_2})\gamma^\mu(1-\gamma_5)\begin{bmatrix}1 & \varepsilon \\ \varepsilon & 1\end{bmatrix}\begin{pmatrix}S_1 \\ S_2\end{pmatrix}$$
$$\underbrace{\hphantom{\begin{bmatrix}1 & \varepsilon \\ \varepsilon & 1\end{bmatrix}}}_{\text{matrix coup}}$$

In section 3 we will consider the case ε very small, when the
coupling is quasi-diagonal. Then we will analyse the "democratic"
case $\varepsilon=1$. Finally we will take a quasi antidiagonal coupling
matrix of the type

$$\text{matrix Coup} = \begin{pmatrix}\varepsilon & 1 \\ 1 & \varepsilon\end{pmatrix}\quad\text{with } \varepsilon \text{ very small.}$$

But before, we analyse the different reactions arising from those
couplings and we take matrix coup = $\begin{pmatrix}1 & \varepsilon \\ \varepsilon & 1\end{pmatrix}$ with no constraint

upon ε. The point is that Z° can couple in three different ways
with the S leptons. We have:

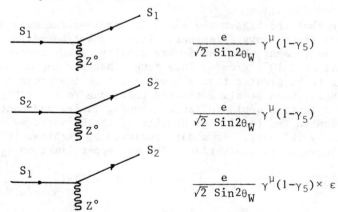

$$\frac{e}{\sqrt{2}\ Sin2\theta_W}\ \gamma^\mu(1-\gamma_5)$$

$$\frac{e}{\sqrt{2}\ Sin2\theta_W}\ \gamma^\mu(1-\gamma_5)$$

$$\frac{e}{\sqrt{2}\ Sin2\theta_W}\ \gamma^\mu(1-\gamma_5)\times\varepsilon$$

In the case analysed by Lee and Weinberg, there was only one reaction to consider during the departure from the chemical equilibrium of the lepton S and the subsequent quenching. Now, due to the three different couplings $S-Z^\circ-S$, we have to take into account no less than eleven different processes. This fact motivated our analysis of the case in which two leptons and not only one are involved. We give a list of the reactions with the corresponding Feynman diagrams and the formula for the averaged product of the cross-section σ and the relative velocity between initial particles v. All the reactions are exothermic.

. First of all we have reactions involving both the S population and the light fermion matter. They are:

*   the annihilation processes

$$S_1\bar{S}_1 \leftrightarrow \ell\bar{\ell} \tag{a}$$
$$S_1\bar{S}_2 \leftrightarrow \ell\bar{\ell} \quad \text{with also} \quad S_2\bar{S}_1 \leftrightarrow \ell\bar{\ell} \tag{b}$$
$$S_2\bar{S}_2 \leftrightarrow \ell\bar{\ell} \tag{c}$$

ℓ stands for a light fermion and $\bar{\ell}$ for its antiparticle. Those are the only reactions responsible for the violation of the total number of S leptons. The corresponding Feynman graphs and formulae for $\langle\sigma v\rangle$ follow.

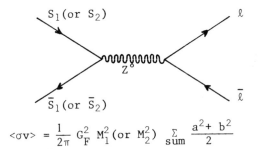

$$\langle\sigma v\rangle = \frac{1}{2\pi} G_F^2 M_1^2 (\text{or } M_2^2) \sum_{\text{sum}} \frac{a^2+b^2}{2}$$

over light fermions with threshold M_1 (or M_2).

The definition of a and b has yet been done in section 1.

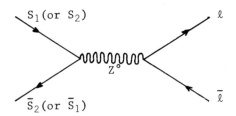

$$\langle\sigma v\rangle = \frac{\varepsilon^2}{8\pi} G_F^2 (M_1 + M_2)^2 \sum \frac{a^2 + b^2}{2}$$

$$\text{threshold } \frac{M_1 + M_2}{2}$$

* the decay of S_1 into S_2 and a pair $\ell\bar{\ell}$

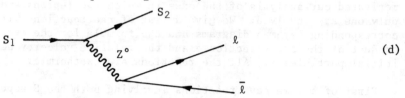

(d)

The decay introduces differences with the one lepton case. The rate is given by

$$\Gamma = \frac{G_F^2}{384\pi^3} \varepsilon^2 M_1^5 \cdot \{1 - 8\alpha^2 - 24\alpha^4 \ln\alpha + 8\alpha^6 - \alpha^8\} \sum \frac{a^2 + b^2}{2}$$

$$\text{threshold } \frac{M_1 - M_2}{2}$$

where α is the ratio M_2/M_1.

* finally the scattering of S upon light matter

$$S_1 + \ell(\text{or } \bar{\ell}) \leftrightarrow S_2 + \ell(\text{or } \bar{\ell})$$

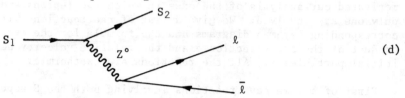

(e)

$$\langle\sigma v\rangle = \frac{G_F^2}{16\pi} \varepsilon^2 S \left(1 - \frac{M_2^2}{S}\right)^2 \left\{\frac{8}{3} + \frac{2}{3}\frac{M_1^2 M_2^2}{S^2} + \frac{M_1^2 + M_2^2}{3S}\right\} \sum \left(\frac{a^2 + b^2}{2} \times \text{helicity}\right)$$

the sum runs over the species of light particles present at time t in the Universe.

S is the square of the center of mass energy of a S_1 colliding upon ℓ we have taken

$$S = M_1^2 + 2M_1 e$$

where e is the mean energy of a light particle at temperature T:

$$e = 3.15 \text{ kT}.$$

This reaction introduces also noticeable modifications with the one lepton case.

. Finally we list processes involving only the S leptons. The general Feynman diagrams:

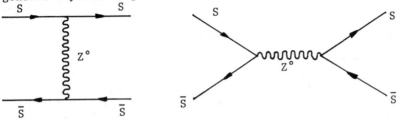

give birth to three reactions:

* $S_1\bar{S}_1 \leftrightarrow S_2\bar{S}_2$ (f)

$$\langle\sigma v\rangle = \frac{1}{\pi} G_F^2 M_1^2 (1 - \frac{M_2^2}{M_1^2})^{1/2} (1 + \varepsilon^2)^2$$

* $S_1\bar{S}_1 \leftrightarrow S_2\bar{S}_1$ (g)

$$\langle\sigma v\rangle = \frac{4G_F^2}{\pi} M_1^2\varepsilon^2 \left[(1-a)(1-b)\right]^{1/2} (1-ab)$$

where $a = (1+\alpha/2)^2$ and $b = (1-\alpha/2)^2$. α has been defined as the ratio M_2/M_1.

* $S_1\bar{S}_2 \leftrightarrow S_2\bar{S}_2$ (h)

we have
$$\langle\sigma v\rangle = \frac{G_F^2}{\pi} S\varepsilon^2(1 - \frac{4M_2^2}{S})^{1/2} \quad \text{where} \quad S = (M_1 + M_2)^2$$

At last we have reactions where a S lepton interacts upon another S lepton (and not an antilepton $\bar{S}$). The general Feynman graphs responsible for them are shown:

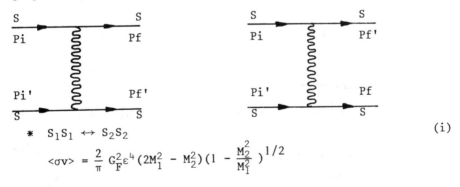

* $S_1S_1 \leftrightarrow S_2S_2$ (i)

$$\langle\sigma v\rangle = \frac{2}{\pi} G_F^2\varepsilon^4 (2M_1^2 - M_2^2)(1 - \frac{M_2^2}{M_1^2})^{1/2}$$

* $S_1S_1 \leftrightarrow S_2S_2$ (j)

$$\langle\sigma v\rangle = \frac{2}{\pi} G_F^2\varepsilon^2 (3M_1^2 - M_2^2)((1-a)(1-b))^{1/2}$$

a and b have yet been defined.

* $S_1 S_2 \leftrightarrow S_2 S_2$ (k)

$$\langle \sigma v \rangle = \frac{G_F^2}{\pi} \varepsilon^2 (S - 2M_2^2)(1 - \frac{4M_2^2}{S})^{1/2}$$

where $S = (M_1 + M_2)^2$.

The equation of evolution of the number density n_1 of lepton S_1 (n_2 for lepton S_2) has to take into account the dilution due to the expansion of Universe and the processes listed before. To each exothermic reaction corresponds the reverse one, responsible for the relaxation of n toward its chemical equilibrium value n^o. The equations write:

$$\frac{dn_1}{dt} = -3 \frac{\dot{R}}{R} n_1 - \Gamma(\text{decay})(n_1 - n_1^o) - \langle \sigma v \rangle \{\text{Reactions a+f+g+2i+j}\}$$

$$(n_1^2 - n_1^{o2}) - \langle \sigma v \rangle \text{ Reactions}\{b+h+k\} (n_1 n_2 - n_1^o n_2^o) -$$

$$- \langle \sigma v \rangle \text{ Reaction e} . n^o (n_1 - n_1^o)$$
$$\text{light lepton}$$

and   (10)

$$\frac{dn_2}{dt} = -3 \frac{\dot{R}}{R} n_2 + \Gamma(\text{decay})(n_1 - n_1^o) + \langle \sigma v \rangle \text{ Reactions}\{f+g+2i+j\}(n_1^2 - n_1^{o2})$$

$$- \langle \sigma v \rangle \text{ Reactions}\{b-h-k\}(n_1 n_2 - n_1^o n_2^o) + \langle \sigma v \rangle \text{ Reaction e } n^o(n_1 - n_1^o)$$
$$\text{light lepton}$$
$$- \langle \sigma v \rangle \text{ Reaction c } (n_2^2 - n_2^{o2}).$$

Those equations are coupled by reactions involving only S or $\bar{S}$ leptons. We emphasize the fact that n_1 refers to the population of S_1 leptons and not $\bar{S}_1$ antileptons. As there is a particle-antiparticle symmetry:

and
$$n_{S_1} = n_{\bar{S}_1} = n_1$$
$$n_{S_2} = n_{\bar{S}_2} = n_2 .$$

Finally we remark that

$$\frac{dn_1 + n_2}{dt} = -3 \frac{\dot{R}}{R} (n_1 + n_2) - \langle \sigma v \rangle (n_1^2 - n_1^{o2}) - \langle \sigma v \rangle (n_2^2 - n_2^{o2}) -$$
$$\text{Reaction a} \qquad \text{Reaction c}$$
$$- 2\langle \sigma v \rangle (n_1 n_2 - n_1^o n_2^o).$$ (11)
$$\text{Reaction b}$$

Processes violating $n_1 + n_2$ (Reactions a, b and c) are the only one to influence the evolution of $n_1 + n_2$. We have solved numerically those equations requiring that

$$\lim_{t \to 0} \frac{n_1 (\text{or } n_2)}{T^3} = \frac{n_1^0 (\text{or } n_2^0)}{T^3} \qquad (12)$$

which is the initial condition.

3. THE DEPARTURE FROM THE CHEMICAL EQUILIBRIUM

We explore now all the range of couplings of S_1-S_2 to the neutral weak Z° vector boson. We start first with a diagonal coupling matrix, we explore the intermediate situation of a "democratic" coupling, and we end with the most interesting physical case: the antidiagonal coupling.

The quasi-diagonal coupling. In that case $S_1(S_2)$ couples preferably to another $S_1(S_2)$ and the mixing interaction between S_1 and S_2 is suppressed by a factor ε. First of all, let ε be equal to zero. We have no mixing at all and the only reactions still present in that case are the annihilations (a) and (c). We remark that both S_1 and S_2 are stable particles. The dileptonic system behaves exactly as two independent neutral heavy leptons and the analysis of Lee and Weinberg is still valid for each particle. The lower mass bound depends on the gap between S_1 and S_2. We have two extreme situations:

. if $M_1 = M_2$. The dileptonic system is degenerated and the final mass density of those leptons is twice that of each particle. Therefore the lower mass bound is raised by a factor $2^{0.54}$ and we get:

$$M_1 = M_2 > (\frac{50 \text{ km/s.Mpc}}{H}) \times 22.0 \text{ GeV} \qquad (13)$$

. if $M_1 \gg M_2$. The S_1 lepton has a present mass density negligeable. All the final mass density is due to S_2. As S_2 annihilates much less rapidly than S_1, the present fossilized population is composed mainly of S_2 neutral leptons. Therefore the lower mass bound is the same as before:

$$M_1 \gg M_2 > (\frac{50 \text{ Km/s.Mpc}}{H}) \times 15.0 \text{ GeV} \qquad (14)$$

Then let ε and the mass gap Δm between S_1 and S_2 be non zero. We have now a rather different situation than before. The main differences come from reactions (d) and (e). S_1 is no more stable: it can decay into S_2 and a $\ell \bar{\ell}$ pair and it can also scatter upon light fermions to give a S_2. Those two processes depend on ε and

Δm. When the two parameters are very small as in fig. 1-b where $\varepsilon = 10^{-4}$ and $\Delta m = 10$ MeV, we still find the pure-diagonal case. The lifetime of 360 hours is exceedingly bigger than the age of the Universe at the time of quenching and as ε is very small, there is no scattering upon light fermions. Each heavy lepton behaves as if it was alone. This configuration has yet been analysed.

If we consider a small gap (10 MeV) and ε equal to 10^{-3}, then 2.10^{-3} as in fig. 2-a and b the only effect is the scattering (e). A lifetime of the order of 20 minutes prevents the decay to be important. The curves show that after a noticeable conversion of S_1 into S_2 there is a quenching. Basically, as long as there are important quantities of light fermions and S_1 leptons, the change of S_1 into S_2 works. But the dilution of the light matter and the decrease of the number density n_1 generate the chemical freezing of S_1. After some time, both S_1 and S_2 populations are fossilized. Notice the decrease of the frozen value of n_1/T^3 as the mixing coupling ε increases. On the other hand if ε is very small as in Fig. 1 (10^{-4}) and if the mass gap is important: $\Delta m = 500$ MeV, the scattering is suppressed but the decay plays a crucial role. S_1 leptons disappear completely after some time of the order of the lifetime of S_1 as is shown in fig. 1-a. Generally, of course, both the decay and the scattering upon light fermions act at the same time. The results lead us to make a few remarks:

. The elimination of S_1 and the subsequent important decrease of its number density is actually a brutal phenomenon in Δm and ε.

. Due to relaxation, when the decay or the scattering become important, the n_1 curve is pushed toward its chemical equilibrium value n_1^0. When those processes play a crucial role, S_1 remains in chemical equilibrium because very fast reactions connect it with light matter.

. Finally when S_1 disappears rapidly, S_2 remains alone and we still find the situation previously analysed by Lee and Weinberg:

$$M_2 > \frac{50 \text{ km/s.Mpc}}{H} \times 15.0 \text{ GeV} \tag{15}$$

Let us consider a "democratic" coupling between the S leptons. The coupling matrix is given by:

$$\frac{e}{\sqrt{2} \, \text{Sin} 2\theta_W} \, (\bar{S}_1, \bar{S}_2) \gamma^\mu (1 - \gamma_5) \begin{bmatrix} 1 & 1 \\ 1 & 1 \end{bmatrix} \begin{pmatrix} S_1 \\ S_2 \end{pmatrix} \tag{16}$$

It turns out that S_1 is coupled very strongly to S_2. The situation can be easily analysed. We start with the extreme configuration $M_1 = M_2$. The cross-sections of reactions (a), (b) and (c) are all

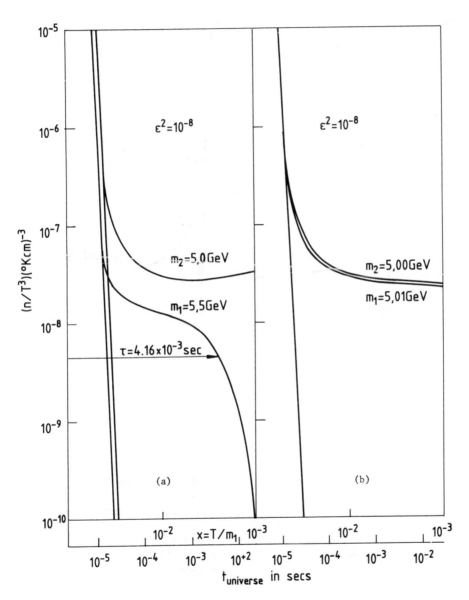

FIG. 1

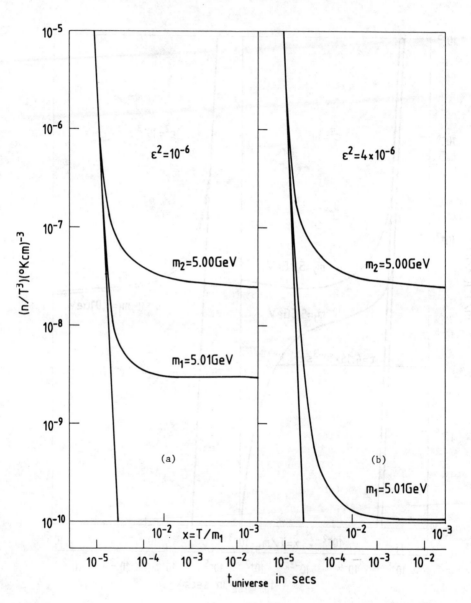

FIG. 2

equal. Therefore the equation describing the evolution of $n_1 + n_2$ writes:

$$\frac{dn_1 + n_2}{dt} = -3\,\frac{\dot{R}}{R}\,(n_1+n_2) - \langle\sigma v\rangle\{\text{reaction}((a)\}\left[(n_1+n_2)^2-(n_1^0+n_2^0)^2\right]$$
$$\text{or (b) or (c)} \tag{17}$$

We remark that $n_1 = n_2 = n/2$, in which n behaves exactly as the number density of the S neutral lepton considered by Lee and Weinberg. Despite the fact that we have actually two heavy neutral leptons, the present mass density of the Universe is the same as if we had only one particle. The lower mass bound remains unchanged.

On the other hand, even for a tiny gap of mass Δm, reactions (d) and (e) are so important that S_1 remains in chemical equilibrium and disappears very rapidly. S_2 evolves lonely and we still find the by-now well-known configuration of one heavy neutral stable particule. Therefore the lower mass bound is always given by:

$$M_2 > \frac{50 \text{ km/s.Mpc}}{H} \times 15.0 \text{ GeV} \tag{18}$$

Finally we present a discussion about the most interesting physical configuration: the anti-diagonal coupling. The interaction between the S-leptons is described by the following coupling matrix:

$$\frac{e}{\sqrt{2}\,\text{Sin}2\theta_W}\,(\bar{S}_1\bar{S}_2)\gamma^\mu(1-\gamma_5)\begin{bmatrix}\varepsilon & 1\\ 1 & \varepsilon\end{bmatrix}\begin{pmatrix}S_1\\ S_2\end{pmatrix} \tag{19}$$

in which ε is a small number eventually equal to zero. S_1 is coupled preferably to S_2. The mixing between the dilepton is strong, but the direct coupling S_1-Z°-S_1 or S_2-Z°-S_2 is suppressed by a ε factor. Therefore the decay (d) and the scattering (e) are predominant reactions owing to the chemical equilibrium of S_1 and its fast elimination as the temperature dropps under its mass M_1. During the quenching of the S_2 population, we have:

$$n_1 = n_1^0 \ll n_2 .$$

Therefore equation (11) simplifies to give:

$$\frac{d}{dt}(n_1+n_2) \simeq \frac{dn_2}{dt} = -3\,\frac{\dot{R}}{R}\,(n_1+n_2) - \underbrace{\langle\sigma v\rangle(n_1^{0^2}-n_1^{0^2})}_{\text{Reaction a}} - \underbrace{\langle\sigma v\rangle(n_2^2-n_2^{0^2})}_{\text{Reaction c}} -$$

$$- \underbrace{2\langle\sigma v\rangle n_1^0(n_2-n_2^0)}_{\text{Reaction b}} \tag{20}$$

$$\frac{dn_2}{dt} = -3\,\frac{\dot{R}}{R}\,n_2 - \underbrace{\langle\sigma v\rangle(n_2^2-n_2^{0^2})}_{\text{Reaction c}} - \underbrace{2\langle\sigma v\rangle n_1^0(n_2-n_2^0)}_{\text{Reaction b}}$$

. the first term refers to the dilution of the S_2 population due
to the expansion of the Universe.

. the second one describes the annihilation (and the reverse
reaction) of a S_2 upon its antiparticle. This process is very
slow because of the ε^2 dependence. For $\varepsilon = 0$ we have in fact no
annihilation of a S_2 upon a $\bar{S}_2$ at all.

. finally, the last term takes into account the annihilation of
S_2 upon S_1 (and the reverse process).

First, we discuss the case $\varepsilon = 0$. The behaviour of n_2 is
shown during the departure from the chemical equilibrium in fig. 3
for various values of the mass gap Δm. It appears that the bigger
the gap the more numerous the present S_2 population. We also
observe that even for a gap of 500 MeV and a mass of 5.0 GeV, the
S_2 mass density largely exceeds the upper limit. Physically, when
the temperature dropps under M_1, the number density of S_1 begins
to fall down. Then, when the temperature dropps under M_2, the S_2
become to disappear via the only allowed reaction:

$$S_2 + \bar{S}_1 \rightarrow \ell\bar{\ell} \qquad \qquad (b)$$

The big problem for S_2 is that there are so few S_1 at that time,
that the annihilation reaction (b) is slowed down noticeably and
even stopped for very large gaps. This is actually a new situation
because we can not only put a limit upon the mass of the S leptons
but also upon the gap between them. The upper bound upon the mass
gap is estimated very optimistically to be:

$$\Delta m = M_1 - M_2 \leq 10\% \, M_2 \qquad \qquad (21)$$

As for the bound upon M_2, we consider the exception to the rule:
$M_1 = M_2$. We find the same result as for the degenerated dilepton
with a pure-diagonal coupling, and we extend it

$$M_2 > (\frac{50 \text{ km/s.Mpc}}{H}) \times 22.0 \text{ GeV} \qquad \qquad (22)$$

Then we analyse the case $\varepsilon \neq 0$. Examples are shown for various
gaps, $M_2 = 5$ GeV and $\varepsilon = 10^{-2}$ in fig. 4.

. For small gaps, nothing is changed. The mixing interaction of
annihilation (b) predominates. The upper limit upon the mass density
of the fossilized S population is yet over-reached for $\Delta m = 500$ MeV
in the example of fig. 4.

. For very large gaps ($\Delta m = 15$ GeV), we observe that the present
mass density of the surviving S_2 does not depend on Δm and decreases
when ε increases. Physically, for a large gap, reaction (b) is

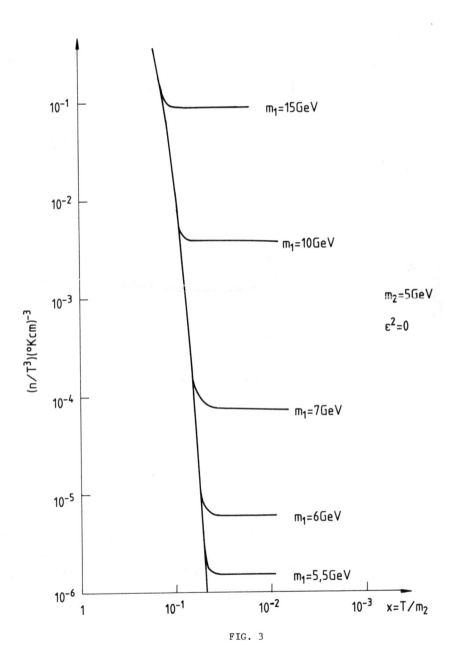

FIG. 3

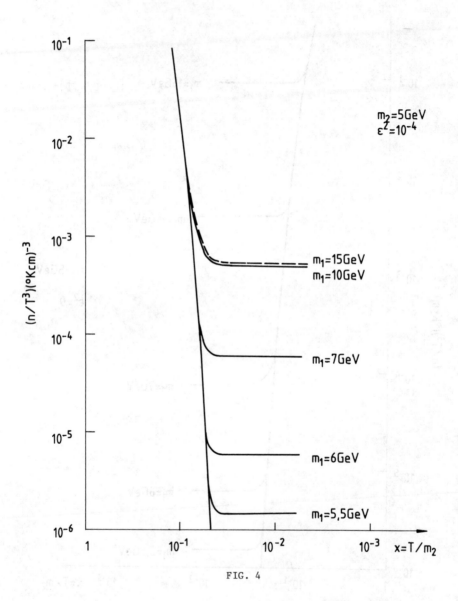

FIG. 4

suppressed and the annihilation (c) acts with an extra factor ε^2 exactly as if the mass M_2 was renormalized to $\varepsilon^{2/3} M_2$.

Finally we give the mass bounds we find from our analysis of the quasi-antidiagonal case:

. If M_2 is larger than

$$(\frac{50 \text{ km/s.Mpc}}{H}) \times \varepsilon^{-2/3} \times 15.0 \text{ GeV} \tag{23}$$

the mass $M_1 (> M_2)$ is free. This is actually the case for very large ε closed to 1 and we are very far from what we call the quasi-antidiagonal case!

. If M_2 is less than the value given by the above formula (and this is the case whenever ε is less than 2.10^{-2}), not only the mass is constrained, but also the gap. We must have:

$$M_2 > \frac{50 \text{ km/s.Mpc}}{H} \times 22 \text{ GeV} \quad \text{and} \quad \frac{M_1 - M_2}{M_2} < 10\% \tag{24}$$

This new result imposes severe constraints on dileptonic theories as we shall see now in section 4.

4. A PHENOMENOLOGICAL MODEL

We apply our results to a class of supersymmetric theories previously considered by John Ellis et al.. In that kind of models they exist three heavy neutral heavy leptons coupling only to the Z° via the formula:

$$\frac{e}{2\sin 2\theta_W} \times Z^\circ_\mu (\tilde{S}^\circ_+ \tilde{S}^\circ_- \tilde{A}^\circ) \begin{bmatrix} \frac{V^2_2 - V^2_1}{2V^2} \begin{bmatrix} 1 & 1 \\ 1 & 1 \end{bmatrix} & \frac{2V_1 V_2}{V^2} \begin{bmatrix} 1 \\ 1 \end{bmatrix} \\ \frac{2V_1 V_2}{V^2} \begin{bmatrix} 1 & 1 \end{bmatrix} & \frac{V^2_1 - V^2_2}{V^2} \end{bmatrix} \gamma^\mu (1 - \gamma_5) \begin{pmatrix} \tilde{S}^\circ_+ \\ \tilde{S}^\circ_- \\ \tilde{A}^\circ \end{pmatrix} \tag{25}$$

We have a mass hierarchy: Mass of $\tilde{A}^\circ$ > Mass of $\tilde{S}^\circ_+$ and $\tilde{S}^\circ_-$.
We define $\tilde{S}^\circ_-$ as the lightest particle. V_1 and V_2 are the respective vacuum expectation values of two different families of higgses $(V^2 = V^2_1 + V^2_2)$.

. The asymmetric vacuum.
We set for example

$$V_2 \simeq V \quad \text{and} \quad V_1 \simeq \varepsilon V$$

where ε is a very small number, possibly equal to zero. If $\varepsilon \neq 0$ $\tilde{S}^\circ_+$ and $\tilde{A}^\circ$ decay and disappear. If $\varepsilon = 0$ there remains $\tilde{S}^\circ_-$ and $\tilde{A}^\circ$ but we have no constraints except that

$$\text{mass of } \tilde{S}^\circ_- > \frac{50 \text{ km/s.Mpc}}{H} \times 30 \text{ GeV} \tag{26}$$

This is a quite high value for the mass of $\tilde{S}^\circ_-$. With a Hubble constant of 75 km/s.Mpc, the $\tilde{S}^\circ_-$ should have to be searched in high energy accelerators with a mass larger than 20 GeV.

. The symmetric vacuum
  We consider now the situation:

$$V_1 \simeq V_2$$

We set

$$\frac{V_2 - V_1}{V} = \varepsilon = \text{very small number}$$

In that case the coupling matrix is quasi-antidiagonal and writes

$$\begin{bmatrix} \dfrac{\varepsilon}{\sqrt{2}} \begin{bmatrix} 1 & 1 \\ 1 & 1 \end{bmatrix} & 1 \\ & 1 \\ 1 & 1 & \sqrt{2}\varepsilon \end{bmatrix} \tag{27}$$

For small ε values the $\tilde{S}^\circ$ leptons annihilate only on the $\tilde{A}^\circ$ still present at the time of their quenching. We can put a limit upon the mass of $\tilde{S}^\circ$:

$$\text{mass of } \tilde{S}^\circ_- > \frac{50 \text{ km/s.Mpc}}{H} \times 27 \text{ GeV} \tag{28}$$

and also upon the mass gap between $\tilde{S}^\circ_-$ and $\tilde{A}^\circ$:

   mass of $\tilde{A}^\circ$ - mass of $\tilde{S}^\circ_-$ < 10% of the mass of $\tilde{S}^\circ_-$

In that case the triplet of leptons ($\tilde{S}^\circ_-$, $\tilde{S}^\circ_+$, $\tilde{A}^\circ$) has to be nearly degenerated. This result imposes severe constraints upon the mass matrix of the theory.

   As a conclusion we remark that the big difference between our analysis and that of Lee and Weinberg concern the quasi-antidiagonal coupling of a dilepton to the Z° vector boson. In that situation, which is not so particular (supergravity imposes V_1 to be closed to V_2), the gap between the dilepton is severely constrained. Therefore, a quasi mass degeneracy of the dilepton is imposed by cosmology.

REFERENCES

* For the study of the departure from chemical equilibrium of the S lepton, see:
  - Lee, B.W. and Weinberg, S., 1977, Phys. Rev. Lett. 39, pp. 165.
  - Dicus, D.A., Teplitz, V.L. and Kolb, E.W., 1977, Phys. Rev. Lett. 39, pp. 168.
  - Gunn, Lee, Lerch, Schramm and Steigman, 1978, Ast. J.

* For the class of supersymmetric models considered in section 4, see:
  - Ellis, J., Ibáñez, L.E. and Ross, G.G., 1982, CERN-TH-3382.

COSMOLOGICAL CONSTANT AND FRIEDMAN UNIVERSES

R. Coquereaux

CERN, 1211 Geneva 23, Switzerland
On leave from Centre de Physique Théorique, CNRS,
13288 Luminy, Marseille, France

A review of the article : "Analytic Discussion of Spatially
Closed Friedmann Universes With Cosmological Constant and Radiation
Pressure" by R.C. and A. Grossmann is given. In this article we
derive explicit formulae for various quantities of interest in the
Universes described in the title ; we also discuss the interpreta-
tion of the cosmological constant in field theories. The analytic
results can be used in order to analyze large structures in the
Universe.

The purpose of this talk is to summarize the content of a
paper written by A. Grossmann and myself, entitled : "Analytic
Discussion of Spatially Closed Friedmann Universes With Cosmological
Constant and Radiation Pressure", published in October 1982 in Ann.
of Physics [1]. In a sense, the title by itself is a good summary
but, the paper being quite long (around 60 pages), it is maybe useful to give here a more detailed account of its content and to go
through several chapters, summarizing the information available to
the astrophysicist interested in cosmological problems.

We consider Friedmann-Lemaître models with positive spatial
curvature, with cosmological term Λ and radiation pressure. The
evolution of the spatial radius of curvature R as function of t
(the cosmic time) is given by

$$\frac{1}{R^2} \left(\frac{dR}{dt}\right)^2 = \frac{C_r}{R^4} + \frac{C_m}{R^3} - \frac{1}{R^2} + \frac{\Lambda}{3} \tag{1}$$

where C_r and C_m are constants characterizing the contribution
of radiation and matter to the energy.

J. Audouze and J. Tran Thanh Van (eds.),
Formation and Evolution of Galaxies and Large Structures in the Universe, 77–83.
© *1984 by D. Reidel Publishing Company.*

The form of Eq. (1) is not particularly appealing ; the equation can be recast into a more tractable form by an appropriate change of variables.

One introduces therefore :
- the characteristic length scale $\Lambda_c = 4/9(Cm^2)$
- the dimensionless radiation parameter $\alpha = C_r \Lambda_c$
- "   "   cosmological constant $\lambda = \Lambda/\Lambda_c$
- "   "   temperature $T = 1/R\sqrt{\Lambda_c}$
- "   "   conformal time $d\tau = dt/R$

Then, one obtains

$$\left(\frac{dT}{d\tau}\right)^2 = \alpha T^4 + \frac{2}{3}T^3 - T^2 + \frac{\lambda}{3} \tag{2}$$

The minus sign in front of T^2 means that we are discussing here only spatially closed Friedmann Universes.

The qualitative behaviour of the solutions of Eq. (2) can be analyzed quite easily by writing it as follows :

$$\left(\frac{dT}{d\tau}\right)^2 + V_\alpha(T) = \frac{\lambda}{3} \tag{3}$$

This is the energy equation for a one-dimensional mechanical system with co-ordinate T, potential $V_\alpha(T)$ and total energy $\lambda/3$. All possible solutions of Eq. (2) are then described by a horizontal line (half-line, or segment) in the $(V_\alpha(T),T)$ plane. Indeed, the kinetic energy $(dT/d\tau)^2$ being non-negative, the associated mechanical system never penetrates under the curve $V_\alpha(T)$; see Figs. 1 and 2.

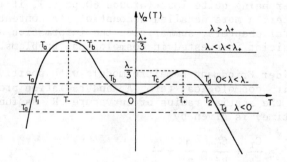

- Figure 1 -

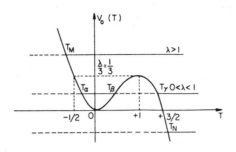

– Figure 2 –

The solutions of a differential equation of the kind

$$(\frac{df}{d\tau})^2 = (\text{polynomial in } f \text{ of degree } \leq 4)$$

are elliptic functions of order two. Therefore, from the analytic
point of view, it is clear (because of Eq. (2)), that $T(\tau)$ is in
all cases an elliptic function of order two. Recall that elliptic
functions generalize the usual trigonometric and hyperbolic functions in the sense that they are meromorphic and doubly periodic
in the complex plane. Usual trigonometric (or hyperbolic) functions are recovered when one of the periods becomes infinite
(this corresponds physically to the case where λ is equal to the
vertical co-ordinate of one of the extrema of the curve $V_\alpha(T)$
(Figs. 1, 2)).

Elliptic functions of order two (which take two times all
complex values in a fundamental periodicity cell) can be conveniently expressed in terms of the Weierstrass functions $\wp$, ξ,
σ ; notice that only $\wp$ is elliptic. Besides, any rational function of an elliptic function is elliptic. One of the purposes of
the article [1] is to give analytic expressions for $T(\tau)$ and related quantities – density function, acceleration function, etc. –
in terms of Weierstrass functions. This study is made in 1) for all
possible Friedman Universes (for all choices of α and λ).

For example, a Universe with $\alpha = 0$ and $\lambda > 1$ and starting
with a big bang is described by the equations (see also Ref. 2))

$$T(\tau) = 6\left[\wp(\tau) + \frac{1}{12} \right]$$

$$\sqrt{\frac{\Lambda}{3}}\, t(\tau) = \text{Ln}\left[\frac{\sigma(\tau_f + \tau)}{\sigma(\tau_f - \tau)} \right] - 2\tau\, \xi(\tau_f) \tag{4}$$

In these expressions, the Weierstrass functions are associated to
a lattice whose periods can be expressed in terms of α and λ.
Here τ_f is "the end of conformal time" (if τ_f is big enough –
at least more than 2π – a very old observer could see the
back of his head!). However, when $\tau \rightarrow \tau_f$, $t \rightarrow \infty$, such a Universe
starting with a big bang expands for ever (as measured in cosmic
time t). Figure 3 shows the behaviour of $T(\tau)$; only the first
part of the curve, from $\tau = 0$ (the big bang) to $\tau = \tau_f$ (the end
of conformal time) is physically meaningful. According to Ref. 3),
the three-dimensional distribution of quasars known to date shows
a singular zone, about 200 Mpc wide, 3000 Mpc away, where no
quasars are observed. For precise values of the cosmological parameters, this empty band, when analyzed in the above-mentioned
Friedmann-Lemaître model, appears like an equatorial band of the
hypersphere S^3. The motivation for our work [1] comes actually
from the analysis made in Ref. 3).

A Friedmann Universe is determined by the two parameters α,
λ ; a further dimensionless parameter τ_0 is needed to specify
a time of observation (the present-day Universe), finally a scale
Λ_c allows us to convert dimensionless quantities into centimetres.
Unfortunately, the measurable quantities do not coincide with the
previous ones and one has to know how to express the former in
terms of the latter. Appropriate changes of variables (with
$\alpha \neq 0$ and $\lambda \neq 0$) are given in Ref. 1).

The red shift function $z = T(\tau)/T(\tau_0) - 1$ considered as a
function of τ (parameter time of emitter) for fixed τ_0 (parameter time of the observer) is also an elliptic function of order
two. Its expression is therefore analytically studied in Ref. 1)
for all possible models.

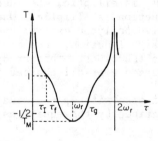

- Figure 3 -

We also study the following problem : suppose that an object X (a quasar) is analyzed by an observer T (the earth) which records its red shift ; let P be another observer, what is the redshift of X as recorded by P? This kind of "red shift transformations" involves not only the relative positions of X, T and S (i.e., the geometry of S^3) but also the dynamics of our Universe (i.e., the Friedman equations). For example, let X be a quasar sitting on an equatorial S^2 sphere of S^3 with respect to a pole P, z_P its red shift as recorded by P and z_T its red shift as recorded by an observer T. Then,

$$\frac{z_T}{z_P} = \frac{\sigma\left[arctg(\frac{tg\ell_p}{cos\beta})\right] \sigma\left[2\tau_0 - arctg(\frac{tg\ell_p}{cos\beta})\right]}{\sigma^2\left[\tau_0 - actg(\frac{tg\ell_p}{cos\beta})\right]} \cdot \frac{\sigma^2(\tau_0 - \frac{\pi}{2})}{\sigma(\frac{\pi}{2})\sigma(2\tau_0 - \frac{\pi}{2})}$$

(5)

Here we restrict ourselves to the case of the Universe already described by Eqs. (4). In this relation, ℓ_p denotes the cosmic latitude of T with respect to the pole P and β is the sighting angle between the pole P and the quasar x ; the Weierstrass σ function is of course associated to the lattice characterized by the value λ of the reduced cosmological constant.

This kind of relation allows one to study large scale structures of space-time ; if there is any regularity in the large scale repartition (or motion) of matter in the Universe, this regularity can only be "seen" by very particular observers (in the same way, the regular elliptic motion of the Halley's comet, which could be seen like an ellipse from above the sun's north pole, looks like a spiraling path against the winter heavens!).

Using Eq. (5) and the values of the cosmological parameters given in Ref. 3), one may recover the observed direction-dependent gap in the repartition of quasars 3).

Friedmann's equation can be, of course, integrated numerically, However, the development of small programmable pocket computers makes now possible the direct computation of interesting quantities by using explicit formulae (this is also a further motivation for our study). The Weierstrass functions are usually not tabulated, for this reason we give also in Ref. 1) several methods for evaluating these functions.

A lattice in the complex plane can be characterized either by its periods ω_r, ω_i or by the so-called Weierstrass invariants g_2, g_3, the relation between these two couples of values being an integral relation. Depending on which couple of values is

known, one or another method should be preferred and we discuss
in Ref. 1) several possibilities.

Tables are also available in microfiche form from the
author for a wide class of models.

The last topic discussed in Ref. 1) is the significance of
the cosmological constant in quantum field theories. The basic
idea is that Λ measures the energy of the vacuum ; by "vacuum"
we mean the lowest state of the field configuration (this is reminiscent of the zero-point energy $\frac{1}{2} \hbar\omega$ of the harmonic oscillator).
We discuss therefore the analogy between Λ and the vacuum energy
of electromagnetism (Casimir effect). Then we analyze the role of
Λ from the point of view of spontaneously broken gauge theories
(in the Weinberg-Salam model and in grand unified theories, GUT).
We also give some information about the analysis made in supersymmetric theories. Our purpose, in the last part of the paper [1]
is not to present new ideas but mainly to guide the reader among
several aspects of this problem.

Another interpretation of the cosmological constant (not
discussed in Ref. 1)) appears if one wants to embody general relativity and non-Abelian gauge theories in a unique formalism of
the Kaluza-Klein type. The basic idea is the following : we assume that we live in a (4+d) dimensional Universe E endowed
with a metric invariant under a group G of isometries. Under
some general assumptions (see, i.e., Ref. 4) for precise details),
E can be written as a local product M S where M is interpreted as space-time and S is an "extra-dimensional world" also
called "internal space" that we do not see because the typical
dimension of S (usually assumed to be compact) is very small
(typically 10^{-32} cm). The scalar curvature R^E of our (4+d)
dimensional Universe E can be written, because of G invariance,
as a sum of contributions involving only M. Roughly speaking, if
one assumes that the shape of our internal space S is constant
when one moves in space-time M, we obtain

$$R^E = R^M(x) - \frac{1}{4} Tr(F_\mu(x) \cdot F^{\mu\nu}(x)) + R^S$$

$R^M(x)$ is the scalar curvature of space-time and describes gravity ;
the second term is the Lagrangian of gauge fields associated with
the group G and generalizes the Maxwell Lagrangian ; the third
term describes the scalar curvature of our internal space and can
be identified with the cosmological constant.

I would like to thank A. Grossmann who kindly accepted to
read the manuscript.

References

1) Coquereaux, R. and Grossmann, A., Ann.Phys. 143 No 2 (1982) 296-356.
2) Kaufman, S.E. and Schucking, E.L., Astrophys.J. 76 (1971) 583.
3) Fliche, H.H., Souriau, J.M. and Triay, R., Astron.Astrophys. 108 (1982) 256-264.
4) Coquereaux, R. and Jadczyk, A., CERN Preprint TH.3483 (1982), to appear in CMP.

II

LARGE STRUCTURES : PANCAKES-CLUSTERS OF GALAXIES

DARK MATTER AND SHOCKED PANCAKES

J.R. Bond[1,2], J. Centrella[3], A.S. Szalay[4,1], and J.R. Wilson[5]

[1] Institute of Astronomy, Cambridge University
[2] Institute of Theoretical Physics, Stanford University
[3] Astronomy Department, University of Illinois
[4] Department of Atomic Physics, Eötvös University,
                                                    Budapest
[5] Lawrence Livermore Laboratories.

We classify massive stable collisionless relics of the Big Bang into three categories of dark matter: hot, with damping mass about supercluster scale; warm, with damping mass of galactic or cluster scale; and cold, with negligible damping. The first objects that form in universes dominated by hot and warm relics are pancakes. Coupled one-dimensional N-body and Eulerian hydrodynamical simulations follow the nonlinear evolution of pancakes, the separation of baryons from dark matter via shock formation and the evolution of the shocked gas by conduction as well as by cooling. We sketch a simple analytic theory based upon the uniformity of pressure over the shocked region which accurately describes our numerical results. Only $\sim$10-20% of the gas cools sufficiently to fragment on sub-galactic scales in neutrino-dominated (hot relic) theories. Cooling is efficient for warm relics. In all cases, the typical fragment size is $\sim 10^9$-10^{10} $M_\odot$.

1. CLASSIFICATION OF DARK MATTER CANDIDATES

Stable collisionless relics of the Big Bang are perhaps the most attractive candidates for the dark matter. Bond and Szalay (1983) have classified the possibilities into three basic types defined by their background velocity dispersion: relics may be hot, warm, or cold. The canonical example of a hot relic is a massive neutrino, with velocity dispersion $6(m_\nu/30eV)^{-1}(1+z)$ km s^{-1}.

J. Audouze and J. Tran Thanh Van (eds.),
Formation and Evolution of Galaxies and Large Structures in the Universe, 87–99.
© 1984 by D. Reidel Publishing Company.

However, any particle which is massive, stable and decouples when
relativistic at an epoch when the temperature of the universe was
$\lesssim T_{qh} \sim 200$ MeV is a hot particle. Any particle which decouples
at a temperature above T_{qh} is warm, with velocity dispersion
$0.085(100/g(T_d))^{1/3}(1 \text{ keV}/m_x)$ km s^{-1}. Here, we call our hypothetical collisionless relic X, and m_x is its mass; $g(T_d)$ is the effective
number of degrees of freedom at the X-decoupling temperature, T_d.
The quark-hadron phase transition temperature, T_{qh}, is the
approximate decoupling boundary between warm and hot since above
T_{qh} the number of relativistic species present is large due to all
the liberated quark-antiquark pairs. At the neutrino decoupling
temperature, $T_d \sim 1$ MeV, g is only 10.75, whereas g $\approx$ 60 just above
T_{qh}. Near electroweak unification at $\sim$100 GeV, g $\sim$ 100 in the
minimal Weinberg-Salam SU(2) ⊗ U(1) ⊗ SU(3) theory. Near grand
unification energies, $\sim 10^{15}$ GeV, g $\sim$ 160 in the minimal Georgi-
Glashow (1974) SU(5) theory. Supersymmetric theories increase g
by only a factor of about two.

The relationship between the density parameter of relativistic
decouplers and their mass is: $\Omega_x = 1.1 \text{ h}^{-2} (100/g(T_d)) \text{ } m_x/1$ keV,
where h is Hubble's constant in units of 100 km s^{-1} Mpc^{-1}. For
g = 10.75, we get the usual 20-10^2 eV mass needed for one species
of massive neutrino to close the universe. The variation corresponds to the range h = 0.5 to 1. For $T_d > T_{qh}$, and for minimal
theories, $m_x \cong 2 \times 10^2 - 2 \times 10^3$ eV is the mass required for $\Omega = 1$.
This mass could be much larger if: (1) there is no large plateau
in g(T) - i.e. no desert; (2) significant entropy generation
occurs after $T_d - \Omega_x$ scales inversely with the entropy amplification factor. The first option could be restricted since the
number of stable relativistic neutrino species at the time of
primordial helium generation cannot be much greater than three to
avoid overproduction (Schwartzman 1969, Olive et al. 1981); this
could translate into a constraint on the number of leptoquark
families. The second option suffers from constraints on the
allowable entropy generation after baryon synthesis.

Hot particles have a damping scale arising from the cons-
tructive effects of gravitational attraction on large scales and
the destructive effects of their random velocity on small scales:
$M_d \cong 3.4 \text{ } m_p^3/m_x^2$ (Bond, Efstathiou and Silk 1980), where
$m_p = 1.22 \times 10^{22}$ MeV is the Planck mass; this is of supercluster
scale. Warm particles damp below the scale

$$M_d \cong 0.11 \text{ } (m_p^3/m_x^2) \text{ } (100/g(T_d))^{4/3} \qquad (1)$$

(Bond, Szalay and Turner 1982, Bond and Szalay 1983), which is
either the scale of galaxies or of clusters depending upon h. If
m_x is very large, due to either (1) or (2), then the damping
scale can be very small, and the particles are effectively cold.
Indeed, we define cold particles to be those with almost no

velocity dispersion, and which thus have $M_d \approx 0$. Collisionless
relics which decouple when they are nonrelativistic are examples.
Preskill, Wise and Wilczek (1982) have recently pointed out that
oscillations of "classical" fields, i.e. of boson vacuum expectation values, can lead to a large time-averaged energy density,
as well as to a rapidly fluctuating part. They claim that spatial
fluctuations in the field have energy density growth identical to
that of nonrelativistic decouplers.

The canonical hot particle is the massive neutrino. Another
candidate is the Majoran, a goldstone boson whose raison d'être
is to generate neutrino masses via spontaneous symmetry breaking
(Georgi, Glashow and Nussinov 1981, Gelmin, Nussinov and Roncadelli
1982). All background $\nu\bar{\nu}$'s would annihilate into a sea of Majorans
- the temperature of which would be higher than that of the background photons. If they are massive, then m $\lesssim$ 10 eV is required
in order to have $\Omega \lesssim 1$. This implies uncomfortably large damping
masses· and, as we shall see, very little gas cooling in a Majoran-
dominated universe. Suggestions for warm relics have included
the gravitinos and photinos of supersymmetric theories, and right-
handed neutrinos. Any of these could also decouple when nonrelativistic, and thus be cold. Heavy neutral leptons of the
sort discussed by Lee and Weinberg (1977), primordial black holes,
and monopoles are other cold relic candidates. If strings form
in phase transitions in the very early universe, and if they
primarily exist as loops of subgalactic dimensions (Kibble 1983),
galaxy formation in string-dominated universes will effectively
follow the cold scenario. The model for classical field
oscillation is provided by the axion (Preskill, Wise and Wilczek
1982).

In all cases in which relics form the dark matter, a remark-
able coincidence is required - namely that Ω_x and Ω_B are not too
dissimilar. We know that Ω_B cannot be too small, or else cooling
on any scales would have been too inefficient. This is an anthropic
argument which rules out extreme variations of Ω_B. If $\Omega \approx 1$ is
required as a consequence of inflation (or simplicity), and m_x
is given by the particle physics, then H_0 would be adjusted so
that the $\Omega_x \sim m_x h^{-2}$ relation is enforced for relativistic
decouplers; the coincidence $\Omega_x \sim \Omega_B$ implies $m_x \sim m_N s^{-1}$, where
$s \sim 10^9$ is the entropy per baryon and m_N is the nucleon mass. In
a simple model of baryon generation, this becomes
$m_x \sim m_N m_p m_{VB}^{-1} \alpha_{GUT} \varepsilon_{CP}$, where m_{VB} is the mass of the intermediate vector boson responsible for baryon generation, α_{GUT} is
the fine structure constant at unification energies, and ε_{CP} is a
CP-violating parameter. Why should such quantities be inter-
related in this manner? The case of nonrelativistic decouplers
requires perhaps an even more stringent restriction upon the
particle physics, namely that the freeze-out temperatures for the
reactions which create X's must be - within some narrow range - a

prescribed fraction of m_x (15 $\lesssim m_x/T_{fx} \lesssim$ 50 for 1 GeV $\lesssim m_x \lesssim 10^{15}$ GeV).

2. COOLING SCALE AND PANCAKES

We have seen that the $m_p{}^3 m_x{}^{-2}$ damping scale applies for hot and warm particles; compare this with the mass scale of stars which is set by the combination $m_p{}^3 m_N{}^{-2} = 1.9 M_\odot$. Another scale at high mass enters into the determination of the fluctuation spectrum: the horizon mass at equipartition between relativistic and nonrelativistic constituents, which occurs at $z_{eq} = 25000 \ \Omega h^2$ when the photon temperature is $T_{\gamma eq} = 5.8 \ \Omega h^2$ eV:

$$M_{Heq} = 0.2 \ m_p{}^3 \ T_{\gamma eq}{}^{-2} \approx 10^{16} \ (\Omega h^2)^{-2} \ M_\odot$$

An initially scale-free density spectrum evolves in the linear phase to one in which there is a sharp damping cutoff at masses smaller than M_d, a strong flattening between M_d and M_{Heq} which then matches onto the original spectral shape above M_{Heq} (Peebles 1982, Bond, Szalay and Turner 1982, Bond and Szalay 1983). It has been conventional to associate the appearance of voids and strings in the galaxy distribution with a large damping cutoff. An important unresolved issue is whether the shoulder below M_{Heq} is sufficient to generate such structure. In any case, for warm and hot particles, the first structures to become nonlinear will be on the scale M_d, will collapse preferentially along one axis, becoming highly asymmetric, and result in shock formation in the central regions (Zeldovich 1970, Sunyaev and Zeldovich 1972). The first structures to collapse in the cold scenario may also be asymmetric and lead to shocks; however, instead of a smooth collective inflow, the shocks may be more localized, arising from cloud-cloud collisions.

Binney (1977), Rees and Ostriker (1977) and Silk (1977) have demonstrated how galaxy masses may be related to a cooling scale. It is instructive to go through this exercise to demonstrate what must be done to get cooling in larger structures - from which galaxies ultimately arise. The Rees and Ostriker (1977) development yields

$$M_{cool} \sim \frac{m_p{}^3}{m_N{}^2} \frac{\alpha^5 m_p}{(m_e m_N)^{\frac{1}{2}}} \frac{\Omega_B}{\Omega} \sim 2 \times 10^{10} \frac{\Omega_B}{\Omega} \cdot M_\odot \ . \tag{2}$$

The ingredients which go into obtaining this scale are as follows. A virialized homogeneous sphere cools via bremsstrahlung faster than free-fall if its temperature satisfies:

$$T/m_e < M \ m_p{}^{-3} \ m_N{}^2 \ (m_N m_e)^{\frac{1}{2}} \ m_p{}^{-1} \ \alpha^{-3} \ \Omega \ \Omega_B{}^{-1} \ .$$

However, the temperature $T \sim M^{2/3}(1+z_t)$ depends not only upon the mass, but also upon the epoch of turn-around, z_t. Indeed, if z_t is too large, Compton cooling should replace bremsstrahlung. In any case, stability can never be regained if T falls in the helium-hydrogen recombination cooling regime. Since the ionization energy of helium is $2\alpha^2 m_e$, and the characteristic temperature for helium recombination is some fraction of this, we obtain the scale M_{cool}. There are 3 ways to increase M_{cool}: (1) raise z_t into the Compton cooling epoch; (2) utilize central condensation so that T can be lower in the central regions since the matter has less far to fall before shocking; (3) stretch the sphere into an oblate configuration so again the gravitational acceleration is less. Effects (2) and (3) operate in pancakes; z_t is constrained by limits on the temperature fluctuations in the microwave background, hence (1) cannot be pushed too far.

3. PANCAKE SHOCK CALCULATIONS

This work is described more fully in Bond, Centrella, Szalay and Wilson (1983), hereafter BCSW. Here, we outline the methods and give the main results. A pancake collapse similar to our runs without conduction has recently been computed by Shapiro, Struck-Marcell and Melott (1983).

3.1 Initial Conditions and the Zeldovich Solution.

We let ξ and x denote comoving Lagrangian and Eulerian spatial variables. We ignore the effects of random velocity dispersions of the collisionless relics since these redshift away as the universe expands, and are small relative to the gravitationally-induced velocities at the time of pancaking (Bond, Szalay and White 1983). For such cold initial conditions, the Zeldovich (1970) solution describes the deviations of the particle positions at time t from their initial values by

$$x(\xi,t) = \xi - b(t)\ \eta(\xi)$$

where $b = (1+z)^{-1}$ in the $\Omega=1$ models we are most concerned with; and $\eta(\xi) = (kb_c)^{-1} \sin{(k\xi)}$ for a single plane wave of comoving wavenumber k and wavelength $L \equiv 2\pi k^{-1}$. A caustic forms when x=0, at $b=b_c$. This formula is exact in the linear regime, and in one-dimension until caustic formation. It also describes the early nonlinear phases of 3-dimensional evolution rather well.

The distribution of the principal eigenvalues, λ_i, of the (anti-) strain tensor $(\partial\eta_i/\partial\xi_j)$ describes the 3-dimensional patterns which first appear in the nonlinear evolution of a density fluctuation spectrum with a damping cutoff. The overdensity is related to these eigenvalues by

$$1 + \delta = (1-b\lambda_1)^{-1} (1-b\lambda_2)^{-1} (1-b\lambda_3)^{-1} . \tag{3}$$

The principal axes, 1,2,3, are always ordered so that $\lambda_1 \geqq \lambda_2 \geqq \lambda_3$. Therefore, by definition, collapse is most rapid along the 1-axis; and caustics, where $\delta \to \infty$, occur at λ_1-maxima. The surfaces of λ_1-maxima are generally curved, with curvature at most of order the damping scale. Different surfaces intersect at points of degeneracy of the 1 and 2-axes, where $\lambda_1 = \lambda_2$. More extreme point-like configurations form where all 3 axes are degenerate. These topological structures are catalogued by Arnold, Shandarin and Zeldovich (1982). In the neighbourhood of every λ_1-maxima surface, the flows are essentially one-dimensional: we expect our 1D simulations to describe the post-caustic evolution of these regions rather well. Directions 2 and 3 may continue undergoing transverse expansion which differs little from the Hubble expansion initially. However, within a few Hubble times of the collapse redshift, transverse flows toward the 1-2 degeneracy lines (strings) may become important. To model this in our 1-D calculations, we utilize an anisotropic expansion factor, $a_2(t) = a(t)(1-b(t)\lambda_2)$, where λ_2 is taken to be constant. This approximation may be quite reasonable since BCSW demonstrate that the tidal force toward strings dominates the pressure gradients opposing such motion, so dark matter and shocked gas should flow toward the strings together, following the Zeldovich solution.

3.2 Evolution Equations.

Neutrinos are followed by direct N-body simulation of their equations of motion: $dva^2/dt = -\nabla\phi$, where $v=dx/dt$ is the comoving peculiar velocity. The gravitational field equations reduce - for these nonrelativistic particles - to Poissons equation, except that the source is the overdensity relative to the background. This couples the N-body code to the gas dynamical code - which utilizes Eulerian hydrodynamics in comoving space. We solve transport equations for the following quantities: (1) Baryon number. (2) Momentum - artificial viscosity is used to treat shocks. (3) Matter energy, including ionic and electronic contributions. Artificial viscosity provides the shock heating. Energy losses arise from bremsstrahlung, Compton cooling, and He and H recombination cooling. Flux-limited conduction is included: transfer primarily occurs via electron-electron collisions, although ion-ion collisions are also incorporated. (4) Ion energy. This transport equation includes ion-ion conduction, and a term describing the relaxation of the ionic temperature towards the electron temperature via ion-electron collisions. It is the ions which are shock-heated and the electrons which cool. The ion and electron temperatures can therefore differ - especially near the shock front. The equation of state is that of an ideal gas with ionization fraction determined by the balance of collisional ionization ($eH \to epe$) with recombination ($ep \to H\gamma$). We do not explicitly calculate the He ionization fraction; it is, however, included in the cooling rate. Complications arise since the pancakes are generally optically thick in their central regions to Ly α and to radiation above the Ly edge. Still, lower energy radiation escapes, allowing continual rapid cooling.

3.3 Pancake Timescales.

Some of the timescales of relevance to the shock problem are plotted as a function of temperature in Fig. 1 for a specific choice of the baryon density – which corresponds to an overdensity of 10^2 above that of the background gas.

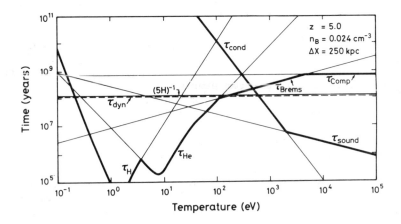

Figure 1.   Characteristic timescales under typical pancake conditions.

The bremsstrahlung and He and H recombination cooling times scale with density as n_B^{-1}, but for given n_B are redshift-independent. Compton cooling is n_B-independent, but varies as $(1+z)^{-4}$. One does not have to go to very large redshift before Compton cooling completely overtakes bremsstrahlung. The conduction timescale across a region of size Δx of uniform density is also plotted; the scaling is $\tau_{cond} \sim n_B (\Delta x)^2 (1+z)^{-2}$. This illustrates that conduction is primarily important in the outer shocked regions. When the sound crossing time across Δx, $\tau_s \sim (1+z)^{-1} \Delta x$, exceeds τ_{cond}, we are in the flux-limited trans-port regime. These cooling and conduction timescales are compared with the dynamical timescale for the collapsing baryon layer, or with $(5H)^{-1} \sim (1+z)^{-3/2}$ – the latter choice is better. (See §3.5).

The major point of Fig. 1 is that once T falls below 100 eV,
cooling is very rapid.

3.4 Shock Evolution.

We began our calculations in the linear regime, $\delta \ll 1$. The
initial phase of the collapse just demonstrates the validity of
the Zeldovich solution - even for the gas, which has a small
pressure. However, near the redshift of caustic formation, z_c -
the collapse redshift - the central density rises rapidly (Eq.3),
and the gas pressure builds up due to the adiabatic compression.
The shock forms approximately at the point where the incoming ram
pressure equals this gas pressure (Sunyaev and Zeldovich 1972),
i.e. near the sonic point. The collisionless relics begin to
separate from the gas after this. In Fig. 2, overdensities of
baryons and neutrinos are given as a function of comoving position
for a pancake of wavelength 25 Mpc which caustics at $z_c=5$. The
universe has $\Omega=1$, $\Omega_B=0.091$ and h=1. The neutrinos form density
spikes at the edges of the well known phase space spirals which
appear in 1D simulations (See, e.g. Bond, Szalay and White 1983).
The gas is confined by the ram pressure, much of it to a region
smaller than 10 kpc. The spread in neutrinos is more than an
order of magnitude greater at these times.

The temperature profile evolution for the 25 Mpc pancake is
shown in Fig. 3. Conduction clearly transports energy from the
hot exterior inward, thereby flattening the temperature gradient.
The profiles drop precipitously toward 1 eV and hydrogen recombination below T $\sim$ 100 eV. By z=4.8, $\sim$9% of the gas has cooled
below this temperature; by z=3, 11% has cooled with conduction,
14% without. It takes conduction time to operate. This is
especially evident in Fig. 4: a 54 Mpc pancake begins with only
$\sim$4% of the gas cold; an inward-eating conduction front breaks
through the center by z=3.8, resulting in essentially a flat
profile by z=3. For small wavelength runs, the temperatures
achieved are lower, conduction is relatively unimportant, and a
large fraction of the gas cools.

Transverse compression results in adiabatic heating of the
outer regions, but very rapid cooling of the inner ones. The
boundary between the two regimes occurs at a mass fraction q=0.15
for our standard L=25 Mpc model, hence does not result in a
significant increase in cooling fraction.

3.5 Semi-analytic Shocks.

The major feature of our numerical runs is that *the pressure
is constant behind the front, and is equal in magnitude to the
ram pressure*. The ram pressure can be determined from the
Zeldovich solution exterior to the shocked region (Sunyaev and

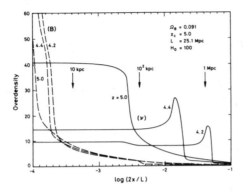

Figure 2. Overdensity of baryons, B, and of neutrinos, ν, -
relative to their backgrounds - against comoving position.

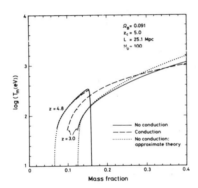

Figure 3. Temperature profile against gas fraction.

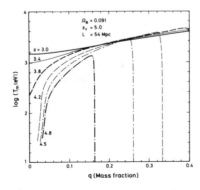

Figure 4. Temperature profile of 54 Mpc - pancake shows the
central breakthrough of the conduction front.

Zeldovich 1972):

$$p(q,z) \approx P_{RAM} \approx 1.9 \times 10^{-14}(1+z)^4 (hL_{10})^2 \Omega_B h^2 f(z) \text{erg cm}^{-3} , \qquad (4)$$

for $q < q_s(z)$, where L_{10} is the wavelength in 10 Mpc units and $q_s(z)$ is the shocked mass fraction at z. The redshift-dependent function $f(z)$ depends upon the initial profile, the transverse motion, and the value of Ω. For $\lambda_2=\lambda_3=0$, f differs greatly from unity only when q_s is large (BCSW): the $(1+z)^4$ term dominates the P_{RAM}-evolution.

In conjunction with Eq.(4), the Rankine-Hugoniot relations for conservation of baryon number, momentum and energy across the front — with incoming quantities determined by the Zeldovich solution — yield a complete theory of the shocked region. For example, the temperature just behind the front is approximately

$$T(q, z_s(q)) \cong 430 (1+z_s) q^2 (h L_{10})^2 \text{ eV} \qquad (5)$$

The neglect of pressure gradients, and also of gravitational potential gradients, which is justified by the ram confinement, reduces our full equations to a single one for T:

$$\left(\frac{\partial \ln T}{\partial \ln a}\right)_q \equiv -\nu(a,T,q) \equiv -\nu_s - \nu_c , \qquad (6)$$

$$\nu_s \equiv -\frac{2}{5}\frac{d \ln P_{RAM}(a)}{d \ln a} , \qquad (7)$$

$$\nu_c \equiv (5H(a) \tau_{cool}(a,T))^{-1} . \qquad (8)$$

For simplicity, we have not included conduction, which would turn Eq.(6) from an ODE into a PDE. Near $a_s(q) = (1+z_s(q))^{-1}$, when the shock is at gas fraction q, $T \sim T_s(a_s/a)^\nu$, hence we term ν the cooling power. The initial condition for Eq.(6) is Eq.(5). For adiabatic cooling, only ν_s is non-zero; using (4), we have $\nu_s \sim 8/5$ for $f \approx 1$. Eq.(6) is easily numerically integrated with all correction factors included. The agreement with our full numerical no-conduction runs is very good, as Fig. 3 demonstrates.

4.   COOLING FRACTIONS AND FRAGMENT SIZES

In Fig. 5, we plot the fraction of gas which has cooled below 100 eV by z=3 for z_c=5 collapses, q_c. Below 100 eV, cooling down to 1 eV occurs rapidly; 100 eV is also the virial temperature for

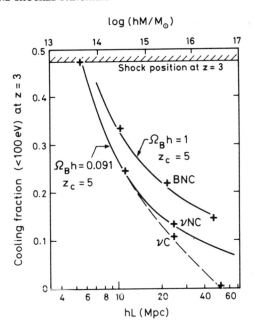

Figure 5. Fraction of cold gas as a function of pancake scale.
Crosses are numerical results: νC and νNC are neutrino-dominated
universes with and without conduction, for which h=1; BNC are pure
baryon universes with no conduction included. The solid and dashed
curves are analytic results; the latter includes conduction effects,
the former does not.

a typical galactic halo. The solid line gives the solution to
Eq.(6) with only bremsstrahlung cooling included, and transverse
Hubble expansion assumed.

$$q_c = A(hL_{10})^{-1} (1 + B \, \Omega_B \, h \, (hL_{10})^2)^{1/3}, \qquad (9)$$

$$A = 0.27 \left(\frac{1+z_c}{6}\right)^{3/10} \left(\frac{4}{1+z}\right)^{4/5} ,$$

$$B = 1.0 \left(\frac{1+z_c}{5}\right)^{1/10} \left(\frac{1+z}{4}\right)^{12/5} \frac{\{1-(\frac{1+z}{1+z_c})^{1/10}\}}{0.04} .$$

A and B are only weakly dependent upon z_c, assuming we are not in
the Compton cooling regime, but are relatively sensitive to z.
In BCSW, we also give an approximate solution with conduction
included (dashed curve). Both expressions agree well with our
full runs.

The comoving size of the cooled region can be computed from the cooled fraction:

$$d \cong 4.7 \, h^{-1} \, q_c \, (hL_{10})^{-1} \text{ kpc },$$

which turns out to be much smaller than the usual 3-dimensional Jeans length. It is more appropriate to consider the longitudinal dimension of the fragmenting region fixed at d; the transverse dimension of the most rapidly growing perturbation is then given by the Sunyaev-Zeldovich length:

$$L_{SZ} \cong 77 \, (1+z)^{-1} \, (q_c L_{10})^{-1} \text{ kpc}.$$

The fragments are thus initially extremely elongated, but can rapidly undergo transverse collapse at the freefall rate due to the efficiency of cooling. The mass of the fragments is

$$M_{SZ} \approx 5 \times 10^8 \, (\Omega_B h)^{-1} \, (hL_{10} q_c)^{-1} \, M_\odot . \tag{10}$$

According to Eq.(9), M_{SZ} is only weakly dependent upon the pancake mass; for $\Omega_B \sim 0.1$, we have $\sim 10^{10} \, M_\odot$ representing the typical fragment scale.

5. DISCUSSION

If we assume the density fluctuations are initially adiabatic, the nature of the dark matter determines how the large scale structure first arises. If cold relics dominate, the theory is a variant of the hierarchical clustering picture developed by White and Rees (1978) for the isothermal picture with dark matter. Peebles (1984) and Primack (1984) discuss the many positive aspects of this picture in this volume. However, it is difficult to see how (1) superclusters and large voids arise, (2) Ω can be one – if indeed it is one. If warm relics dominate, cooling is no problem, but again (1) and (2) do not come naturally. Further, dwarf irregulars and ellipticals would have to arise via fragmentation in the warm scenario, and if M_d is of cluster scale, so would all galaxies.

In neutrino-dominated models, (1) and (2) could naturally follow. On the other hand, building a theory with $\sim 90\%$ of the gas too hot to immediately condense on galaxies could be a problem. The X-ray emitting gas in rich clusters cannot have a mass much larger than ~ 2 times that in the galaxies of the clusters (Ku et al. 1982). However, since rich clusters may arise at the points of 3-fold eigenvalue degeneracy, and their gas has $T \sim 7$ keV,

which is significantly higher than that of pancake gas, we should
not identify the two sorts of hot gas. Intergalactic gas with
$T > 10$ eV and one-tenth the critical density cannot be ruled out
(Sherman 1982). Indeed, it may be desirable to have most of the
baryons unclustered on galactic scales, since there is apparently
a gap between the baryon density required at the primordial nucleosynthesis epoch to generate consistent light element abundances
(Gry 1984, Schramm 1984) and that inferred from luminous matter.
Thus, the neutrino theory may well survive this cooling problem.

REFERENCES

Arnold, V.I., Shandarin, S.F. and Zeldovich, Ya.B. 1982, Geophys.
    Astrophys. Fluid Dynamics 20, 111.
Binney, J. 1977, Ap. J. 215, 483.
Bond, J.R., Efstathiou, G. and Silk, J. 1980, Phys. Rev. Lett. 45,
    1980.
Bond, J.R. and Szalay, A.S. 1983, Ap. J., in press.
Bond, J.R., Szalay, A.S. and Turner, M.S. 1982, Phys. Rev. Lett.
    48, 1636.
Bond, J.R., Centrella, J., Szalay, A.S. and Wilson, J.R. 1983,
    M.N.R.A.S., to be published.
Bond, J.R., Szalay, A.S. and White, S.D.M. 1983, Nature 301, 584.
Gelmini, G.B., Nussinov, S. and Roncadelli, M. 1982, preprint
    MPI-PAE/Pth 37182.
Georgi, H. and Glashow, S.L. 1974, Phys. Rev. Lett. 32, 438.
Georgi, H., Glashow, S.L. and Nussinov, S. 1981, Nuc. Phys. B
    193, 297.
Gry, C. et al. 1984, this volume, p. 279.
Kibble, T.W. 1983, private communication.
Ku, W.H.M. et al. 1982, M.N.R.A.S. 202.
Lee, B.W. and Weinberg, S. 1977, Phys. Rev. Lett. 39, 165.
Olive, K.A., Schramm, D.N., Steigman, G., Turner, M.S. and Yang,
    J. 1981, Ap. J. 246, 557.
Peebles, P.J.E. 1982, Ap. J. 258, 415.
Peebles, P.J.E. 1984, this volume, p. 185.
Preskill, J., Wise, M.B. and Wilczek, F. 1982, Harvard preprint
    HUTP-82/A048.
Primack, J.R. and Blumenthal, G.R. 1984, this volume, p. 163
Rees, M.J. and Ostriker, J. 1977, M.N.R.A.S. 179, 541.
Schramm, D.N. and Freese, K. 1984, this volume, p. 197.
Schwartzman, V.F. 1969, Sov. Phys. JETP Lett. 9, 184.
Shapiro, P.R., Struck-Marcell, C. and Melott, A.L. 1983, preprint.
Sherman, R.D. 1982, Ap. J. 256, 370.
Silk, J. 1977, Ap. J. 211, 638.
Sunyaev, R.A. and Zeldovich, Ya.B. 1972, Astron. Astrophys. 20,
    189.
Zeldovich, Ya.B. 1970, Astron. Astrophys. 5, 84.

PANCAKES, HOT GAS AND MICROWAVE DISTORTIONS

A.S. Szalay[1], J.R. Bond[2,3] and J. Silk[4,5]

1. Dept. Atomic Physics, Eotvos University, Budapest.
2. Institute of Astronomy, University of Cambridge.
3. Inst. of Theoretical Physics, Stanford University.
4. Astronomy Dept., University of California, Berkeley.
5. Institut d'Astrophysique, Paris.

The pancake picture with massive neutrinos has a nonlinear mass scale for dark matter different from that inferred from the galaxy autocorrelation function. In the first part of this paper, we suggest ways in which the correlation function for mass density may differ from that of galaxies. In the second part, we predict the magnitude of the Zeldovich-Sunyaev effect arising from pancakes. Electrons in the hot gas created by pancake shocks can upscatter photons in the microwave background radiation, causing spectral distortions. Angular differences in these distortions lead to temperature fluctuations which are on the edge of observability, and can be used as a test of the pancake scenario.

I. CORRELATION FUNCTION AND PANCAKES

Recently, 3-dimensional N-body simulations of galaxy clustering with adiabatic initial conditions were carried out by several groups (Clypin and Shandarin 1982, Frenk et al. 1983, Centrella and Melott 1983). The fluctuation spectra used had sharp cut offs at around 40 Mpc, as predicted by calculations of linear fluctuation growth in a neutrino-dominated universe (Bond and Szalay 1981, 1983; Peebles 1982). The presence of voids and filaments, the percolation properties (Shandarin 1983) and even the shape of the correlation function found in these experiments agree with observations.

The correlation function, $\xi(r)$, is calculated from the distribution of the idealized, collisionless particles. On those scales for which the overdensity is nonlinear, the

101

correlation function is found to steepen and to be given by a
single power law. This overcomes the Peebles (1974) objection
to the pancake model, namely that a power law on small scales
is unlikely. Indeed, in these simulations, this property is
used to fix the present time as the moment at which the slope
of ξ equals -1.8.

Even though the correct shape is obtained, an additional
problem arises with the overall amplitude of ξ. If nonlinearity
is required on pancake scales, then it should be approximately
equal to the nonlinear scale of galaxy clustering, as Fall
(1981) and Peebles (1980) emphasized for baryon-dominated
universes; Bond, Efstathiou and Silk (1980) and Peebles (1982)
pointed out that neutrino-dominated universes also have this
problem. The simulations have made this argument precise and
acute, as White et al. (1984) extensively discuss in this
volume. A simple algorithm for 'mocking' dissipation - by
identifying particles at the highest local densities with
galaxies - makes this discrepancy even more pronounced. White
et al. also find that nonlinearity is predicted to occur at low
redshift, hence galaxies would form too late.

These two difficulties originate as a result of equating
the galaxy-galaxy correlation function with the correlation
function of the collisionless particles. Is this justified?
In the adiabatic picture with a large damping cutoff, galaxies
arise only after large structures have collapsed, shocked,
and their central regions have cooled to $\sim 10^4$K; this cold gas
then fragments into small clouds which give rise to galaxies as
Bond et al. (1984) , hereafter BCSW, detail in this volume.
The cooled fraction is a sensitive function of mass scale and
collapse epoch, and can be anywhere in the range of 1-20% of
the baryon mass for neutrino-dominated models. Most of the
gas is far too hot to condense on galactic scales. Since the
process of galaxy formation is so sensitive to the details
of the gas dynamics, we can imagine a number of ways in which
the correlation function of the dark matter may differ from
that of galaxies. Examples are as follows:

1. *Effects of Reheating*: Early pancakes form from low
temperature (0.1K) gas; they easily cool in their central
regions, emitting radiation which will photoionize gas in
regions which have not yet pancaked. Stars, quasars etc.
which form after the first pancakes have collapsed will also
contribute to this ionizing flux. If these sources are maintained
during subsequent pancake collapses, line cooling becomes
ineffective (Sunyaev, private communication), which lowers the
cooling rate. If *no cooling* occurs, central regions are
adiabatically compressed as long as the gas pressure is less
than the incoming ram pressure. The central density in a late

pancake could then be very much smaller than in a similar early
pancake, so sooling would *remain ineffective,* making galaxy
formation unlikely. (If the gas can be heated above 10^5K,
rapid cooling will definitely not occur; such temperatures are
difficult to attain with typical photo-ionizing sources (but
see point 5). In some circumstances, the inhibition of line
emission may be prove sufficient to avoid rapid cooling).

2. *Local Variations in Cooling*: Even in early pancakes, the
cooled fraction is a sensitive function of the length scale,
$q_c \sim L^{-1}$ (BCSW, Eq. 9), and so will vary from pancake to pancake
and even from place to place within one pancake. The one to
one relation between dark matter and galaxies is therefore
broken. Similarly, the dependence of q_c on collapse redshift
also modulates the galaxy formation rate.

3. *Effects of Conduction*: Variations in the cooled fraction
are further amplified by thermal conduction; the cold layer can
even be completely erased for very large pancakes (BCSW),
thereby preventing galaxy formation in those regions.

4. *Structure of Tight Clusters*: Many tightly bound clumps
resembling rich clusters appear in the simulations, especially
at intersections of pancakes. They arise from transverse flows
in pancakes and filaments, and can be associated with some of
the catastrophes discussed by Arnold et al. (1982). These
catastrophes start as simple pancakes, with the more complicated
topology arising later as different collisionless streams
interpenetrate. Shock formation in the simple pancakes will
prevent the gas from forming the more complicated topology by
screening the later flows, which accrete onto the already
shocked region. Though galaxies are collisionless, the dissipation which has already occurred makes their streaming pattern
different than that of the dark matter; their catastrophes
will also differ. This is in spite of both components being
subject to transverse motion. Furthermore, the infall velocity
for a spherically symmetric collapse is considerably higher
than for a flat system of the same mass. This means that
cooling will also be less effective in these regions than in
pancakes. These two effects can lead to a lesser concentration
of galaxies in tight clusters than would be predicted from the
dark matter distribution.

5. *Explosions*: As discussed by Ostriker and Cowie (1981), a
chain reaction of explosions can occur in the intergalactic
medium. If such processes are initiated in pancakes, the
shocks will not be spherically symmetric due to the inhomogeneity of the pancakes. They will proceed faster towards the
density gradient. By itself, this will modify the relation
between dark matter and galaxies. Any metal injection would

aid enormously in cooling hot gas. On the other hand the kinetic energy from the explosions provides another heating mechanism for the IGM. For our purposes, it is more successful than heating by photo-ionization, since the temperatures attained can be far in excess of 10^5K, making cooling difficult.

In the current version of the pancake picture, there are three different sites for galaxy formation: pancakes, filaments and rich clusters. The contribution from each site to the number of galaxies and to their autocorrelation function is not well known. The relation of ξ for galaxies to ξ for dark matter and, indeed, to that for clusters must be clarified before we may regard the correlation function argument as a definitive. test for the neutrino-dominated picture.

II. CBR DISTORTIONS FROM HOT PANCAKE GAS

The upscattering of the microwave background photons by hot electrons is described by the Kompaneets equation for the photon distribution. Its solution is frequency independent in the Rayleigh-Jeans part of the spectrum, and the spectral distortions are given by the CBR temperature fluctuation (Zeldovich and Sunyaev 1969).

$$dT/T = -2y = -2\int n_e \sigma_T (kT_e/mc^2) \, dr \tag{1}$$

The integral is taken along the line of sight, T_e, n_e, and m are the electron temperature, density and mass, and σ_T is the Thomson cross section.

The distortions themselves are difficult to measure. Our goal is to predict angular fluctuations arising from the statistical dispersion of this line of sight integral, since this is observable. It is a well known property of the Einstein-de Sitter cosmology that the comoving arclength d at redshift z is related to angular scale (for small θ) by

$$d(\theta,z) = 2c\theta/[H_o g(z)] = 10 \, (\theta/5.7') \, h^{-1}g(z) \, \text{Mpc}, \tag{2}$$

$$g(z) = 1 - (1+z)^{-\frac{1}{2}} \tag{3}$$

Radio experiments of small beam size probe only small separations; they measure integral quantities along the line of sight. Typical experiments involve either beam-switching through some angle, with the beam pair swept around a part of the sky, or imaging of a given region.

Pancakes are isobaric in their shocked regions, hence $n_e T_e \sim P$ is constant, so our line integral becomes a simple sum over all pancakes along the line of sight:

$$y_T = \sum_n y_n = \sum_n 1/2 \ P_n \ (\sigma_T/mc^2) \ (x_n/\cos i_n) \tag{4}$$

Here P_n is the ram pressure of the nth pancake given by BCSW, Equation (4). The pancakes are generally tilted to the line of sight, having an inclination angle i_n, and proper width $x_n/\cos i_n$. For face-on pancakes this is the width of the shocked region at the redshift of observation. An expression which is valid in the early phases of shock development and proves adequate for our purposes is :

$$x = 1.3 \ (1+z)^{-1} \ L_{10} \ (\tau - 1)^{3/2} \ Mpc, \tag{5}$$

where $L_{10} = L/10$ Mpc is the comoving wavelength of the collapsing region, $\tau = \pi q/\sin \pi q = (1+z_p)/(1+z)$, z_p is the redshift of pancaking and q is the mass fraction just entering the shock at z. Typically, $x \sim 0.05L$. The edge-on pancakes have $x/\cos i \sim L$, hence give much larger effects. The contribution from each pancake is:

$$y = 1.08 \times 10^{-9} \ \Omega_B h^2 L_{10}^2 \ (1+z_p)^3 \ F(q)/\cos i \tag{6}$$

$$F(q) = 5 \ \pi^2 q^2 \ (\tau - 1)^{3/2}/[\tau^3 (1 - \tau \cos \pi q)]$$

$F(q)$ includes explicitly the ram pressure evolution as in Sunyaev and Zeldovich (1972) or Bond et al. (1983). For simplicity, it assumes that all pancakes continue transverse expansion at the Hubble rate until the present. We estimate the error in this approximation later.

Along the line of sight there will be many pancakes, and each will have a different collapse redshift, length scale, orientation, etc. To calculate the contribution in detail requires the knowledge of the appropriate probability distributions. Lacking this, we use a simple model to estimate the expected magnitude of the effect. We assume all pancakes form at the same epoch, z_p, with the same length, and are evenly spaced in comoving length along the line of sight. The number of pancakes along this line (N) is a random variable with mean

$$\langle N \rangle = (600/hL_{10}) \ [1 - (1+z_p)^{-\frac{1}{2}}] = (600/hL_{10}) \ g(z_p) \tag{7}$$

up to factors of order one which depend upon the precise distribution of spacing. We assume the statistics of N are Poisson, with $\sqrt{N}$ fluctuations. We further assume that the inclination angle is statistically independent of the other

variables, and is randomly oriented. Then the dispersion of
y will depend on the relative dispersion in the number of
pancakes along the line of sight ($\sim 1/\sqrt{N}$) and on the relative
dispersion of the value <1/cos i> averaged over the N pancakes.
Even though 1/cos i has a very skewed distribution (truncated
at e = L/x $\sim$ 10-20, corresponding to the maximal edge-on width
L), <1/cos i> tends to a Gaussian with the relative dispersion
$\sim 1/\sqrt{N}$, due to the central limit theorem. The numerical values
are mildly dependent on e:

$$<1/\text{cos i}> = A(1 \pm \sigma/\sqrt{N}) = \begin{array}{l} 2.5(1 \pm 1.0/\sqrt{N}), \text{ for e = 10} \\ 2.9(1 \pm 1.5/\sqrt{N}), \text{ for e = 20} \end{array} \qquad (8)$$

The average for F would involve a statistical mean over the
probability distribution of collapse redshift and length scale,
were these available. In our simple model we find by numerical
integration the value of <F(q)> to be about 1.0 for z_p= 3 and
slowly changing to 1.1 at z_p= 2 and 0.7 at z_p= 7. Switching
the transverse expansion off yields an increase by 6 in the
value of <F>. We obtain by multiplying these numbers an
approximate expression for the mean spectral distortion and its
dispersion. We quote the standard deviation:

$$dT/T = \sqrt{N}\sqrt{(1+\sigma^2)}\, 2.16\text{x}10^{-9}\Omega_B h^2 L_{10}{}^3 (1+z_p)^3 \; <F> \; <1/\text{cos i}> \qquad (9)$$

Angular fluctuations in the temperature will arise as a consequence of the variance about the mean as we go from one line
of sight to another, as long as the two directions are
statistically independent. This will be valid provided their
angular separation θ is much greater than the angular scale
associated with the transverse dimension of a pancake, which
follows from Equation (1):

$$\theta \gtrsim 1° \gtrsim \theta_L; \quad \theta_L = 5.73' \; hL_{10}/g(z) \qquad (10)$$

We have assumed that the transverse and initial longitudinal
scales are about the same. In Table 1, we give more accurate
evaluations of dT/T as a function of L and z_p for Ω = 1, Ω_B = 0.1,
and h = 1 for these large-angle fluctuations.

L / z_p	1.0	2.0	3.0	5.0	7.0
20	7.1E-07	3.0E-06	7.1E-06	2.1E-05	4.5E-05
40	4.0E-06	1.7E-05	4.0E-05	1.2E-04	2.5E-04
60	1.1E-05	4.7E-05	1.1E-04	3.3E-04	7.0E-04
80	2.3E-05	9.6E-05	2.3E-04	6.9E-04	1.4E-03
100	4.0E-05	1.7E-04	4.0E-04	1.2E-03	2.5E-03

Table 1. The dependence of dT/T on large (>1°) angular scales
on the pancake length scale L in Mpc, and on the redshift of
pancaking z_p.

Since pancakes form a cell-like structure in the universe, they are highly correlated at small separations. In this case consider a 'beam-switch' differential measurement. We can only get a contribution from the pancake if it intersects with one beam, but not with the other. The only obvious way to achieve this is that we have either a pancake edge-on, nearly tangential to the line of sight, or we have a branching point of three pancakes. The total averaged contribution will be proportional to $d(\theta,z)/L$, therefore on small angles we expect dT/T to be rising as θ up to the large angle limit.

A review of the present observational situation was given by Partridge (1981), further references therein. The overall limits on dT/T for scales from a few arc minutes to a few degrees are about 10^{-4}. On the other hand as recently shown by Danese et al. (1983) there is no hope of detecting premeval fluctuations below the 10^{-4} level due to the contributions from unresolved radio sources. However, by imaging techniques, there is a nonvanishing chance, that one may see signals at 3σ or higher above the rms background, with a rather sharp boundary: a clear signature of a pancake.

Even though our numbers are order of magnitude estimates, it is clear that the dT/T coming from hot gas is very close to present upper limits. With a more sophisticated treatment one can obtain strong limits on the characteristic scale and on the epoch of galaxy formation in the pancake picture. The inclusion of transverse contraction leading to filaments and clusters will be important. The estimates presented here can already be used to rule out extremely large pancakes, which form at early epochs.

We acknowledge useful discussions with George Efstathiou, Craig Hogan, Rashid Sunyaev and Simon White.

REFERENCES

Arnold, V.I., Zeldovich, Ya. B. and Shandarin, S.F. 1982, Geophys. Astrophys. Fluid Dynamics 20, 111.
Bond, J.R., Centrella, J., Szalay, A.S., and Wilson, J.R. 1983 , M.N.R.A.S. to be published.
Bond, J.R., Centrella, J., Szalay, A.S., and Wilson, J.R., 1984, this volume, p. 87.
Bond, J.R., Efstathiou, G., and Silk, J. 1980, Phys.Rev. Lett. 45, 1980.
Bond, J.R., and Szalay, A.S. 1981, Proc. Neutrino 1981, Hawaii, 1,59.
Bond, J.R., and Szalay, A.S., 1983, Ap. J. to be published.
Centrella, J., and Melott, A., 1983, Ap.J. to be published.

Clypin, A.A., and Shandarin, S.F., 1982, I. Appl. Math. prepr. No. 136, Moscow.

Danese, L., DeZotti, G., and Mandolesi, N., 1983, Astron. Astrophys. 121, 114.

Fall, M.S., 1981, Rev. Mod. Phys. 51, 21.

Frenk, C., White, S.D.M., and Davis, M., 1983, Ap. J. to be published.

Ostriker, J.P., and Cowie, L.L., 1981, Ap. J. Lett. 243, L127.

Patridge, B., 1981, Proc. Int. School Cosmology, Erice, Italy, P.121, edited by B.J.T. Jones, Reidel, Dordrecht.

Peebles, P.J.E., 1974, Ap. J. Lett., 189, 51.

Peebles, P.J.E., 1980, The Large Scale Structure of the Universe, Princeton University Press, Princeton.

Peebles, P.J.E., 1982, Ap. J. 258, 415.

Shandarin, S.F., 1983, Pisma Astr. Zh. 9, 795.

Sunyaev, R.A., and Zeldovich, Ya. B., 1972, Astron. Astrophys., 20, 189.

White, S.D.M., Frenk, C., and Davis, M., 1984, this volume, p. 117.

Zeldovich, Ya. B., and Sunyaev, R.A., 1969, Astrophys. Space Sci. 4, 301.

SPECTRUM OF THE COSMIC BACKGROUND RADIATION

N. Mandolesi
Istituto TE.S.R.E. - CNR, Via de' Castagnoli, 1 - 40126
Bologna Italy

ABSTRACT

Distortions of the spectrum of the cosmic background radiation from a Planckian shape may be expected for a variety of physical reasons. Presently available experimental data are compared.

1. INTRODUCTION

The standard hot model of the Universe predict a cosmic background radiation (CBR) with a precise black body spectrum. Small distorsions from an exact Planckian shape are indeed expected in order to account for the observed structure of the Universe. The detection of those small deviations from a black body spectrum and their features would then be highly informative on the growth and evolution of the primordial fluctuations responsible for the Universe populated by galaxies and clusters of galaxies.

Experimentally the shape of the CBR spectrum has been investigated since its discovery[1]. The available observational

J. Audouze and J. Tran Thanh Van (eds.),
Formation and Evolution of Galaxies and Large Structures in the Universe, 109–115.
© 1984 by D. Reidel Publishing Company.

situation is consistent with no significant distortions from a
Planck spectrum in the Rayleigh-Jeans region[2], while at milli-
meter and submillimeter wavelengths some deviations from a
black body spectrum are reported by Woody and Richards[3].

2.   MEASUREMENTS OF THE CBR

Measurements of the temperature of the CBR may be distinguished according to the techniques used in the observations
and crudely divided into three groups:

i) microwave observations in the Rayleigh-Jeans region of
   the spectrum; the instruments used are microwave radiometers with low side lobes antennas Dicke-Switched superhe-
   terodyne receivers; the measurements are usually carried
   out at high altitude sites in order to minimize atmospheric effects.

ii) direct infrared observations at the peack and in the Wien
    part of the spectrum; the instruments used employ typical
    infrared technology with incoherent detectors; in this
    frequency range atmospheric problems are so severe that
    observations are made possible only at balloon or rockets
    altitudes or with satellites.

iii) Interstellar molecules; direct ground based measurements
     of the CBR at the peak (millimeter and submillimeter wave-
     lengths) are prohibited by the high atmospheric emissivi-
     ty; indirect ground based observations are instead allowed
     by interstellar molecules (mainly CN and CH) which have
     rotational transitions at millimeter and submillimeter
     wavelengths[4]. The exitation temperature is obtained by
     measuring the relative intesities of optical absorption
     lines from ground and exited rotational States.

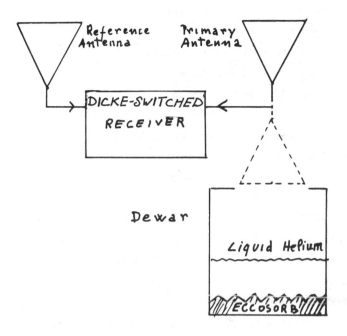

Fig. 1 - Typical layout of a microwave radiometer for measuring
         CBR temperatures

In what follows the experimental technique used by microwave radiometers will be discussed in some detail; direct and
indirect infrared techniques are examined in review papers by
Weiss[5] and Danese and De Zotti[7].

The concept of a microwave experiment for measuring the
temperature of the CBR is simple: compare the power received
by the radiometer (Fig. 1) when its primary antenna is looking
at the sky with that received when the antenna is directed into
a cold load used as an absolute calibrator.

In terms of antenna temperature, proportional to the power
received by the radiometer one gets:

$$T_{OUT} = (T_{CBR} + T_{ATM} + T_{GAL} + T_{GROUND} + T_{UNEXPECTED})_{SKY} -$$
$$- (T_{CAL} + T_{W-CAL})_{CL} \qquad (1)$$

where the subscripts have the following meaning:

OUT   = output signal

CBR   = cosmic background radiation

ATM   = atmospheric emission

GAL   = galactic contribution

GROUND   = ground contribution

UNEXPECTED = euphemism for instrumental contribution

CAL   = absolute calibrator

W-CAL   = contribution from warm parts of calibrator

CL   = cold load

By carefully evaluating or measuring of the terms of eq.
(1) T_{CBR} can be deduced and converted in brightness temperature
if the observing wavelength is too short to follow the Rayleigh
-Jeans radiation law.

In addition the experiment design is such to reduce and
minimize the extraneous sources of radiation and carefully accounts for the residual values. For instance the atmospheric
emission is strongly reduced by having a high altitude observing site, while the residual is determined by zenith scans; the
galactic contribution is minimized by choosing a high altitude
site and the observing period and the residual is measured by
galactic scans. Low side lobes antennas (usually corrugated
horns) are used to minimize the ground pick-up. As source of
absolute calibration a slab of eccosorb (product with emissivity near to unity at microwaves) immersed in a bath of liquid
helium is used ($T_{OL} \sim 3.8$ K at 3,800 m s.l.m.).

An accurate design of the liquid-helium deward and laboratory tests provide the extra contribution T_{W-CL}.

3.   SPECTRUM OF THE CBR

The measurements in the low frequency range ($\lambda > 0.35$ cm), all made before 1968, used different experiment designs and are of varying quality; thus they are inadequate to set stringent limits on distortions.

Valuable infrared observations have been made by Woody and Richards[7] with a balloon borne experiment over the wavelength range from 0.025 to 0.40 cm; their data show some evidence for a non-thermal spectrum.

CBR measurements with interstellar molecules have been made in the early seventies[6]; fresh improved measurements could be extremely important.

New measurements of the CBR spectrum have been carried out at low frequencies[2].

The observations were made at 5 wavelengths (0.33, 0.9, 3.0, 6.3, 12.0 cm) using radiometers with scaled corrugated horn antennas; the atmospheric emission was continuously monitorized at an additional wavelength (3.2 cm).

A single liquid-helium-cooled absolute reference was used for all 5 radiometers; the contemporary measurements were performed from a high, dry site (University of California's White Mountain Research Station at 3,800 m a.s.l.) on the nights of July 5 and July 6, 1982.

The results of this multifrequency experiment are shown in Fig. 2 together with the previous ground based, CN and Woody and Richards[3] infrared data.

4.   CONCLUSIONS

The Fig. 2 shows that the new results[2] are in good agreement with the previous ones, while the accuracy is on average a factor of $\sim$ 2 better when compared with other low frequency results.

A standard statistical analysis gives the following results:

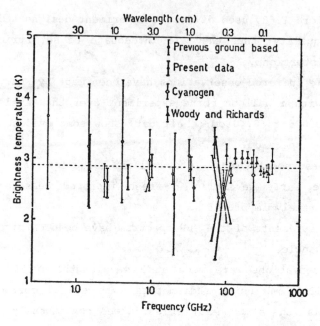

Fig. 2 - Spectrum of the CBR

i) Best fit black body temperature

T (K)

2.78 ± 0.10   New data only[2)]
2.74 ± 0.05   All ground based measurements
2.90 ± 0.03   All available data

ii) Fit to a Bose-Einstein spectrum with chemical potential μ
    and density parameter Ω

T(K)	μ	Ω	
2.77 ± 0.06	$(1.1 ^{+3}_{-1.1}) \times 10^{-3}$	0.1	Ground based measurements
2.77 ± 0.06	$<7 \times 10^{-3}$	1	Ground based measurements
2.92 ± 0.03	$(5.2 ^{+3}_{-2.8}) \times 10^{-3}$	0.1	All available data
2.92 ± 0.03	$(1.4 ± 0.9) \times 10^{-2}$	1	All available data

iii) Fit to a comptomized spectrum with a comptomization parameter u

T(K)   u

2.90 ± 0.03 $(2 \pm \frac{7}{2}) \times 10^{-3}$ All available data

In conclusion there is no significant evidence of deviations from a Planck spectrum in the Rayleigh-Jeans region.

Ground based measurements, however, yield a CBR temperature lower than that reported by Woody an Richards (2.96 K), the
difference being significant at $\sim 3\,\sigma$ level. If such difference
is interpreted as an excess at the peak the data entail:
$\mu < 4 \times 10^{-3}$ ($\Omega = 0.1$) or $\mu > 7 \times 10^{-3}$ ($\Omega = 1$) and $u < 0.01$
($1\,\sigma$ limits).

More precise spectral measurements, already planned at
both wavelengths ranges, microwave and infrared, could lead
hopefully to a direct determination of the irregularities of
the early Universe.

REFERENCES

1) Penzias, A.A. and Wilson, R.W.: 1965, Ap. J., 142, 419.

2) Smoot, G., De Amici, G., Friedman, S.D., Witebsky, C.,
   Mandolesi, N., Partridge, R.B., Sironi, G., Danese, L.,
   and De Zotti, G.: 1983, Submitted to Phy. Rev. Letters.

3) Woody, D.P., and Richards, P.L.: 1981, Ap. J., 248, 18.

4) Thaddeus, P.: 1972, Am. Rev. Astron. Astrophys., 97, 973.

5) Weiss, R.: 1980, Am. Rev. Astron. Astrophys., 18, 489.

6) Danese, L., and De Zotti, G.: 1977, Rivista del Nuovo Ci-
   mento, 7, 277.

7) Woody, D.P., and Richards, P.L.: 1979, Phy. Rev., Letters,
   42, 925.

IS THE UNIVERSE MADE OF MASSIVE NEUTRINOS?

Simon D.M. White, Carlos S. Frenk and Marc Davis

Department of Astronomy and Space Sciences Laboratory,
University of California, Berkeley, CA 94720, U.S.A.

ABSTRACT The hypothesis that the mass density of the Universe is
dominated by massive neutrinos leads to a relationship between
the cosmological parameters Ω and H and the characteristic length
scale of density fluctuations in the neutrino distribution at
recombination. N-body simulations can be used to perform the
mild nonlinear extrapolation required to compute a scale for the
present neutrino distribution which can be compared directly with
the clustering scale of galaxies. If nonlinear structures in the
neutrino distribution formed significantly before the present
day, the neutrino scale substantially exceeds that of the
galaxies for any acceptable set of cosmological parameters. However, formation of structure before a redshift of 4 is required
to account for high redshift quasars. The discrepancy is particularly marked in open universes and is exacerbated when it is
recognised that in the standard picture galaxies form only from
high density material within pancakes.

1 INTRODUCTION

Reported measurements of a non-zero neutrino rest mass (1,2)
a couple of years ago stimulated a revival of interest in the
possibility that neutrinos are the dominant component of the mass
density of the present Universe. In addition to identifying the
missing mass, this hypothesis can explain why constraints on the
baryon density of the Universe based on analyses of cosmic
nucleosynthesis disagree with dynamical estimates of the total
density. It can also resolve the apparent conflict between
observational limits on the small-scale anisotropy of the microwave background and the "natural" assumption that inhomogeneities

117

J. Audouze and J. Tran Thanh Van (eds.),
Formation and Evolution of Galaxies and Large Structures in the Universe, 117–123.

in the early universe are predominantly in the most rapidly
growing adiabatic mode (3). If the mass of the electron neutrino
is really ~ 30 eV as suggested by some experiments, then in the
standard Big Bang cosmology, neutrinos must provide a substantial
fraction of the total cosmic density.

A great virtue of the massive neutrino hypothesis is that in
its application to cosmology it has only one uncertain parameter,
the mass of the most massive stable neutrino. In conventional
models this mass is directly related to observable cosmological
parameters through the equation

$$m_{30} = 3.2 \ \Omega \ h^2 \ \theta^{-3} \tag{1}$$

where m_{30} is the neutrino mass in units of 30 eV, Ω is the
density parameter, h is Hubble's constant in units of 100
km/s/Mpc, and θ is the microwave background temperature in units
of 2.7 °K. Because neutrinos decouple from the rest of the
universe at the relatively early epoch when kT $\simeq$ 2 MeV, the
neutrino population can be treated as a noninteracting collisionless but self-gravitating gas in calculations of the evolution of
its density distribution on large scales. This situation is
sufficiently simple that the evolution can be treated with a
degree of rigour and confidence that is unusual in studies of
this type. As a result certain predictions of the massive
neutrino hypothesis are more easily falsifiable than is usual in
theories for the formation of large scale structure. One firm
prediction is that the phase-space density of neutrinos can nowhere exceed their phase-space density in the early universe; the
latter quantity can be calculated directly with very few uncertainties (4). This may exclude the identification of neutrinos
with the missing mass which appears to be present in certain
dwarf spheroidal satellites of the Milky Way (5,6,7). A second
firm prediction is that free streaming of the neutrinos after
they decouple but before they become non relativistic will wipe
out all small scale fluctuations in their density distribution
(8). Fourier components with wavelengths less than

$$\lambda = 41 \ m_{30}^{-1} \ \theta^{-1} \ \text{Mpc} \tag{2}$$

are very strongly suppressed, the amplitude of waves of length λ
being reduced by an order of magnitude. Bond and Szalay (9) give
a transfer function which can be used to calculate the power
spectrum of density fluctuations present at recombination from
any assumed fluctuation distribution at very early times. The
suppression of short wavelength perturbations means that nonlinear structure in a neutrino-dominated universe must form by
some variant of Zeldovich's pancake scheme in which coherent
structures of supercluster size separate from the expansion and
then fragment to form galaxies.

In view of the large scales predicted by equations (1) and (2) an obvious question is whether the massive neutrino hypothesis is consistent with the relatively small scale on which the observed galaxy distribution becomes nonlinear; this scale may be characterised by the distance, r_o, at which its autocorrelation function is equal to unity. Recent work confirms that $r_o = 5 h^{-1}$ Mpc with relatively small uncertainties (10,11). By comparing the predictions of linear theory with observation, Peebles (12) concluded that it is very difficult to make the massive neutrino hypothesis consistent with observation. The main uncertainty in his conclusions stems from the assumptions he needed to estimate the epoch of galaxy formation and to take account of the non-linear effects which are important on scales where the quantitative properties of the galaxy distribution are well known. This uncertainty can be eliminated by simulating the nonlinear evolution of the neutrino distribution from the known linear initial conditions. On scales of a few Mpc the nonlinearity is quite mild and systematic effects resulting from the technical limitations of N-body simulation methods can be kept at a relatively low level.

Last year we carried out a series of simulations of the growth of structure in a universe where the density distribution at early times possessed a coherence length (13). Identifying this length with the appropriate scale in a neutrino-dominated universe we were able to scale the nonlinear clustering in our Einstein-de Sitter models to fit observations of the galaxy distribution for values of h near 0.8. Our elation at this success was tempered by a number of uncomfortable problems. Firstly the age of the universe implied by this scaling is only 8.3 billion years and is uncomfortably short. Secondly we compared the autocorrelation function of the particles in our models with that of galaxies rather than of neutrinos in the real Universe. In the pancake theory galaxies form only in the regions of space where the neutrinos are densest and their correlation length should therefore exceed that of the neutrinos (see below). Finally we reproduced the galaxy autocorrelation function successfully only during a brief period quite soon after nonlinear structures first collapsed. The observation of quasars at red-shifts approaching 4 is inconsistent with the identification of this epoch with the present day. These difficulties are all related to the requirement that the model ratio of initial coherence length to final clustering scale should match the value inferred from the observed galaxy distribution and equations (1) and (2). In what follows we explore them further and find the discrepancies to be sufficiently large to cast considerable doubt on neutrino-dominated cosmologies of the kind extensively discussed in the recent literature.

2 SCALING N-BODY MODELS TO A NEUTRINO-DOMINATED UNIVERSE

The N-body models discussed here and in our earlier paper
(13) followed the evolution of the distribution of 1000 particles
using a direct N-body integrator written by S.J. Aarseth. The
particles were initially placed at those grid points of a cubic
mesh which lay within a spherical boundary, and the whole system
was given the uniform expansion required to make its total energy
zero. We then perturbed the position and velocity of all the
particles with a superposition of small amplitude waves of random
direction and a restricted range of wavelengths. This procedure
was designed to set up initial conditions corresponding to a
specific spectrum of linear density fluctuations at early times.
Subsequent evolution was followed explicitly by the code and
results for any given initial spectrum were obtained by averaging
simulations which started from different statistical realisations
of it. Our early experiments used an initial power spectrum which
was flat between upper and lower wavelength cutoffs. This is a
poor representation of the actual power spectrum expected for a
population of massive neutrinos and so there is some uncertainty
in how the experiments should be scaled to represent a neutrino-
dominated universe. We adopted a scaling based on Peebles'
analysis of the density autocorrelation function at early times
(12). We have now performed experiments starting with fluctuation spectra based on Bond and Szalay's (9) neutrino transfer
function which confirm that this choice was appropriate.

In order to compare the galaxy distribution with theoretical
models it is desirable to choose a measure of clustering which is
as insensitive as possible to the details of small scale clustering, and which can be determined accurately from galaxy surveys.
Such a scale is the length, R, determined from the integral
relation,

$$R^{-3} \int_{o}^{R} \xi(r)r^2 dr = 1, \qquad (3)$$

where $\xi(r)$ is the autocorrelation function of the galaxy distribution. The integral on the left-hand side of this equation is
closely related to the quantity J_3 discussed by Peebles
(12). Recent observations give $R = 5 \pm 0.7$ h^{-1} Mpc (10), where
the uncertainty is based on comparison of results for different
areas of the sky. In a neutrino-dominated universe the dimensionless ratio of the present galaxy clustering scale to the present
neutrino scale may thus be expressed in terms of purely cosmological parameters as

$$R/\lambda = 0.39 \pm 0.06 \; \Omega \; h \; \theta^{-2} \; . \qquad (4)$$

In our N-body simulations we know the value of λ which is built
into the initial conditions and we can calculate R, and thus R/λ,
for the particle distribution as a function of time. If we
assume that the distribution of galaxies follows that of the
neutrinos (see below) the model value of this ratio determines
the value of $h\theta^{-2}$ required for agreement with observation. This
quantity must clearly lie within the observationally allowed
range if the model is to be acceptable at the particular time
considered.

A further constraint arises because the models must be
allowed to evolve long enough for galaxies to form and to age to
their presently observed state. In a neutrino-dominated universe
no galaxy-sized objects can form until the large-scale neutrino
distribution has gone nonlinear. In our simulations we identify
the onset of galaxy formation with the time at which 1% of the
elements in our initial grid pattern have undergone collapse.
Later times can then be identified by the expansion factor,
$1 + z_{GF}$, since the onset of galaxy formation. To be consistent
with observations of galaxies at high redshift, z_{GF} should exceed
2, and to be consistent with the existence of high redshift
quasars it should exceed 4. This is an extremely conservative
prescription because galaxy formation cannot occur simultaneously
in all regions. In our simulations we find that by $z_{GF} = 2$ only
30% of the initial volume elements have undergone collapse.

In Figure 1 we show the values of $h\theta^{-2}$ obtained by comparing
R/λ in N-body models of neutrino clustering with equations (4).
$h\theta^{-2}$ is shown as a function of z_{GF} for each ensemble of simulations; note that in any particular ensemble Ω is specified and
cannot be adjusted. A first clear inference is that neutrino-
dominated universes with Ω significantly less than one can be
definitely ruled out. For example, with $\Omega = 0.2$ and $z_{GF} = 2.5$ a
Hubble constant of about 400 km/s/Mpc is required for "agreement"
with observation, showing a discrepancy of a factor of at least 4
between the large coherence length of the initial conditions and
the small observed clustering scale of galaxies (cf. equation 4).
For $\Omega = 1$ the situation is considerably better, but it is still
difficult to obtain an acceptable model. If the power spectrum of
neutrino fluctuations in the early universe is taken to be of the
"constant curvature" form,

$$|\delta_k|^2 \sim k^n \, T(k), \quad n = 1, \tag{5}$$

where $T(k)$ is the transfer function of Bond and Szalay (9), then
$h \geqslant 1$ is required to match length scales if $z_{GF} = 2.5$; this
implies an age of less than 7 billion years for the Universe and
so is in conflict with the ages of globular star clusters.

The scaling discrepancy between a neutrino-dominated

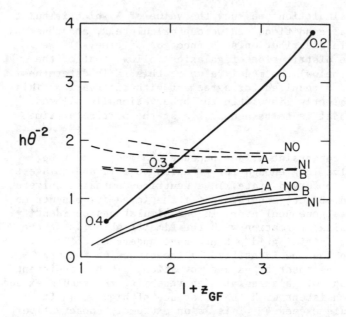

Figure 1.Constraints on the cosmological parameters of a
neutrino-dominated universe from the requirement that its clustering scale match that observed for galaxies. Each line represents results from one ensemble of N-body simulations and shows
the required value of $h\theta^{-2}$ as a function of the expansion factor
since the first appearance of collapsed structures. The steep
line marked 0 refers to an ensemble of open universes and is also
labeled with the values of Ω corresponding to several different
expansion factors. The clustering in these low density models is
indistinguishable on large scales from that in the corresponding
flat models. All other curves in the figure refer to Einstein-de
Sitter models; full lines give the scaling requirements when all
particles are used to estimate the clustering length, while
broken lines give the corresponding requirements when only particles in regions with $\rho > 2\langle\rho\rangle$ are assumed to be "visible". The
lines marked A and B refer to the two ensembles of our earlier
paper (13) and the lines marked N0 and N1 refer to ensembles with
initial conditions corresponding to equation (5) with n=0 and n=1
respectively.

universe and observation is actually worse than appears from this
comparison of galaxy clustering with the predictions of our
simulations. In a neutrino-dominated universe galaxies are
expected to form only in regions of high neutrino density where
the baryonic material can shock, cool and fragment. As a result
there should be no galaxies in low density regions which may,
nevertheless, contain a significant fraction of the neutrinos. In

such a situation the correlation length, R, for the galaxies is
expected to exceed the corresponding scale for the neutrinos. To
illustrate this using our models, we assumed that the only "visible" particles would be those in regions where the local density
was at least twice the overall mean. On small scales this assumption increased the amplitude of the correlation function by about
the square of the factor by which the total particle number decreased. As a result the clustering length increased considerably
and agreement with the observed scale required h > 1.6 in a flat
universe with n = 1. Although this particular procedure is
clearly quite arbitrary, it shows that a realistic treatment of
galaxy formation is likely to worsen the scaling difficulties of
neutrino-dominated universes. It seems that for any plausible
cosmological parameters the coherence length of a neutrino-
dominated universe is too large to be consistent with the relatively small scale on which we see nonlinear structure in the
present galaxy distribution. The discrepancies are large enough
that the chances of finding some means of salvaging the neutrino
picture look very poor. Finally we note that if there are m
flavours of stable neutrino with nearly equal mass, the scaling
constraints are essentially the same as in a universe with density Ω/m which is dominated by a single neutrino species. Such
multiple neutrino models are strongly ruled out by observation.

This research was supported by the United States Department
of Energy under contract number DE-AT03-82ER40069.

REFERENCES

1. Lyubimov,V.A.,Novikov,E.G.,Nozik,V.Z.,Tretyakov,E.F. &
    Kozik,V.S.1980 Phys.Lett.B94,pp 266-268.
2. Reines,F.,Sobel,S.W. & Pasierb,E.1980 Phys.Rev.Lett.45,
    pp 1307-1311.
3. Doroshkevich,A.G.,Khlopov,M.Yu.,Sunyaev,R.A.,Szalay,A.S. &
    Zel'dovich,Ya.B.1980 B.Ann.N.Y.Acad.Sci.375,pp 32-42.
4. Tremaine,S.D. & Gunn,J.E.1979 Phys.Rev.Lett.42,pp 407-410.
5. Aaronson,M.1983 Astrophys.J.,266,L11-L15.
6. Faber,S.M. & Lin,D.N.C.1983 Astrophys.J.,266,pp L17-L20.
7. Lin,D.N.C. & Faber,S.M.1983 Astrophys.J.,266,pp L21-L25.
8. Bond,J.R.,Efstathiou,G. & Silk,J.I.1980 Phys.Rev.Lett.45,
    pp 1980-1984.
9. Bond,J.R. & Szalay,A.S.1983 Astrophys.J., in press.
10. Davis,M. & Peebles,P.J.E.1983 Astrophys.J., in press
11. Bean,A.J.,Efstathiou,G.,Ellis,R.S.,Peterson,B.A. & Shanks,T.
    1983 Mon.Not.R.astr.Soc.,in press.
12. Peebles,P.J.E.1982 Astrophys.J.,258,415-424.
13. Frenk,C.S.,White,S.D.M. & Davis,M.1983 Astrophys.J.,in press.

THE PREGALACTIC UNIVERSE AND THE FORMATION OF LARGE SCALE STRUCTURE

Craig J. Hogan

California Institute of Technology

ABSTRACT: We summarize a number of reasons for believing that pregalactic sources of energy, such as quasars or protogalaxies, may have significantly influenced the development of the large scale galaxy distribution. A brief review is presented of specific mechanisms by which this may have occurred and their observational consequences. It is possible that the large scale structure may be primarily due to such activity, rather than the direct effects of primordial adiabatic or isothermal fluctuations.

The astrophysical production of radiative or hydrodynamical free energy from nuclear burning or gravitational accretion leads to a wide variety of processes which significantly affect the motion and distribution of matter within galaxies. Such complicated nonlinear processes as shock waves, magnetic dynamos, jets, turbulent flows, galactic winds and fountains, etc. are taken for granted on small scales $\lesssim 10^{11} M_\odot$. Generally one does not ask what is the "fundamental reason" for this activity, or inquire about the details of the initial conditions. Instead, one tries to understand what the dominant process is in any given situation, in order to construct a realistic simple model of what is going on. Of course, some situations are so complicated that this effort fails, as in the problem of star formation.

On the very largest scales, we have a quite different situation. Here the universe is accurately described by a simple homogeneous and isotropic Friedmann model, and the only force involved is gravitation. These essential features of the very large scale structure of the universe depend crucially on "initial conditions" (or at least on events which occurred very early, before the net baryon charge was generated), and are quite unaffected by the complications on small scales.

125

J. Audouze and J. Tran Thanh Van (eds.),
Formation and Evolution of Galaxies and Large Structures in the Universe, 125–131.
© *1984 by D. Reidel Publishing Company.*

But there is a particularly fascinating transition regime on scales $\simeq 10^{15-16} M_\odot$, the largest scales where nonlinear inhomogeneity appears in the clustering of galaxies. We do not know whether this structure is due primarily to small perturbations in large scale cosmological initial conditions on these scales, or to astrophysical activity of the type which is observed on smaller scales (1). I will adopt the term "neogenic" to refer to the hypothesis that the large scale inhomogeneous structure in the galaxy distribution is primarily due to free energy released by comparatively recent events occurring at say $z \lesssim 10^5$, and that it is not significantly affected in a direct way by fluctuations on the same scale originating at earlier times. In this view the direct effects of inhomogeneous structures originating in the earlier universe occurred on such a small scale that they have been erased by subsequent cosmological evolutionary processes which can be understood without reference to the cosmological initial conditions. The opposite hypothesis (and the most popular), which I will call "paleogenic," is that the galaxy distribution is due mainly to direct effects of primordial fluctuations on the same scale as the presently observed inhomogeneity, and that the action of quasars or similar neogenic energy sources may be safely neglected (at least in first approximation) in formulating a theory of large scale structure.

Of course, there is room for middle ground between the two views, if it turns out that neither primordial fluctuations nor pregalactic evolutionary processes can be safely neglected on these scales. But it is simpler to talk about it as a simple dichotomy, and indeed it would seem to be a fortuitous coincidence if both effects, while having totally unrelated causes, should have comparable amplitude. In this paper I summarize some reasons for preferring a neogenic scenario, and describe some simple models of pregalactic activity which may be able to explain the present day large scale inhomogeneity. I also describe some observational strategies for choosing between the various models. In concentrating on the neogenic view I do not wish to imply that there are very conclusive reasons for preferring it; it is just that the paleogenic scenarios are amply reviewed in this volume and elsewhere.

It is surprising that these issues in extragalactic astronomy are intimately related to current research topics in high energy physics. If the paleogenic view is correct, then the structure must be caused by fluctuations introduced into the metric at very early times indeed -- $t \lesssim 10^{-35}$sec -- during an early phase in the universe's history when observed scales could have been in causal contact. This is possible in principle (for example in "inflationary" universes) but only before the net baryon number was created, the reason being that the baryons introduce an absolute comoving scale and thereby lead to the classical particle horizons of Friedmann universes. (Baryon generation occurs at very high temperatures, which is why we require an early time). The dimensionless amplitude $\delta\varphi$ of the binding-energy fluctuations (which ultimately form the binding energy of galaxy clusters in paleogenic scenarios) can be calculated in inflationary universes from the fundamental Lagrangian adopted for matter (2,3,4). To lead to a reasonable paleogenic scenario the amplitude must lie in the range $\delta\varphi \sim 10^{-6}$ to 10^{-5} (depending on the composition of the "missing mass".) The neogenic hypothesis asserts that the primordial $\delta\varphi \lesssim 10^{-6}$ This probably implies that that some seed structure which triggers all the astronomical activity must be generated in one of the more recent phase transitions, say at electroweak symmetry breaking ($T \simeq 100\,\text{GeV}, t \simeq 10^{-9}\text{sec}$) or at quark confinement (~ 0.2 to $1\,\text{GeV}, t \simeq 10^{-5}\text{sec}$).

Several astronomical coincidences support the "naturalness" of the neogenic scenario. The first is well known: because gravity is scale free, some nongravitational process must have selected the present day non-linear mass scale (defined as the total mass around a typical galaxy in excess of the mean), $M_0 \sim 10^{15} \Omega h^{-1} M_\odot$. It is a suggestive coincidence that M_0 approximately coincides with the mass of matter within a Hubble length at the epoch t_{eq} when the matter and radiation densities are equal, the "radiation pressure Jeans mass" $M_{j\gamma} = 10^{16} (\Omega h^2)^{-2} M_\odot$. Indeed $M_{j\gamma}$ is known to be the largest scale over which a nongravitational force (in this case radiation pressure) can have affected the evolution of density perturbations (5), so it seems practically certain that this scale, which is imprinted at $z \lesssim z_{eq} = 10^{4.5} \Omega h^2$, has something to do with the scale of galaxy clustering. In the paleogenic scenarios, $M_{j\gamma}$ imprints a characteristic mass by passively modifying a preexisting spectrum of primordial fluctuations, through damping or some other scale dependent modification of the growth rate of fluctuations. In the neogenic scenarios, nongravitational forces play a more active role, and we would expect that the largest scale of inhomogeneity ought naturally to be the largest scale over which nongravitational forces can effectively transfer momentum in an expansion time (and thereby alter the pure gravitational equations of motion). In some neogenic scenarios (described below) this scale is still precisely the characteristic radiation Jeans scale $M_{j\gamma}$ which appears in the paleogenic models.

Another scale is also well known but less frequently commented upon: a characteristic binding energy $(v/c)^2 \simeq 10^{-6}$ or velocity $v \simeq 300\,\text{km/sec}$ also exists in the galaxy distribution (6) and must be imprinted nongravitationally. It is this property of the clustering which naturally leads us to take the neogenic view. The characteristic velocities of galaxies are about the same as velocities *within* galaxies which we know are not of primordial origin, such as the motion of hot gas in the interstellar medium or in galactic winds (whose energy balance relies on heating by supernovae), or matter in disks which must have undergone considerable dissipation in a gaseous phase, and thus probably have a binding energy determined in some way by the cooling function of dilute gas. It is also suggestive that the crossing-times of stars in galaxies and galaxies in groups and clusters span a continuous range of order $t \simeq 10^{8-10} y$, commensurate with the evolutionary timescales of stars or efficient Eddington-limited sources of radiation (e.g., QSO's), and that the binding energy densities on each scale $(v/c)^2 \rho$ are commensurate with each other and with the mean cosmological energy density of starlight.

These order of magnitude coincidences are fortuitous in the paleogenic scenarios (indeed, the most natural situation in a paleogenic model would be an obvious discontinuity in the various dynamical quantities between systems of paleogenic origin and those of neogenic origin), but they arise naturally if one assumes that astrophysical sources of energy are primarily responsible for the structure. From this point of view the paleogenic scenario therefore has what particle physicists like to call a "fine-tuning problem:" there is no particular reason why the amplitude of perturbations on the scale $M_{j\gamma}$ should be just sufficient to collapse at an epoch comparable to a stellar evolution time. In the neogenic picture, the time at which this structure appears ought naturally to be commensurate with the evolution time of the energy sources. The paleogenic view is of course still quite plausible; the fluctuations are specified by a single number, the dimensionless amplitude $\delta\varphi$, and it could very well be that an amplitude of the correct magnitude appears in (say) a supersymmetric model of the new inflation (3,10), as the ratio of a fundamental mass to the Planck mass. But the preceding comments still cause us to worry about exactly how large a scale we have to go to before we can safely ignore evolutionary effects caused by the release of energy.

Thus, although the very large scale uniformity of the universe must be a consequence of highly symmetric conditions at early times, there is no compelling evidence that primordial fluctuations are of direct relevance to the observed galaxy clustering, since we know that matter on these scales must have been rearranged by nongravitational forces in any scenario, and that the energy to do this without paleogenic fluctuations is available without recourse to sources of free energy significantly different from those we observe. Proof of significant primordial fluctuations would require evidence of inhomogeneity on much larger scales than $M_{j\gamma}$; such proof could in principle come from observation of the angular spectrum of microwave background anisotropy (7). The paleogenic model would predict that temperature fluctuations at large angular scale ϑ should have the characteristic amplitude of the scale-free initial perturbations, $\delta T/T \simeq \delta\varphi$, independent of ϑ (8); while the neogenic model would predict that the angular fluctuations, if they are observed at all, are produced by discrete sources of radiation and anisotropies at large ϑ ought to hold to the characteristic behavior of two-dimensional white noise perturbations in brightness temperature, $(\delta T/T)$ or ϑ^{-1} (9). At present no convincing evidence exists in either direction.

We now turn to a brief discussion of the mechanisms which can produce large scale structure from pregalactic stellar energy. Even if the neogenic view is correct, there is still reason to hope that we may be able to develop a more convincing simple model than we have been able to do for star formation, because there are very few plausible mechanisms for moving matter around on scales as large as $\simeq 10^{15-16} M_\odot$. Furthermore, the mechanisms are quite different in character and hence unlikely to be of comparable importance -- in other words, it is likely that the largest-scale inhomogeneity was generated by whatever single nongravitational force had the largest range, and is therefore likely to be described as a simple process.

The first of these mechanisms to be considered utilized explosive energy from supermassive ($\simeq 10^5 M_\odot$) stars (11). An explosiveenergy source has also recently been proposed with a more detailed treatment of gas dynamics and cooling (and without restrictive assumptions about the nature of the exploding object) by Ostriker and Cowie (12) and Ikeuchi (13). Their basic idea is "explosive amplification"; an exploding galaxy or quasar can create a mass of bound material greater than its own mass. The kinetic energy of explosion is converted by shocks into heat; gradients in gas pressure lead to compression, cooling, and collapse of matter in a shell surrounding the original explosion. The characteristic scale of structure l created is determined by equating Hubble velocity lH with sound speed $v_s \simeq (p/\rho)^{\frac{1}{2}}$. To order of magnitude, the temperature required in the gas is equal to the virial temperature in the final bound systems; for present-day clustering with $v \sim 300-1000 \, km/sec$, this temperature is about $(v/c)^2 m_{proton} \simeq 10^{7-8} K$. This process is therefore restricted to occur at redshifts $z \lesssim 7$, because at higher redshift ionized gas can cool in less than an expansion time by Compton cooling on the microwave radiation, so that energy is radiated before it can be usefully converted into motion and the scale of collapsing shells is much less than v_s/H. At redshifts $z \lesssim 7$, explosive amplification proceeds until the medium is too hot to cool in a free-fall time. A sphere of hydrogen gas in the relevant density range in hydrostatic equilibrium with its own gravity can cool in a free-fall time if its mass is less than $\simeq 10^{12} M_\odot$ or its radius in less than $\simeq 75 \, kpc$ (14-18), and so one can argue that galaxies appear quite naturally in this picture.

Promising as these models are, it is well to bear in mind several difficulties. It is not certain that shocks would remain coherent on a sufficiently large scale to explain the observed galaxy clustering structure. Also, massive, tightly bound systems such as the Coma cluster have a virial temperature so high ($>10^8 K$) that the corresponding proto-cluster gas cloud could not have radiated its binding energy in a free-fall time; we would not expect to find such bound systems in the universe today, because they would have "bounced" rather than collapsed. Finally, the objects which explode probably radiate about 1% of their rest mass as visible light (which is more than they release in the explosions). The fact that we do not see it means that there is a significant dust obscuration out to a redshift $z \sim 5$, which degrades the optical light into what is now the submillimeter region. The energy density of this radiation is probably not much smaller than the microwave background, and we know that it must have been released inhomogeneously on a scale of order $5 \, h^{-1}$ Mpc, or about 3 arc minutes in angular size; thus it is somewhat surprising that submillimeter astronomers have not seen anisotropy of cosmological origin. With IRAS and other new far-infrared technology, a systematic search for the radiation in different wavebands may be made.

Carr and Rees (19) have discussed various other "neogenic" scenarios. One important point made by them is that it is relatively easy to affect the efficiency of galaxy formation with a comparatively small amount of energy, because the temperature required to raise the Jeans mass to $\simeq 10^{11} M_\odot$ in dilute gas at the mean cosmological density at recent epochs is in the UV range, $\sim 10^5 K$. Thus it may be possible to introduce apparent large scale structure in the galaxy distribution without having to move the corresponding mass around, using only $\simeq 10$ eV/gas atom instead of the $\simeq 1$ keV it would take to produce a genuine mass perturbation. If this type of effect has been important it might explain why virial estimates of the mean cosmological density consistently give $\Omega < 1$ (see e.g.(6)), in spite of our prejudice that Ω ought to be $1 \pm \varepsilon$, based on fairly general cosmological fine-tuning arguments applied to the initial conditions (20). If the galaxies have an unrepresentative autocorrelation function which is greater than that of the background matter, it would help to explain their unexpectedly low velocities.

If pregalactic energy sources formed quite early ($z \gtrsim 100$; $t \lesssim 10^7 \mathrm{y}$), then another neogenic scenario is possible (5,21), which is a straightforward generalization of conventional paleogenic "isothermal" fluctuations. One assumes that small protogalaxies or QSO's are created by some unspecified evolutionary processes on smaller scales, which we need not be concerned about. The model then aims to show that their radiation would generate fluctuations on large scales which later lead to galaxy clustering. Thus the hope is that the final step in the formation of structure can be described by such a simplified model without reference to what occurred earlier on smaller scales, which may well have involved a complicated system similar to the interstellar medium of our Galaxy. In the model, random fluctuations in luminosity per mass lead to perturbations in the radiation energy density; the resulting flux of radiation drags matter (in the form of interstellar photoionized gas) along with it by Thomson drag; and this leads to density perturbations, up to masses of order $M_{j\gamma}$. One attractive feature is that a precise linear calculation is possible, similar to those describing the evolution of perturbations in the pre-recombination big bang but with terms added to represent the sources of radiation. (The spectrum of the added radiation is not important because the dominant opacity is still Thomson scattering.) In this way one can determine exactly how much inhomogeneity must appear in the large scale radiation field to create a given matter density perturbation on the observed scales, and fairly rigorous predictions can be made about the amplitude, angular spectrum and polarization of the resulting microwave background anisotropy, which are not sensitive to the various parameters of the model, such as the size of the protogalaxies or the energy density of radiation they release. We expect the peak amplitude to be on an angular scale of about one or two degrees, and to be greater than about $\delta T / T \simeq 5 \times 10^{-6}$; and the polarization anisotropy on this scale is about 1×10^{-5}. At larger ϑ, $\delta T / T \propto \vartheta^{-1}$, independent of the parameters. Fortunately, this prediction seems well within the range of present or planned experiments to measure the anisotropy (22,23,24). Although the starting conditions of this model are rather unsatisfyingly complicated, it does provide a natural explanation of why galaxies, if they began to form early enough, should now be clustered on as large a scale as they are, and also accounts for the binding energy of clusters in a natural way.

This work was supported by a Bantrell Fellowship at Caltech.

REFERENCES

1.   Ostriker, J.P., in "Astrophysical Cosmology," eds. H.A. Brück, G.V. Coyne, and M.S. Longair, (Proceedings of the Vatican Study Week, 1982),p.473.

2.   Hawking, S.W., Phys. Lett. 115B, p. 295 (1982).

3. Bardeen, J.M., Steinhardt, P.J. and Turner, M.S., Preprint, University of Chicago (1983).

4. Guth, A. and Pi, S.-Y., Phys. Rev. Lett. 49, 1110 (1982); Starobinski, A.A., Phys. Lett. 117B, p. 175 (1982).

5. Hogan, C.J., Mon. Not. R. astr. Soc., 202, p. 1101 (1983).

6. Davis, M. and Peebles, P.J.E., Ap. J.,267, p.465 (1983).

7. Hogan, C.J., Kaiser, V. and Rees, M.J., Phil. Trans. R. Soc. Lond., A307, p. 97 (1982).

8. Peebles, P.J.E., Ap. J. Lett., 263, p. L1 (1982).

9. Hogan, C., Ap. J. Lett., 256, p. L33 (1982).

10. Steinhardt, P., Proc. 1982 Nuffield Workshop; Ellis, J., Nanopoulos, D.V., Olive, K.A. and Tamvakis, K., preprint CERN (1983).

11. Doroshkevich, A.G., Zeldovich. Ya.B. and Novikov, I.D., Sov. Astr. 11, p. 233 (1967). (orig. Astr. Zh. 44, p. 295).

12. Ostriker, J.P. and Cowie, L.L., Ap. J. Lett., 243, p. L127 (1981).

13. Ikeuchi, S., P.A.S.P., 33, p. 211 (1981).

14. Rees, M.J. and Ostriker, J.P., Mon. Not. R. astr. Soc., 179, p. 451 (1977).

15. Silk, J., Ap. J. 211, p. 638 (1977).

16. Binney, J., Ap. J. 215, p. 483 (1977).

17. Gunn, J.E., in "Astrophysical Cosmology" (Vatican Study Week, 1982), p. 557.

18. Faber, S., in "Astrophysical Cosmology" (Vatican Study Week, 1982), p.191.

19. Carr, B.J. and Rees, M.J., Preprint, University of Cambridge (1983).

20. Guth, A. 1983, Phil. Trans. Roy. Soc. London, A307, p. 141 (1982).

21. Hogan, C. and Kaiser, N., Ap. J. (1983), in press.

22. Wilkinson, D.T., Phil. Trans. Roy. Soc. London, A307, p.55 (1982).

23. Partridge, R.B., in "The Origin and Evolution of Galaxies," ed. B.J.T. Jones and J.E. Jones, p.121 (Reidel, 1983).

24. Fabbri, R., Guidi,I.& Natale, V., Phil. Trans. Roy. Soc. London, A307, p.67 (1982).

LARGE SCALE INHOMOGENEITIES AND GALAXY STATISTICS

Richard Schaeffer

Service de Physique Théorique,
CEN-Saclay, 91191 Gif-sur-Yvette Cedex, France

The density fluctuations associated with the formation of large scale cosmic pancake-like and filamentary structures can be evaluated (1) using the Zeldovich approximation (2) for the evolution of non-linear inhomogeneities in the expanding universe. At the scale introduced by these non-linear density fluctuations due to pancakes, the standard scale-invariant ($\sim r^{-\gamma}$) correlation function $\xi(r)$ is modified. The corresponding J_3 integral

$$J_3(d) = \int_0^d r^2 \, \xi(r) \, dr \quad ,$$

is then found to be dominated by the standard short range correlation function

$$J_3(d) \sim \frac{1}{3-\gamma} \left(\frac{r_o}{d} \right)^{\gamma-3}$$

for small $d (<10h^{-1}\text{Mpc})$ and by the pancake contribution

$$J_3(d) \sim 0.41 \, fd^3 \left(\frac{\lambda}{2d} \right)^{1/3} \left[1 - c \left(\frac{2d}{\lambda} \right)^{1/3} \right]$$

for large $d (>10h^{-1}\text{Mpc})$.

Here f is the pancake filling factor, λ the coherence length and $c<1$ a parameter. All these constants can be adjusted to the existing data for galaxy counts. They can also, in principle, be obtained from the initial fluctuation spectrum. This form of J_3 is obtained from the universal form of the pancake singularity and is thus model independent (but of course, the parameters f, λ

133

J. Audouze and J. Tran Thanh Van (eds.),
Formation and Evolution of Galaxies and Large Structures in the Universe, 133–135.
© *1984 by D. Reidel Publishing Company.*

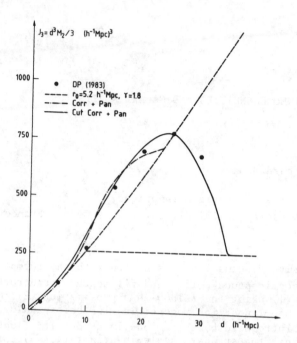

Fig.1 - J_3 integral (full circles) deduced by Davis and Peebles (1983) from the CFA sample as compared to the integral (dashed curve) calculated from the correlation function that reproduces the CFA data at $d \leq 10 \ h^{-1}$Mpc. The pancake contribution can be added to the short range power law contribution assuming that the latter extends to all scales (dash-dotted curve) or that the latter vanishes for $d_c > 10 \ h^{-1}$Mpc (full curve).

and c are not). It provides a statistical test for the existence of large scale inhomogeneities. The application to several recent three dimensional data sets shows that despite large observational uncertainties over the relevant scales, characteristic features that can be attributed to pancakes may be present in most, but not all, of the various galaxy samples. In particular, adjusting (Fig.1) the parameters to the CFA sample (3) leads to

$$\lambda c^{-3} \approx 70 \ h^{-1} \text{Mpc and fc} \approx 0.22 \ .$$

From the limit $c \leq 1$, a maximum coherence length $\lambda < 70h^{-1}$Mpc can be obtained and the filling factor is constrained to $0.22 \leq f \leq 1$.

The modification of the higher order moments (>2) of the number of galaxies in a cell of size d due to pancakes can be calculated in a similar way, and depends on the same parameters λ and f. The determination of these moments from the galaxy data at scales relevant for pancakes (10 to 40h Mpc) will provide additional constraints on λ and f. It will be an important discriminant of intrinsic large scale structures.

REFERENCES

(1) Schaeffer, R. and Silk, J., to be published.
(2) Zeldovich, Ya. B. 1970, Astron. Astrophys. 5, pp.84-89.
(3) Davis, M. and Peebles, P.J.E., to be published.

CLUSTERS OF GALAXIES AS INDICATORS OF GALAXY ORIGIN

Piotr Flin

Jagiellonian University Observatory
ul. Orla 171, 30-244 Kraków, Poland.

ABSTRACT
In this paper I am discussing various difficulties
occuring while the clusters of galaxies are investigated. The special emphasis is put on the brightness
profiles of galaxy clusters, ellipticities and position
angles of clusters. The distribution of position angles
of galaxies in the Local Supercluster is also discussed.

1. INTRODUCTION

It is commonly accepted that structures existing now
in the Universe reflect the physical processes which
have led to their formation. Theories of galaxy origin
predict phenomena and relations among cluster parameters, which can be tested, in principle, by observations. Therefore, the detailed studies of structures
in the Universe performed even in one range of the
electromagnetic spectrum, e.g. the optical one, could
be useful in eliminating some theories.

In this paper I would like to present some results
connected with investigations of galaxy clusters, pointing out difficulties connected with their interpretation. On the other hand, an observer dealing with data
and trying to interpret them, always meets with mixture of effects, those survived from the very origin
of structures and the others connected with evolution.
Moreover, the same phenomenon could be explained

137

J. Audouze and J. Tran Thanh Van (eds.),
Formation and Evolution of Galaxies and Large Structures in the Universe, 137–146.
© 1984 by D. Reidel Publishing Company.

theoretically either by origin processes or by evolution.

2. SECONDARY MAXIMUM IN THE BRIGHTNESS PROFILE

This effect may be used as a typical example of a phenomenon having different theoretical explanations /for
review see: Flin (1982a) , Trevese and Vignato (1982) /.
Due to the fact that some authors are claiming that
this phenomenon, when observed, is caused either by
the departure from assumed spherical symmetry of the
cluster or by subclustering, I decided to check the
correctness of these statements. Dressler sample (1980)
of Abell clusters served as an observational basis.
From total number of 55 clusters the subsample consisting of 30 clusters was extracted. All the clusters, in
which galaxies are extending over the distance 2 Mpc
/assuming H_0 = 50 km / sec • Mpc belong to that sub-
sample/

    For each cluster iterative procedure was applied
and cluster center was determined. The iterative procedure of finding the maximal projected density of galaxies is the same, as described in the previous paper
/Flin (1982 b)/, but with modification /this modification
changes some numerical values given in the previous paper, but does not change the conclusion/. The cluster
center was fixed when:
1. the number of galaxies is maximal in the innermost
   ring with radius 0.25 Mpc,
2. the procedure is numerically stable.
Keeping the cluster center fixed, the covariance ellipse
was fitted in classical manner /Trumpler and Weaver
1953/. Later on the counts of galaxies in elliptical
rings were made and brightness profiles were constructed
/fig. 1/, assuming equal weight to each galaxy.

    The distances of observed secondary maximum were
compared with the distribution of galaxies in cluster.
I distinguished two types of subgroups:
1. laying in narrow position angle from the cluster
   center, and
2. elongated subgroups forming ring-like structures.
The results of these analyses show that taking into
account the axial symmetry instead of previously assumed spherical symmetry does not eliminate the occurence of secondary maximum. Moreover, some of these
observed secondary maxima are caused by "ring-like"
structures.

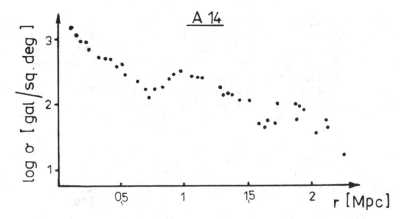

Fig. 1. The brightness profile of galaxy cluster.

In few cases the existence of secondary maximum cannot
be attributed to any evident subgroup. Sometimes, for
weak phenomenon its appearance depends on the cluster
center coordinates. Deviations from the smooth distribution, regarded as secondary maxima are greater
than those due to statistical fluctuations, when their
measure is $N^{1/2}$.

But two points ought to be stressed in connection
with brightness profiles: firstly, in some clusters
when secondary maximum is observed in optical region,
the brightness profiles observed in X-ray range are
quite smooth /Jones (1982) /, secondly, assumed axial
symmetry sometimes causes the existence of "new" secondary maximum, which is different from that observed
when spherical symmetry of cluster is assumed.

3. THE ELLIPTICITIES AND POSITION ANGLES OF CLUSTERS

In several papers Carter and Metcalfe (1980) , Binggeli (1982) , Burgett (1982) ellipticities of galaxy
clusters were investigated. Further, the analysis of
the distribution of ellipticities and /or the comparison of cluster position angle with position angle
of the brightest member galaxy was carried out.

The method of covariance ellipse with modification
increasing the obtained ellipticities was developed by
Carter and Metcalfe (1980) ,and was also used by Burgett
(1982) .Binggeli (1982) fitted the straight line by using
the ortogonal least squares method and determined the

the sums of galaxy distances from two axes. These methods as well as the fit of classical covariance ellipse
give different values of obtained ellipticities and position angles. Furthermore, while the same method is applied, the values of discussed parameters depend on
taking into account unweighted coordinates of galaxies
/equal weight to each galaxy/ or luminosity weighted
coordinates / Burgett (1982) /. This discrepancy is not
important when a sample of clusters is investigated
for statistical analysis of cluster properties. The obtained values of clusters ellipticities are important
while compared with theoretical predictions. E.g. Binney and Silk (1979) proposed a model giving numerical
values of structure ellipticities in different moments.
Therefore, I think, that the principal difficulty occuring here is as follows:which values of ellipticities
ought to be compared with the model? Moreover,the change
of ellipticities with radial distance from the cluster
center was noticed for several clusters /Peach (1982),
Burgett (1982) /.

    As an example, I am presenting three such rela-
tions from my studies. In fig. 2, e - denotes ellipticity of cluster, r - radial distance from the cluster
center in Mpc. Analyzing the change of ellipticities
the problem of errors becomes very important. Carter
and Metcalfe (1980) and also Binggeli (1982) applied
Monte Carlo runs. Presently numerical simulations are
performed with more realistic assumption than cluster
spherical symmetry and King distribution of galaxies
within cluster. Burgett (1982) compared obtained ellipticities at different, but close each other distances from the cluster center. In my study obtained error
$\Delta e \approx 0.05$ is the mean of differences of ellipticities
counted at the same distances from cluster centers determined as the mean arithmetic of member galaxy coordinates /first step of iteration/ and the adopted center with maximal projected density. The difference of
obtained ellipticities at the same radial distances   r
from the cluster center depends also on the distance   r,
which is quite understandable. The number of galaxies
is increasing with increasing r, so Δe are dimishing.
Differences of ellipticities counted for consecutive
steps of iteration are much smaller than the quoted
above value Δe. The change of ellipticity with radial
distance is observed for different cluster centers during iteration. Taking into account these mentioned
above evidences it ought to be accepted that this effect
is statistically significant. Also, from the quoted
by Burgett (1982) values and their errors the same con-

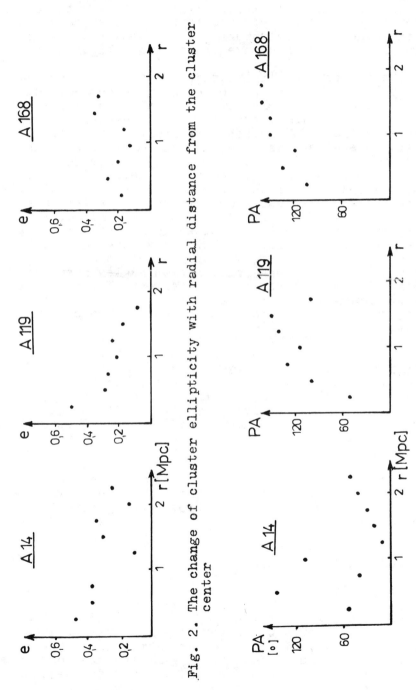

Fig. 2. The change of cluster ellipticity with radial distance from the cluster center

Fig. 3. The change of cluster position angle with radial distance from the cluster center

clusion can be drawn out. Therefore, the observed effect of the change of cluster ellipticities with radial
distance from the centers should be regarded as real.
The influence of subclustering can be observed mainly
as the instantaneous change of counted ellipticities.

The further complication during cluster investigation, besides the problems of cluster center determination, is occuring; cluster position angle is also
changing with radial distance from the center, as can
be seen in figure 3. In connection with this effect
the analysis of errors is even more complicated than
in the case of ellipticity - radial distance relation.
Monte Carlo runs were applied by Binggeli (1982) , Burgett obtained Δ PA in the same manner as Δe. I obtained
ΔPA in the way previously described for Δe determination. The same effect, that is the dependance of Δ PA
on r, which has the same explanation as relation Δe
versus r, is appearing. In both relations the values
Δe and ΔPA are reversely proportional to cluster ellipticity e. The change of position angles seems to
be also the real phenomenon. The effect of subclustering
is manifesting itself through abrupt change of position
angle, much stronger than the change of cluster ellipticities.

In testing different theories of galaxy origin
the important question arises:at which distance from
the cluster center the cluster position angle ought to
be taken,in order to compare it with position angle of
the brightest member galaxy, or with position angle of
neighbouring cluster. Binggeli (1982) chose 2 Mpc, which
seems reasonable from observational point of view. The
cluster ellipticities counted at this distance are, in
general, insensitive to the cluster center coordinates.
From the theoretical point of view one can expect that
at this radial distance the density is much smaller
than in the cluster core, and evolutionary effects are
also smaller.

4. ALIGNMENT OF GALAXIES

4.1. Abell clusters

In some clusters, mainly L-type clusters the position
angles of major axes of member galaxies are not distributed randomly /for review see Flin (1982a,)(1983) /.
It means that angular momenta of galaxies are correlated.
This effect is very important information for testing

galaxy origin, or... galaxy evolution. There are not common agreements about the significancy of evolutionary processes which could change the angular momenta of galaxies belonging to clusters. I believe that enviromental influence to galaxies is not enough effective for changing position angles of many galaxies in clusters. If so, galaxy position angles are not changed from the very origin. The discussion of enviromental influence to galaxies is recently reviewed by White (1982) and the problems of gravitational interaction of galaxies are discussed by Farouki and Shapiro (1981). The precize discussion of different effect causing the observed alignment based on numerical simulation is performed by MacGillivray et al. (1982a) and MacGillivray and Dodd (1982a,b).

4.2. The Local Supercluster

Recent investigation of the distribution of ellipticities and position angles of spiral and irregular galaxies performed by MacGillivray at al. (1982b) permit them to find "marginally significant tendency for galaxies to be aligned along to plane of Local Supercluster". This result was obtained using Hawley and Peebles method (1975). Also earlier investigations show rather similar results. The only contradictory result was obtained by Jaaniste and Saar (1977). MacGillivray at al. (1982b) suggested an error in Jaaniste and Saar spherical trygonometry.

However the situation is much more complicated Flin and Godłowski (1983). Jaaniste and Saar paper is internally inconsistent. The description of the basic figure is different than the figure itself. Moreover, derived and applied formulae do not correspond to the manner of versors defining. Because their method is very interesting and looks very promising we decided to re-analyze it. From detailed analysis of their paper it is clear that from possible solutions they selected only one. As a result of such approach the celestial position of galaxy has a very strong influence to the results. Due to mentioned previously choice of the solution, angular momenta of galaxies more parallel to the supergalactic plane are selected on the northern hemisphere, more perpendicular - on the southern hemisphere. These a priori expected effects also depend on the line-of-sight galaxy inclination. The preliminary result of our analysis is showing that more galaxies are laying on the southern hemisphere, but they are close to the supergalactic plane. It cannot be excluded that this is causing the result obtained by Jaaniste and Saar.

    On the other hand there is a suspicion that the
Hawley and Peebles method is prefering such result, as
obtained by MacGillivray at al. (1982b) . If this sus-
picion will materialize, we shall consider MacGillivray
et al. result as random distribution of position angles.
There is a paper in preparation, in which the same sample
of galaxies /galaxies taken from UGC and ESO / Uppsala
Survey / is analyzed using both methods, which ought to
give consistent result.

    MacGillivray at al. (1982b) concluded, that their
result is supporting the scenario, in which subunits
are formed after the protocluster. This point of view
seems to be confirmed by Binggeli (1982) and not confirmed by Helou and Salpeter (1982). The former author
investigated the distribution of position angles of neighbouring clusters, the latter authors studied 40 galaxies
in the Virgo cluster, where for each galaxy the direction
of angular momentum was determined. The sample of 306
galaxies in the Local Supercluster was analyzed by Yama-
gata et al. (1981). For these galaxies the distinction
of "S" and "Z" spiral arm was made. They concluded, that
their result is consistent either with perpendicular
model or with random distribution.

5. CONCLUSION

The aim of this paper was to show what optical studies
of galaxy clusters could tell us about the origin of
the structures in the Universe. It is clear that such
investigations permit us rather to eliminate, than to
support some models of cluster /galaxy/ origin.

    The attention ought to be paid to the fact, whether
the observed relations among cluster parameters exactly
follow those predicted by the particular theory of galaxy origin. In the case of alignment of galaxies in
clusters such careful discussion is presented by Helou
and Salpeter (1982). Moreover, different possible spurious effect must be taken into account, when from the
analysis of two-dimensional data someone would like to
draw out conclusion concerning three dimensional structures. The above mentioned papers presenting the results
of MacGillivray et al. (1982a) and MacGillivray and Dodd
studies (1982a,b) ought to be regarded as excellent discussion of these problems. I believe that contradiction
among different investigations is rather due to the scar-
city of observational data and difficulties with their
interpretations, and therefore cannot be regarded as

principal impossibility of the usage of clusters as
indicators of galaxy origin. The detailed studies of
clusters which are now in progress permit us to eliminate some theories, or rather else classes of theories,
of galaxy origin. Therefore, I think that the present
status in this subject is difficult, but not hopeless.

ACKNOWLEDGEMENTS

I am very grateful to Dr Jean Audouze and the Organizing
Commeettee for financial support, enabling me to attend
this Meeting.

REFERENCES

Binney,J., Silk,J., 1979, Mon.Not.Roy.astr.Soc. 188,
          p. 273
Binggeli,B., 1982, Astron. Astroph. 107, p. 338
Burgett,W.S., 1982, Master Thesis, University of Okla-
          homa
Carter,D., Metcalfe,N., 1980, Mon.Not.Roy.astr.Soc. 191,
          p. 325
Dressler,A., 1980, Astroph,J. Suppl. 42, p. 565
Farouki,R., Shapiro,S.L., 1981, Astroph.J. 243, p. 32
Flin,P., 1982a, in "The Origin and Evolution of Gala-
          xies" /ed. V. de Sabbata/ World
          Scientific, Singapore, p. 63
Flin,P., 1982b, in Proceedings of the European Collo-
          quium", "Clusters of Galaxies" /eds.
          D. Gerbal and A. Mazure/, Editions
          Frontieres, Gif sur Yvette
Flin,P., 1983, in "The Origin and Evolution of Galaxies"
          /eds. B.J.T. Jones and J.E. Jones/
          D. Reidel Publ. Co., Dordrecht,
          p. 329
Flin,P., Godłowski,W., 1983 in preparation
Hawley,D.L., Peebles, P.J.E., 1975, Astr.J. 80, 477
Helou,G., Salpeter,E.E., 1982, Astroph.J. 252,p.75
Jaaniste,J., Saar, E.E., 1977, Tartu Obs. Preprint A-2
Jones,C., 1982, private information
MacGillivray,H.T., Dodd,R.J., McNally,B.V., Lightfoot,
          J.F., Corwin,H.G., Heathcote,S.R.
          1982a, Astroph.Sp.Sci. 81, p. 231
MacGillivray,H.T., Dodd,R.J., McNally,B.V., Corwin,H.G.,
          1982b, Mon.Not.Roy.astr.Soc. 198,
          p. 605
MacGillivray,H.T., Dodd,R.J., 1982a, Astroph.Sp.Sci 83,
          p. 127, 1982b, Astroph.Sp.Sci 83,
          p. 373

Peach,J.V., in Proceedings of the European Colloquium
        "Clusters of Galaxies" /eds. D. Gerbal and
        A. Mazure/, Editions Frontieres, Gif sur
        Yvette
Trevese,D., Vignato,A., 1982, Astron.Astroph. 110, p.
        238
Trumpler,R.J., Weaver,H.F., 1953, Statistical Astronomy,
        University of California Press, Berkeley
        and Los Angeles
White,S.D.M., 1982, in "Morphology and Dynamics of Ga-
        laxies" /eds. L. Martinet and M. Mayer/,
        Twelfth Advanced Course of the Swiss Society
        of Astronomy and Astrophysics, Sauverny,
        Geneva Obs., p. 289
Yamata,T., Hamabe,M., Iye.M., 1981, Tokyo Observatory
        Annals XVIII, p. 164

III

MASSIVE HALOS

ARE MASSIVE HALOS BARYONIC?

Dennis J. Hegyi

Department of Physics
University of Michigan
Ann Arbor, Michigan

ABSTRACT

The problems with massive halos being composed of baryonic matter are discussed. Specifically, a halo composed of either gas, snowballs, dust and rocks, low mass stars, Jupiters, dead stars or neutron stars is shown to be unlikely. Halos could be composed of black holes less than 100 M_0 if they, unlike the stars in this mass range, are extremely efficiently accreting or primordial. At present, however, particles from supersymmetric theories appear to offer the most interesting possibilities as the constituents of halos.

I. INTRODUCTION

Spiral galaxies are surrounded by halos, large amounts of sub-luminous or non-luminous matter. These halos are approximately spherical in shape and may extend out to distances as far as ten times the optical radius of a spiral galaxy.

The supporting evidence for halos is quite compelling. Using dynamical arguments based on the rotation curves of spiral galaxies, it is possible to accurately determine the halo mass as a function of galactic radius. Also, a number of independent arguments require that halos be approximately spherical. Based on the information available about halos, it is not difficult to show that halos contain about 10-100 times the mass in the disks of spiral galaxies, and consequently, contain a significant fraction of the cosmological mass density.

149

J. Audouze and J. Tran Thanh Van (eds.),
Formation and Evolution of Galaxies and Large Structures in the Universe, 149–162.
© *1984 by D. Reidel Publishing Company.*

In contrast to the definite statements that can be made
regarding the existence of halos, very little can be said about
the exact nature of the halo mass. At present, it appears that
the most direct way to determine the composition of the halo mass
is to show what halos cannot contain.

In the present investigation we argue that halos are not
composed of baryons. Our approach will be to show the problems
associated with the following types of baryonic matter: gas,
snowballs, dust and rocks, Jupiter-like objects, low mass stars,
dead stars and neutron stars. It appears very difficult to avoid
the problems that we shall present if halos are baryonic. We
shall discuss a model in which it is claimed, a primordial halo
composed of gas can be converted into Jupiters, and show that it
is not self-consistent.

Though not baryonic, black holes are a possible constituent
of halos. If halos are composed of black holes they must be
extremely efficiently accreting or primordial. Aside from the
possibility of efficiently accreting black holes, we expect the
cosmological baryonic abundance to be low at the time of
nucleosynthesis. We shall briefly discuss the current situation
regarding the observed nuclear abundances in terms of
cosmological production in a low baryon density universe.

One of the earliest discussions of massive halos surrounding
spiral galaxies was given by Hohl (1,2). He found his models of
spiral disks to be unstable with respect to the growth of long
wavelength modes, and as a result, the disks tended to develop
into bar-shaped structures within about two revolutions. Hohl
was able to stabilize his models by adding a fixed central force
which he identified with a halo population of stars and the
central core of the galaxy. Kalnajs (3), considering only exact
solutions for infinitely thin spiral disks, explored ways of
stabilizing the initially cool rotational state. Perhaps his
most interesting result was that by embedding the spiral disk in
a uniform density halo, stability could be obtained.

The possibility that spiral galaxies might be surrounded by
massive halos was emphasized by Ostriker and Peebles (4). Using
a 300-star galactic model they studied the instability of spiral
structure to the development of bar-like modes. The onset of
instability was reached when t, the ratio of the kinetic energy
of rotation to the total gravitational energy, increased to a
value ~0.14. From a literature survey, the authors concluded
that for systems ranging from fluid MacLaurin spheroids to flat
galactic systems with 10^5 stars, the critical value for the onset
of instability appears to be t ≈ 0.14. Two different ways were
suggested to stabilize the spiral structure, a hot disk
population with radial orbits and a hot spherical halo. From a

variety of arguments, it is now known that the halo mass distribution is spherical.

The strongest observational evidence supporting the existence of massive halos is dynamical. The rotation curve of a galaxy must satisfy the criterion that in equilibrium the inwardly directed gravitational force must balance the outwardly directed centrifugal force. Rotation curves of galaxies have been obtained by both optical and radio techniques (5-11). Data obtained on more than 50 spiral galaxies reveal symmetric rotation curves which support the equilibrium condition

$$M_r = \frac{K}{G} v^2 r \ .$$ (1)

where M_r is the mass within radius r, K is a constant ranging from $2/\pi$ for a thin disk to unity for a spherically symmetric mass distribution, G is the gravitational constant, and v is the circular rotational velocity at galactic radius r. The observations show that v is a constant independent of r, and, as may be seen from eq. (1), $M_r \propto r$.

Beyond about 50 Kpc it is difficult, typically, to observe rotation curves, and binary galaxies (12,13) have been used to sample the halo mass distribution at large radii. Unfortunately there are a variety of selection effects which binary galaxies are subject to and it has not yet been possible to untangle these effects sufficiently to unambiguously interpret the results (14).

As already mentioned, several arguments have been used to show that the mass distribution of halos is spherically symmetric. The persistence of warps in spiral disks (15,16), star counts (17), and the scale height of stars perpendicular to spiral disks (18,19), all indicate relatively spherical halos, i.e. with aspect ratios close to unity.

II. EVIDENCE AGAINST NONBARYONIC HALOS

Much of the discussion about nonbaryonic halos has been presented elsewhere by Hegyi and Olive (20). Here we shall summarize parts of that discussion and amplify other parts. Before starting, however, we define a "standard halo" which we shall need to evaluate a variety of properties of baryonic halos. For this halo, $M_r \approx 10^{12} M_\odot$ in a radius of 100 kpc.

First we consider a halo made of gas. In a cold gaseous halo, particles moving on radial orbits would quickly collide with other gas particles and collapse on a gravitational timescale $\tau_c = (3\pi/32G\rho)^{1/2} \approx 5 \times 10^8$ yrs. Since halos must persist for 10^{10} years, they must be in hydrostatic equilibrium and they must be hot. Our standard halo, if it were gaseous, requires an equilibrium temperature of $T_{EQ} \approx 2 \times 10^{6}$°K which is sufficiently hot to violate the upper limits on the X-ray background by a factor of 20. The X-ray emissivity is sensitive to Ω_{Halo}, the fraction of the critical density contained in halo. We use $\Omega_{Halo} > 0.05$ (21).

A halo of snowballs will not be stable on a cosmological timescale. Snowballs, consisting primarily of hydrogen, are distinguishable from Jupiters because they are bound electrostatically. It turns out that the binding energy of a hydrogen molecule to solid hydrogen is sufficiently small so that it easily escapes, even when the temperature of the snowball is at 3°K, the temperature of the present cosmic background radiation. In fact, halos must have formed when the temperature of the cosmic background radiation was over 7°K; since halos are composed of non-interacting particles and cannot evolve to higher densities, they must have formed when the density of the universe had a density about equal to the present density of halos.

The argument against a halo of snowballs requires two steps. Based on laboratory measurements on solid hydrogen, its vapor pressure at 3°K has been found to be about 9×10^{-12} mm (22). This is high enough so that it is possible to show that there is no equilibrium between the solid and gaseous phase of hydrogen. The second part of the discussion involves the rate at which molecules evaporate to reach equilibrium. The time for evaporation (23) of a H_2 molecule (molecules rather than atomic hydrogen, will leave the snowball preferentially because their binding energy is lower) is

$$t_{ev} \sim [\nu_0 \, e^{-b/kT}]^{-1} . \tag{2}$$

The evaporation time is the inverse of the product of two terms: a Boltzmann factor which is the probability of a system attaining the escape energy, and an attempt frequency, the number of times per second that the system strikes the barrier. The reader is referred to (20) for more details. Here we report that at 3°K, the evaporation time per molecule is less than 10^{-8} seconds.

Next we consider a halo composed of dust and rocks, i.e. metals. A halo made of metals would contain a factor of about 50 times the mass of the disk of a spiral galaxy. The factor of 50 arises as the ratio of $\Omega_{Halo}/\Omega_{Disk} > .05/.001 \approx 50$. The problem is that if even a very small fraction of the halo mass mixed with the disk it would lead to a large metal abundance in the disk. Since the halo is believed to have formed before the disk and since there are disk stars with metal abundances $Z \sim 10^{-5}$, this implies that less than about one part in 5×10^6 of the halo mixed with the disk gas. It is difficult to believe that the halo could be composed of metals without contaminating the disk at such a low level.

The next possibility that we consider is a halo composed of low mass stars or Jupiters (24), that is, objects which are gravitationally bound with $m \lesssim 0.08 \, M_{\odot}$ which do not have high enough central temperatures to support nuclear burning. By making observations of the surface brightness of the halo, it is possible to set limits on the mass in low mass stars. If a connection can be established between the nuclear burning stars $(M > 0.8 M_{\odot})$ and the Jupiters, then by establishing constraints on the luminous portion of the initial mass function, constraints are simultaneously set on the non-nuclear burning portion of the initial mass function.

To connect the luminous and non-luminous parts to the halo init al mass function, a single power law relation has been assumed. The justification for this assumption is that the physics which affects the lower mass limit for nuclear burning is independent of the physics which governs gravitational collapse and it would be a considerable coincidence if these two mass scales coincided. Nuclear burning depends on the fine structure constant, α, and the strength and range of strong interactions while the physics of gravitational collapse depends on α and the gravitational constant, G. Since the assumption that the halo initial mass function is a single power law is the strongest assumption in this manuscript, we shall return to this subject to present other supporting evidence and discuss the substantial problems that must be overcome to seriously consider a radically different initial mass function, namely a halo of Jupiters with negligible mass in nuclear burning stars.

As we shall show, the mass-to-light ratio, M/L, of a halo is a function of the slope of the initial mass function, x, and the lowest mass condensation which forms gravitationally, m_{min}, also known as the Jeans mass. The initial mass function is defined by

$$\phi_m = Am^{-(1+x)} , \tag{3}$$

where ϕ_m is the number of stars per unit mass per unit volume of the halo. In general, A and x will depend on the mass range considered. The total mass density in stars and Jupiters, ρ_m, is

$$\rho_m = \int_{m_{min}}^{m_G} m \, \phi_m \, dm , \tag{4}$$

which, using eq. (3) may be found to be,

$$= \frac{A}{1-x} \left[m_G^{1-x} - m_{min}^{1-x} \right] . \tag{5}$$

Here, m_G is the mass of a giant which is taken to be $m_G \approx 0.75\ M_\odot$ and for the present argument, we neglect the small fraction of the mass contained in more massive objects.

Using the initial mass function, the luminosity density of the halo, ρ_L, may be seen to be,

$$\rho_L = \int_{m_0}^{m_G} L_m\ \phi_m\ dm + L_G . \tag{6}$$

For

$$L_m = c\ m^D , \tag{6a}$$

$$\rho_L = \frac{Ac}{D-x} \left[m_G^{D-x} - m_0^{D-x} \right] + L_G , \tag{7}$$

where L_m is the luminosity of a star of mass m, and c and D are constants chosen for a particular spectral band. The lower limit of integration, m_0, in eq.(6), the lower limit for nuclear burning, has been taken to be $m_0 \approx 0.08\ M_\odot$ (25). The quantity, L_G, is the light due to giants. Since observational constraints are available in the I and K Johnson spectral bands for the halo of the edge-on spiral galaxy NGC 4565 we shall evaluate ρ_L in these bands. The data of Gunn and Tinsley (26) in the range $0.08\ M_\odot$ to $0.8\ M_\odot$ have been fit with the power law in eq.(6a). For the luminosity in the I band, $L_{m,I}$, $c = 1.49 \times 10^{-3}$ and $D = 2.71$ and, correspondingly, for $L_{m,K}$, $c = 3.12 \times 10^{-2}$ and $D = 2.11$ where mass is expressed in solar units and in each spectral band, the luminosity equals unity for a zero magnitude star.

To express the contribution of giants to the surface brightness we have used the method described in Tinsley (27). Since Tinsley discussed a metal abundance $Z = .01$, we corrected the Tinsley models using the calculations of Sweigart and Gross (28). Fitting the later calculations (for $m = .7\ M_\odot$, $Y = .30$) for the change in main sequence lifetime as a function of Z, the correction to the lifetime was found to be $\propto \exp[28.6Z - .286]$, that is, increasing Z increased main sequence lifetimes. Also, it may be seen that this factor is equal to unity for $Z = 0.01$. For these calculations we have used $Z = 10^{-5}$, a value appropriate to halo stars. Lifetimes for smaller metal abundances are not changed appreciably.

To calculate M/L for the halo of NGC 4565, we shall use $M/L = \rho_m/\rho_L = \sigma_m/\sigma_L$, where σ_m and σ_L are the projected mass and luminosity density. It is necessary to evaluate the projected halo mass density in terms of the 21 cm rotational velocity 253 km/s (29) and the maximum extent of the halo, R_{max}. This may be seen to be

$$\sigma_m = \frac{v^2}{2\pi G}\frac{1}{r}\tan^{-1}\sqrt{\left(\frac{R_{max}}{r}\right)^2 - 1} \tag{8}$$

at galactic radius r. The distance to NGC 4565 is unlikely to be larger than 24 Mpc, and since the rotation curve has been observed out to 11.6 ', $R_{max} = 81$ Kpc. Using eq.(8) and eq.(7), it may be seen that M/L for the halo is only a function of x and m_{min}.

We now turn to the observational data on the surface brightness of the halo of NGC 4565. Data taken with the annular scanning photometer (30) in the Kron I band has been discussed by Hegyi (31), see Figure 1. That data has been transformed to the Johnson system and expressed in solar units. A least squares fit to that data using the functional form $\sigma_L = a/r + b$ has been performed. (This functional form assumes that R_{max} is large compared to r so that the $\tan^{-1}$ function in eq. 8 reduces to $\pi/2$.) A 2σ lower limit to σ_m/σ_L expressed in solar units in the Johnson I band is

$$M/L_I > 60 \ M_\odot/L_{\odot,I} \ . \tag{9}$$

Observations in the K band have been made by Boughn, Saulson, and Seldner (32) using a chopping secondary. Their 2σ lower limit is

$$M/L_K > 38 \ M_\odot/L_{\odot,K} \ . \tag{10}$$

We shall now determine whether the available observational and theoretical constraints on x and m_{min} can accomodate the limits on M/L in eqs. (9) and (10). The strongest constraints on x, derived from the observation of spectral features (26) and the initial mass function in the solar neighborhood (33) require x $\leqslant$ 1 at the 2σ level. Also there is no data in conflict with x $\leqslant$ 1. Photometric data ranging from globular clusters to elliptical galaxies can be fit using the weaker constraints x $\leqslant$ 1.35, by a single free parameter, the metal abundance (34,35).

Constraints on m_{min}, the smallest mass to collapse gravitationally (36,37,38), have a lower limit of $\geqslant$ 0.007 $M_\odot$. A more recent calculation (39) in which new reactions to form

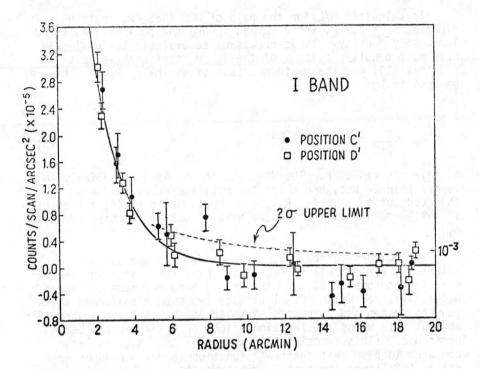

Figure 1. The measured surface brightness of the halo of
NGC 4565 versus galactic radius. Positions C' and D' are two
symmetric scanning positions. The curve fitted to the data is
the de Vaucouleur's surface brightness law and the 2σ upper limit
to the data is labelled. [1 count/scan/arc sec$^2 \times 10^{-5}$ is
25.34 mag I_{Kron}.]

molecular hydrogen are considered, requires $m_{min} > 0.004\ M_\odot$.
That result was found for optically thin clouds. An
equally forceful position has been presented in which it is
argued that the first objects to form have $m_{min} \geqslant 1500\ M_\odot$.

If we choose $m_{min} = .004\ M_\odot$ and find x to satisfy the I and
K band NGC 4565 observations, we find x > 1.6 and 1.7
respectively. On the other hand, if we choose x ⩽ 1 and try to
find the allowed range for m_{min}, we find no solution. It is not
possible to put enough mass in the halo for this x without
violating the surface brightness observations. For x = 1.35, we
find $m_{min} < 2 \times 10^{-4}$ at least a factor of twenty below the
calculated lower limit on m_{min}. These are the problems if one
chooses to consider a single power law initial mass function and
a halo of stars and/or Jupiters.

There are some observations which have a bearing on our
assumption of whether the initial mass function is a single power
law below the nuclear burning cutoff. Probst and O'Connell (41)
argue that the initial mass function in the solar neighborhood
does not even rise as steeply as a single power law for stellar
masses less than $0.1 M_0$. Instead the slope turns over, meaning
that there is little mass contained in stars with $m < 0.1 M_0$.
Since these results are based on stars with solar metal
abundance, the conclusions are strengthened for stars which have
lower metal abundances and which cannot cool as effectively.

Though we have argued that it seems reasonable to use a
power law for the slope of the initial mass near $0.08 M_0$ and that
any possible gravitational condensation of smaller mass would
adhere to the same power law, let us now consider the possibility
that only Jupiters formed. As a prototypical model, we shall
consider the model presented at this conference by Professor
Rees. In that model, a Jean's mass at recombination, $10^5-10^6 M_0$,
cools and forms a very thin disk of thickness equal to the Jean's
length of a $10^{-3} M_0$ condensation, that is, a Jupiter.
Subsequently the disk fragments contributing 10^8-10^9 Jupiters to
the formation of a halo of Jupiters.

There appear to be two large-scale instabilities which the
disk must avoid if Jupiters are to form: the tendency of the
disk to form a bar, and the instability of a cool disk to form
massive condensations which are a significant fraction of the
total disk mass (42). We shall disucss the second instability
using the Toomre stability criterion.

The basic kinematic criterion for stability is that the time
for a blob of material to orbit the disk, t_{orb}, should be longer
than the time for a pressure wave or sound wave to cross the
disk, t_s. Writing $t_{orb} \sim r/v$ and $t_s \sim r/c_s$, we have

$$t_{orb} > t_s \tag{11}$$

leading to

$$r/v > r/c_s \tag{12}$$

or

$$v < c_s . \tag{13}$$

This is the condition that, for stability, the orbital velocities
be less than the individual particle velocities. Adding the
dynamics, namely, in equilibrium, the following condition for
circular motion must be satisfied,

$$v^2/r = GM/r^2 . \tag{14}$$

For a disk with mass per unit area, σ, $M \sim \pi\sigma r^2$, then substituting for M in eq. (14) and multiplying by r, we have

$$v^2 = \pi G\sigma r . \tag{15}$$

Substituting this result into eq. (13) leads to

$$\pi G\sigma r < c_s^2 \tag{16}$$

The speed of sound is $c_s^2 \sim KT/m_p$. Also, from the Jeans mass condition we have

$$GM_J/r_J \sim KT/m_p . \tag{17}$$

where M_J is the Jeans mass and r_J is the Jeans length. Substituting eq. (17) into eq. (16), it may be seen that

$$\pi G\sigma r < c_s^2 \sim GM_J/r_J \tag{18}$$

or

$$\pi\sigma r r_J < M_J . \tag{19}$$

If we write the thickness of the disk, t, in terms of the radius of the disk, r, then $t = \epsilon r$. With $t \simeq r_J$ and $M \sim \pi\sigma r^2$, we have

$$\epsilon M < M_J . \tag{20}$$

From the numbers required by the model, that is, dividing a $10^5 \ M_\odot$ object into $10^{-3} \ M_\odot$ objects or 10^8 Jupiters, it may be seen that the ratio t/r required for a disk of thickness equal to the Jeans length of a Jupiter is $\sim \sqrt{10^{-8}} = 10^{-4}$. Using this value for ϵ on the left hand side of eq. (20) yields $\sim 10 \ M_\odot$, while the desired Jeans mass is $10^{-3} \ M_\odot$. The inequality is not satisfied by a factor of 10^4. That is, such thin disks are unstable and form $\sim 10 \ M_\odot$ objects, not Jupiters. An alternative interpretation is that a disk which is hot enough for stability is too hot to allow low mass gravitational condensations to develop.

The halo cannot be made of stars which have an initial mass greater than 2 $M_\odot$. Such stars either evolve to white dwarfs with mass $\approx 1.4 \ M_\odot$ (43) or to neutron stars which also, coincidently, have masses $\approx 1.4 \ M_\odot$. Taylor and Weisberg (44) have found two neutron stars with masses of 1.4 $M_\odot$ to within 1% and all other neutron star mass determinations are consistent with 1.4 $M_\odot$. Consequently, any star with initial mass greater than 2 $M_\odot$ must lose 40 per cent of its mass during evolution. The ejected mass cannot be hot because of previous arguments and it cannot cool and fall in the disk because there is too much mass to be contained. Also, since a significant fraction of the mass of the evolved stars, > 10%, might be expected to be converted into helium and metals during evolution, problems similar to those raised by metallic halos could be present.

Though black holes do not have a well defined baryon number, we shall briefly consider them because if halos are not baryonic, they are evidently either composed of black holes or some weakly or very weakly interacting particles (see review by Joel Primack in this volume).

It appears unlikely that many black holes in the mass range 1-50 $M_\odot$ formed in the halo. Stars in this mass range eject a considerable fraction of their mass. Unless the black holes can accrete virtually all their ejecta, problems similar to those with metallic halos arise. Black holes which are more massive than 100 $M_\odot$ appear to be excluded by new observations (45), though they need to be confirmed. Thus, halos could be composed of black holes in the mass range ~50-100 $M_\odot$ (46) or they could be primordial.

Arguing by eliminating specific baryonic forms of matter is not the most persuasive way to argue that halos are not baryonic, but, unfortunately, we are unable to present a forceful positive argument eliminating baryons directly. In this context, it is worth considering the constraints that primordial nucleosynthesis places on baryonic halos, though we admit that there are strong assumptions implicit in the nucleosynthesis calculations.

In this context, we shall take the simplest point of view, namely, that all the dark matter in halos and rich clusters is either all baryonic, or not baryonic and see which conclusion, if any, the nuclear abundances favor.

A lower limit to the mass fraction of the closure density in baryons, Ω_b, may be obtained from the luminous matter in galaxies and could be as low as .001. The thermal X-ray fluxes from clusters of galaxies yield higher baryon abundances but do not exclude $\Omega_b \sim .001$. On the other hand if all the dark matter were baryonic, the mass content of halos and rich clusters would require a lower limit for the baryonic abundances to be, $\Omega_b \gtrsim 0.1$.

The deuterium abundance of ~1×10^{-5} by mass does not favor either high or low baryon abundances. There are problems with both ranges. However, the deuterium abundance may not be well known (see Audouze this volume). The He[4] abundance is presently observed to be in the range Y ~ .22-.25 (47). Since an observed helium abundance is an upper limit on the primordial abundance, and since $\Omega_b \gtrsim .1$ requires Y > .26, the helium observations favor a low baryon abundance. The observed Li^7 abundance (48) is consistent with two abundance ranges, Ω_b ~ .001-.003 and Ω_b ~ .01-.02. It appears inconsistent with $\Omega_b \gtrsim .1$. Taken together, the abundance data favors a low baryon abundance (49). A key test of the cosmological baryon abundance will be a new measurement of the primordial helium abundance which is independent of the possible systematic effects in the present spectroscopic measurements.

I would like to thank G. William Ford, Martin Rees, Alar Toomre and Scott Tremaine for useful discussions.

REFERENCES

1. Hohl, F. 1977, NASA TR R-343.

2. Hohl, F. 1975, in IAU Symposium No. 69, "Dynamics of Stellar Systems, ed. A. Hayli (Dordrecht, Neth: Reidel), pp. 349.

3. Kalnajs, A.J. 1972, Ap. J. 175, pp. 63.

4. Ostriker, J.P and Peebles, P.J.E. 1973, Ap.J. 186, pp. 467.

5. Rogstad, D.H. and Shostak, G.S. 1972, Ap. J. 176, pp. 315.

6. Roberts, M.S. and Rots, A.H. 1973, Astr. Ap. 26, pp. 483.

7. Haschick, A.D. and Burke, B.F. 1975, Ap. J. (Letters) 200, pp. L137.

8. Roberts, M.S. 1975 in IAU Symposium No. 69, "Dynamics of Stellar Systems, ed. A. Hayli (Dordrecht, Neth: Reidel), pp. 331.

9. Sancisi, R. 1977, IAU Symposium No. 77, "Dynamics of Stellar Systems, ed. A. Hayli (Dordrecht, Neth: Reidel).

10. Krumm, N. and Salpeter, E.E. 1979, A.J. 84, pp. 1138.

11. Rubin, V.C., Ford, Jr., W.K., and Thonnard, N. 1978, Ap. J. (Letters) 225, pp. L107.

12. Turner, E.L. 1976, Ap. J. 208, pp. 304.

13. Peterson, S.D. 1979, Ap. J. 232, pp. 20.

14. Rivolo, A.R. and Yahil, A. 1981, Ap. J. 251, pp. 477.

15. Saar, E.M. 1978 in IAU Symposium 84, "The Large-Scale Characteristics of the Galaxy", ed. W.B. Burton, p. 513.

16. Tubbs, A.D. and Sanders, R.H. 1979, Ap. J. 230, pp. 736.

17. Monet, D.G., Richstone, D.O. and Schechter, P.L. 1981, Ap. J. 245, pp. 454.

18. Van der Kruit, P.C. 1981, Ast. Ap. 99, pp. 298.

19. Rohlfs, K. 1982, Astr. Ap. 105, pp. 296.

20. Hegyi, D.J. and Olive, K.A., to be published 1983, Physics Letters.

21. Faber, S.M. and Gallagher, J.S. 1979, Ann. Ref. Astr. Ap. 17, pp. 135.

22. Johnson, V.J. 1960, "A Compendium of the Properties of Materials at Low Temperature (Phase I)", U.S. Air Force.

23. Hollenbach, D. and Salpeter, E.E. 1971, Ap. J. 163, pp. 155.

24. Dekel, A. and Shaham, J. 1979, Astro. Ap. 74, pp. 186.

25. Straka, W.C. 1971, Ap. J. 165, pp. 109.

26. Tinsley, B.M. and Gunn, J.E. 1976, Ap. J. 203, pp. 52.

27. Tinsley, B.M. 1976, Ap. J. 203, pp. 63.

28. Sweigart, A.V. and Gross, P.G. 1978, Ap. J. Suppl. 36, pp. 405.

29. Krumm, N. and Salpeter, E.E. 1979, A.J. 84, pp. 1138.

30. Hegyi, D.J. and Gerber, G.L. 1977, Ap. J. (Letters) 218, L7.

31. Hegyi, D.J. in Proceedings of the Moriond Astrophysics Meeting (ed. J. Andouze, P. Crane, T. Gaisser, D. Hegyi, and J. Tran Thranh Van) Frontiers, 1981, pp. 321.

32. Boughn, S.P., Saulson, P.R. and Seldner, M. 1981, Ap. J. (Letters) 250, pp. L15.

33. Miller, G.E. and Scalo, J.M. 1979, Ap. J. Suppl. 41, pp. 513.

34. Aaronson, M., Cohen, J.G., Mould, J. and Malkan, M. 1978, Ap. J. 223, pp. 824.

35. Frogel, J.A., Persson, S.E. and Cohen, J.G. 1980, Ap. J. 240, pp. 785.

36. Low, C. and Lynden-Bell, D. 1976, M.N.R.A.S. 176, pp. 367.

37. Rees, M.J. 1976, M.N.R.A.S. 176, pp. 483.

38. Silk, J. 1982, Ap. J. 256, pp. 514.

39. Palla, F., Salpeter, E.E., and Stahler, S.W. (preprint).

40. Tohline, J.F. 1980, Ap. J. 239, pp. 417.

41. Probst, R.G. and O'Connell, R.W. 1982, Ap. J. (Letters) 252, L69.

42. Toomre, A. 1964, Ap. J. 139, pp. 1217.

43. Chandrasekhar, S. 1935, M.N.R.A.S. 95, pp. 207.

44. Taylor, J.H. and Weisberg, J.M. 1982, Ap. J. 253, pp. 908.

45. Lin, D.N.C. and Faber, S.M. 1983 (preprint).

46. Carr, B.J., Bond, J.R. and Arnett, W.D. 1983 (preprint).

47. Pagel, B. 1982, Phil. Trans. of R.S. London 307, pp. 19.

48. Spitz, M. and Spitz, F. 1982, Nature 297, pp. 483.

49. Olive, K.A., Schramm, D.N., Steigman, G., Turner, M.S., Yang, J. 1981, Ap. J. 246, pp. 557.

WHAT IS THE DARK MATTER?
IMPLICATIONS FOR GALAXY FORMATION AND PARTICLE PHYSICS

Joel R. Primack[*] and George R. Blumenthal[†]

[*]Board of Studies in Physics and Santa Cruz Institute
for Particle Physics, [†]Lick Observatory, Board of Studies
in Astronomy and Astrophysics, University of California,
Santa Cruz, CA 95064, U.S.A.

ABSTRACT We discuss three arguments that the dark matter which dominates the present universe is not baryonic - based on excluding
specific baryonic models, deuterium abundance, and the absence of
small-angle fluctuations in the microwave background radiation.
If the dark matter consists of elementary particles, it may be
classified as hot (free streaming erases all but supercluster-
scale fluctuations), warm (free streaming erases fluctuations
smaller than galaxies), or cold (free streaming is unimportant).
We consider scenarios for galaxy formation in all three cases. We
discuss several potential problems with the hot (neutrino) case:
making galaxies early enough, with enough baryons, and without too
much increase in $M_{tot}/M_{\ell um}$ from galaxy to rich cluster scales.
The reported existence of dwarf spheroidal galaxies with relatively
heavy halos is a serious problem for both hot and warm scenarios.
Zeldovich ($n = 1$) adiabatic initial fluctuations in cold dark matter
(axions, or a heavy stable "ino") appear to be lead to observed
sizes and other properties of galaxies, and may also yield large
scale structure such as voids and filaments.

I. INTRODUCTION

There is abundant observational evidence that dark matter (DM)
is responsible for most of the mass in the universe (1). Dark
matter is detected through its gravitational attraction in the
massive extended halos of disk galaxies and in groups and clusters
of galaxies of all sizes. It is appropriate to call this matter
"dark" because it is detected in no other way; it is not observed
to emit or absorb electromagnetic radiation of any wavelength.
Matter observed in these latter ways we will call "luminous". Here

J. Audouze and J. Tran Thanh Van (eds.),
Formation and Evolution of Galaxies and Large Structures in the Universe, 163–183.
© 1984 by D. Reidel Publishing Company.

we consider the nature of the dark matter.

II. THE DM IS PROBABLY NOT BARYONIC

There are three arguments that the DM is not "baryonic", that is, that it is not made of protons, neutrons, and electrons as all ordinary matter is. As Richard Feynman has said in other contexts, one argument would suffice if it were convincing. All three arguments have loopholes. The arguments that DM $\neq$ baryons are as follows:

A. Excluding Baryonic Models (2)

The dark matter in galaxy halos cannot be gas (it would have to be hot to be pressure supported, and would radiate); nor frozen hydrogen "snowballs" (they would sublimate); nor dust grains (their "metals", elements of atomic number ≥ 3, would have prevented formation of the observed low-metallicity Population II stars); nor "jupiters" (how to make so many hydrogen balls too small to initiate nuclear burning without making a few large enough to do so?); nor collapsed stars (where is the matter they must have ejected in collapsing?).

The weakest argument is probably that which attempts to exclude "jupiters": arguments of the form "how could it be that way?" are rarely entirely convincing.

B. Deuterium Abundance (3)

In the early universe, almost all the neutrons which "freeze out" are synthesized into ^{4}He. The fraction remaining in D and ^{3}He is a rapidly decreasing function of η, the ratio of baryon to photon number densities. The presently observed D abundance (compared, by number, to H) is $(1-4) \times 10^{-5}$. Since D is readily consumed but not produced in stars, 10^{-5} is also a lower limit on the primordial D abundance. This, in turn, implies an upper limit $\eta \leq 10^{-9}$, or

$$\Omega_b \leq 0.035 \, h^{-2} \, (T_o/2.7)^3, \tag{1}$$

where Ω_b is the ratio of the present average baryon density ρ_b to the critical density

$$\rho_c = 3H_o^2/8\pi G = 1.9 \times 10^{-29} \, h^2 \, g \, cm^{-3} = 11 \, h^2 \, keV \, cm^{-3}, \tag{2}$$

H_o is the Hubble parameter (the subscript o denotes the present epoch), and observationally $h \equiv H_o \, (100 \, kms^{-1}Mpc^{-1})^{-1}$ lies in the range $\frac{1}{2} \leq h \leq 1$. The total cosmological density $\Omega \equiv \rho_{tot}/\rho_c$ is very difficult to determine observationally, but it appears to lie in the range $0.1 \leq \Omega \lesssim 2$. Cosmological models in which the universe passes through an early de Sitter "inflationary" stage, predict Ω very close to unity.

In a baryon dominated universe ($\Omega \simeq \Omega_b$), the deuterium bound, Eq. (1), is consistent only with the lower limit on Ω, and then only for the Hubble parameter at its lower limit. An Einstein-de Sitter or inflationary ($\Omega = 1$) or closed ($\Omega > 1$) universe cannot be baryonic.

C. Galaxy Formation

In the standard cosmological model, which we will adopt, large scale structure forms when perturbations $\delta \equiv \delta\rho/\rho$ grow to $\delta \gtrsim 1$, after which they cease to expand with the Hubble flow. Let us further assume that perturbations in matter and radiation density are correlated (these are called adiabatic perturbations, since the entropy per baryon is constant; these are the sort of perturbations predicted in grand unified models). Then photon diffusion ("Silk damping") erases perturbations of baryonic mass smaller than (4)

$$M_{Silk,b} \approx 3 \times 10^{13} \, \Omega_b^{-\frac{1}{2}} \, \Omega^{-\frac{3}{4}} \, h^{-\frac{5}{2}} \, M_\Theta \, . \tag{3}$$

Thus galaxies ($M_b \lesssim 10^{11-12} M_\Theta$) can form only after the "pancake" collapse of larger-scale perturbations (5). Perturbations δ in a matter dominated universe grow linearly with the scale factor

$$\delta \propto a = (1+z)^{-1} = T_0/T \tag{4}$$

where $z = (\lambda_0 - \lambda)/\lambda$ is the redshift and T is the radiation temperature. In a baryonic universe, δ grows only between the epoch of hydrogen recombination, $z_r \simeq 1300$, and $z \simeq \Omega^{-1}$. It follows that at recombination $\delta T/T \approx \delta\rho/3\rho \geq 3 \times 10^{-3}$ for $M \gtrsim M_{Silk}$, which corresponds to fluctuations on observable angular scales $\theta > 4'$ today. Such temperature fluctuations are an order of magnitude larger than present observational upper limits (6).

The main loophole in this argument is the assumption of adiabatic perturbations. It is true that the orthogonal mode, perturbations in baryonic density which are uncorrelated with radiation (called isothermal perturbations), do not arise naturally in currently fashionable particle physics theories where baryon number is generated in the decay of massive grand unified theory (GUT) bosons, since in such theories $\eta = n_b/n_\gamma$ is determined by the underlying particle physics and should not vary from point to point in space. But galaxies originating as isothermal perturbations do avoid both Silk damping and contradiction with present $\delta T/T$ limits.

A second loophole is the possibility that matter was reionized at some $z \gtrsim$, by hypothetical very early uv sources. Then the fluctutions in $\delta T/T$ at recombination associated with baryonic proto-pancakes could be washed out by rescattering.

Despite the loopholes in each argument, we find the three arguments together to be rather persuasive, even if not entirely compelling. If it is indeed true that the bulk of the mass in the universe is not baryonic, that is yet another blow to anthropocentricity: not only is man not the center of the universe physically (Copernicus) or biologically (Darwin), we and all that we see are not even made of the predominant variety of matter in the universe!

III. THREE TYPES OF DM PARTICLES: HOT, WARM & COLD

If the dark matter is not baryonic, what is it? We will consider here the physical and astrophysical implications of three classes of elementary particle DM candidates, which we will call hot, warm, and cold. (We are grateful to Dick Bond for proposing this apt terminology.)

Hot DM refers to particles, such as neutrinos, which were still in thermal equilibrium after the most recent phase transition in the hot early universe, the QCD deconfinement transition, which presumably took place at $T_{QCD} \sim 10^2$ MeV. Hot DM particles have a cosmological number density roughly comparable to that of the microwave background photons, which implies an upper bound to their mass of a few tens of eV. As we shall discuss shortly, free streaming destroys any perturbations smaller than supercluster size, $\sim 10^{15} M_\theta$.

Warm DM particles interact much more weakly than neutrinos. They decouple (i.e., their mean free path first exceeds the horizon size) at $T > T_{QCD}$, and consequently their number density is roughly an order of magnitude lower, and their mass an order of magnitude higher, than hot DM particles. Perturbations as small as large galaxy halos, $\sim 10^{12} M_\theta$, could then survive free streaming. It was initially suggested that, in theories of local supersymmetry broken at $\sim 10^6$ GeV, gravitinos could be DM of the warm variety (7). Other candidates are also possible, as we will discuss.

Cold DM consists of particles for which free streaming is of no cosmological importance. Two different sorts have been proposed, a cold Bose condensate such as axions, and heavy remnants of annihilation or decay such as heavy stable neutrinos. As we will see, a universe dominated by cold DM looks remarkably like the one astronomers actually observe.

It is of course also possible that the dark matter is NOTA –

none of the above! A perennial candidate, primordial black holes, is becoming increasingly implausible (8-10). Another possibility which, for simplicity, we will not discuss, is that the dark matter is a mixture, for example "jupiters" in galaxy halos plus neutrinos on large scales (3).

IV. GALAXY FORMATION WITH HOT DM

The standard hot DM candidate is massive neutrinos (3-5), although other, more exotic, theoretical possibilities have been suggested, such as a "majoron" of nonzero mass which is lighter than the lightest neutrino species, and into which all neutrinos decay (11). For definiteness, we will discuss neutrinos.

A. Mass Constraints

Left-handed neutrinos of mass $\lesssim 1\,\mathrm{MeV}$ will remain in thermal equilibrium until the temperature drops to $T_{\nu d}$, at which point their mean free path first exceeds the horizon size and they essentially cease interacting thereafter, except gravitationally (12). Their mean free path is, in natural units ($h = c = 1$), $\lambda_\nu \sim [\sigma_\nu n_{e^\pm}]^{-1}$ $\sim [(G_{wk}^2 T^2)(T^3)]^{-1}$, and the horizon size is $\lambda_h \sim (G\rho)^{-\frac{1}{2}} \sim M_{P\ell} T^{-2}$, where the Planck mass $M_{P\ell} \equiv G^{-\frac{1}{2}} = 1.22 \times 10^{19}\,\mathrm{GeV} = 2.18 \times 10^{-5}\,\mathrm{g}$. Thus $\lambda_h/\lambda_\nu \sim (T/T_{\nu d})^3$, with the neutrino decoupling temperature

$$T_{\nu d} \sim M_{P\ell}^{-\frac{1}{3}} G_{wk}^{-\frac{2}{3}} \sim 1\,\mathrm{MeV}. \tag{5}$$

After T drops below $1\,\mathrm{MeV}$, e^+e^- annihilation ceases to be balanced by pair creation, and the entropy of the e^+e^- pairs heats the photons. Above $1\,\mathrm{MeV}$, the number density n_{ν_i} of each left-handed neutrino species (counting both ν_i and $\bar\nu_i$) is equal to that of the photons, n_γ, times the factor 3/4 from Fermi vs. Bose statistics; but e^+e^- annihilation increases the photon number density relative to that of the neutrinos by a factor of 11/4.[1] Thus today, for each species,

$$n_\nu^o = \frac{3}{4} \cdot \frac{4}{11} n_\gamma^o = 109 \left(\frac{T_\gamma}{2.7K}\right)^3 \mathrm{cm}^{-3}. \tag{6}$$

Since the cosmological density

$$\rho = \Omega\rho_c = 11\,\Omega h^2\,\mathrm{keV\,cm}^{-3}, \tag{7}$$

it follows that

$$\sum_i m_{\nu_i} < \rho/n_\nu^o \leq 100\,\Omega h^2\,\mathrm{eV}, \tag{8}$$

where the sum runs over all neutrino species with $m_{\nu_i} \lesssim 1\,\text{MeV}$.[2]
Observational data imply that Ωh^2 is less than unity (3). Thus
if one species of neutrino is substantially more massive than
the others and dominates the cosmological mass density, as for
definiteness we will assume for the rest of this section, then a
reasonable estimate for its mass is $m_\nu \sim 30\,\text{eV}$.

At present there is apparently no reliable experimental evidence for nonzero neutrino mass. Although one group reported (15)
that $14\,\text{eV} < m_{\nu_e} < 40\,\text{eV}$ from tritium β end point data, according to
Boehm (16) their data are consistent with $m_{\nu_e} = 0$ with the resolu-
tion corrections pointed out by Simpson. The so far unsuccessful
attempts to detect neutrino oscillations also give only upper limits
on neutrino masses times mixing parameters (16).

B. Free Streaming

The most salient feature of hot DM is the erasure of small
fluctuations by free streaming. It is easy to see that the minimum mass of a surviving fluctuation is of order $M_{P\ell}^3/m_\nu^2$ (17,4).

Let us suppose that some process in the very early universe –
for example, thermal fluctuations subsequently vastly inflated,
in the inflationary scenario (18) – gave rise to adiabatic fluctuations on all scales. Neutrinos of nonzero mass m_ν stream
relativistically from decoupling until the temperature drops to
m_ν, during which time they will traverse a distance $d_\nu \approx \lambda_h(T = m_\nu)$
$\sim M_{P\ell}\, m_\nu^{-2}$. In order to survive this free streaming, a neutrino
fluctuation must be larger in linear dimension than d_ν. Correspondingly, the minimum mass in neutrinos of a surviving fluctuation
is $M_{J,\nu} \sim d_\nu^3\, m_\nu\, n_\nu(T = m_\nu) \sim d_\nu^3\, m_\nu^4 \sim M_{P\ell}^3\, m_\nu^{-2}$. By analogy with Jeans'
calculation of the minimum mass of an ordinary fluid perturbation
for which gravity can overcome pressure, this is referred to as
the (free-streaming) Jeans mass. A more careful calculation (4,19)
gives

$$d_\nu = 41\,(m_\nu/30\,\text{eV})^{-1}\,(1+z)^{-1}\,\text{Mpc}, \tag{9}$$

and

$$M_{J,\nu} = 1.77\,M_{P\ell}^3\, m_\nu^{-2} = 3.2 \times 10^{15}\,(m_\nu/30\,\text{eV})^{-2}\,M_\Theta, \tag{10}$$

which is the mass scale of superclusters. Objects of this size
are the first to form in a ν-dominated universe, and smaller scale
structures such as galaxies can form only after the initial collapse of supercluster-size fluctuations.

C. Growth of Fluctuations

The absence of small angle $\delta T/T$ fluctuations is compatible with this picture. When a fluctuation of total mass $\sim 10^{15} M_{\odot}$ enters the horizon at $z \sim 10^{4}$, the density contrast of the radiation plus baryons δ_{RB} ceases growing and instead starts oscillating as an acoustic wave, while that of the neutrinos δ_{ν} continues to grow linearly with the scale factor $a = (1+z)^{-1}$. Thus by recombination, at $z_r \approx 1300$, $\delta_{RB}/\delta_{\nu} < 10^{-1}$, with possible additional suppression of δ_{RB} by Silk damping (depending on the parameters in Eq. (3)). This picture, as well as the warm and cold DM schemes, predicts small angle fluctuations in the microwave background radiation just slightly below current observational upper limits (6).

In numerical simulations of dissipationless gravitational clustering starting with a fluctuation spectrum appropriately peaked at $\lambda \approx d_{\nu}$, the regions of high density form a network of filaments, with the highest densities occurring at the intersections and with voids in between (5,20-22). The similarity of these features to those seen in observations (23,24) is certainly evidence in favor of this model.

D. Potential Problems with ν DM

A number of potential problems with the neutrino dominated universe have emerged in recent studies, however. (1) From studies both of nonlinear (22) clustering ($\lambda \lesssim 10$ Mpc) and of streaming velocities (25) in the linear regime ($\lambda > 10$ Mpc), it follows that supercluster collapse must have occurred recently: $z_{sc} \lesssim 0.5$ is indicated (25), and in any case $z_{sc} < 2$ (22). But then, if QSOs are associated with galaxies, their abundance at $z > 2$ is inconsistent with the "top-down" neutrino dominated scheme in which superclusters form first: $z_{sc} > z_{galaxies}$. (2) Numerical simulations of the nonlinear "pancake" collapse taking into account dissipation of the baryonic matter show that at least 85% of the baryons are so heated by the associated shock that they remain ionized and unable to condense, attract neutrino halos, and eventually form galaxies (25a). (3) The neutrino picture predicts (26) that there should be a factor of ~ 5 increase in M_{tot}/M_{lum} between large galaxies ($M_{tot} \sim 10^{12} M_{\odot}$) and large clusters ($M_{tot} \gtrsim 10^{14} M_{\odot}$), since the larger clusters, with their higher escape velocities, are able to trap a considerably larger fraction of the neutrinos. Although there is indeed evidence for a trend of increasing M_{tot}/L with M_{tot} (1,27), when one takes into account the large amount of x-ray emitting gas in rich clusters (28) one finds comparable $M_{tot}/M_{lum} \sim 14$ for galaxies with large halos and for rich clusters (29,30). (M_{lum} here includes matter luminous in x-ray as well as optical wavelengths, in contrast to luminosity L that includes only the latter.) (4) Both theoretical arguments (31) and data on Draco (32,33) imply that dark matter dominates the gravitational potential of dwarf spheroidal galaxies. The phase-space constraint (34) then sets a lower limit (33) $m_\nu > 500$ eV, which is completely incompatible with the cosmological constraint Eq. (8). (Note that for neutrinos as the DM in spiral galaxies, the phase space constraint implies $m_\nu > 30$ eV.)

These problems, while serious, may not be fatal for the hypothesis that neutrinos are the dark matter. It is possible that galaxy density does not closely correlate with the density of dark matter, for example because the first generation of luminous objects heats nearby matter, thereby increasing the baryon Jeans mass and suppressing galaxy formation. This could complicate the comparison of nonlinear simulations (22) with the data. Also, if dark matter halos of large clusters are much larger in extent than those of individual galaxies and small groups, then virial estimates would underestimate M_{tot} on large scales and the data could be consistent with M_{tot}/M_{lum} increasing with M_{lum}. But it is hard to avoid the constraint on z_{sc} from streaming velocities in the linear regime (25) except by assuming that the local group velocity is abnormally low. And the only explanation for the high M_{tot}/L of dwarf spheroidal galaxies in a neutrino-dominated universe is the rather ad hoc assumption that the dark matter in such objects is baryons rather than neutrinos. Of course, the evidence for massive halos around dwarf spheroidals is not yet solid.

V. GALAXY FORMATION WITH WARM DM

Suppose the dark matter consists of an elementary particle species X that interacts much more weakly than neutrinos. The Xs decouple thermally at a temperature $T_{Xd} \gg T_{\nu d}$ and their number density is not thereafter increased by particle annihilation at temperatures below T_{Xd}. With the standard assumption of conservation of entropy per comoving volume, the X number density today n_X^0 and mass m_X can be calculated in terms of the effective number of helicity states of interacting bosons (B) and fermions (F), $g = g_B + (7/8)g_F$, evaluated at T_{Xd} (35). These are plotted in Fig. 1, assuming the "standard model" of particle physics. The simplest grand unified theories predict $g(T) \approx 100$ for T between 10^2 GeV and $T_{GUT} \sim 10^{14}$ GeV, with possibly a factor of two increase in g beginning near 10^2 GeV due to N = 1 supersymmetry partner particles. Then for T_{Xd} in the enormous range from ~1 GeV to ~T_{GUT}, $n_X^0 \sim 5 g_X cm^{-3}$ and correspondingly $m_X \simeq 2\Omega h^2 g_X^{-1}$ keV (36), where g_X is the number of X helicity states. Such "warm" DM particles of mass $m_X \sim 1$ keV will cluster on a scale $\sim M_{p\ell}^3 m_X^{-2} \sim 10^{12} M_\odot$, the scale of large galaxies such as our own (7,37,38).

What might be the identity of the warm DM particles X? It was initially (7) suggested that they might be the $\pm\frac{1}{2}$ helicity states of the gravitino $\tilde{G}$, the spin 3/2 supersymmetric partner of the graviton G. The gravitino mass is related to the scale of

supersymmetry breaking by $m_{\tilde{G}} = (4\pi/3)^{\frac{1}{2}} m_{SUSY}^2 m_{P\ell}^{-1}$, so $m_{\tilde{G}} \sim 1$ keV corresponds to $m_{SUSY} \sim 10^6$ GeV. This now appears to be phenomen-
ologically dubious, and supersymmetry models with $m_{SUSY} \sim 10^{11}$ GeV
and $m_{\tilde{G}} \sim 10^2$ GeV are currently popular (39). In such models, the
photino $\tilde{\gamma}$, the spin $\frac{1}{2}$ supersymmetric partner of the photon, is
probably the lightest R-odd particle, and hence stable. But in
supersymmetric GUT models $m_{\tilde{\gamma}} \sim 10 m_{\tilde{g}}$, and there is a phenomenological lower bound on the mass of the gluino $m_{\tilde{g}} > 2$ GeV (40). The
requirement that the photinos almost all annihilate, so that they
do not contribute too much mass density, implies that $m_{\tilde{\gamma}} \gtrsim 2$ GeV
(14,41), and they become a candidate for cold rather than warm
dark matter.

A hypothetical right-handed neutrino ν_R could be the warm
DM particle (42), since if right-handed weak interactions exist
they must be much weaker than the ordinary left-handed weak interactions, so $T_{\nu_R d} \gg T_{\nu d}$ as required. But particle physics provides
no good reason why any ν_R should be light.

Thus there is at present no obvious warm DM candidate elementary particle, in contrast to the hot and cold DM cases. But
our ignorance about the physics above the ordinary weak interaction
scale hardly allows us to preclude the existence of very weakly
interacting light particles, so we will consider the warm DM case,
mindful of Hamlet's prophetic admonition

There are more things in heaven and earth, Horatio,
Than are dreamt of in your philosophy.

A. Fluctuation Spectrum

The spectrum of fluctuations δ_ν at late times in the hot DM
model is controlled mainly by free streaming; $\delta_\nu(M)$ is peaked
at $\sim M_{J,\nu}$, Eq. (10), for any reasonable primordial fluctuation
spectrum. This is not the case for warm or cold DM.

The primordial fluctuation spectrum can be characterized by
the magnitude of fluctuations as they just enter the horizon. It
is expected that no mass scale is singled out, so the spectrum is
just a power law

$$\delta_{DM,H} = \left(\frac{\delta\rho_{DM}}{\rho_{DM}}\right)_H = \kappa \left(\frac{M_{DM}}{M_0}\right)^{-\alpha} \tag{11}$$

Furthermore, to avoid too much power on large or small mass scales
requires $\alpha \approx 0$ (43), and to form galaxies and large scale structure
by the present epoch without violating the upper limits on both
small (6) and large (44) scale (quadrupole) angular variations in
the microwave background radiation requires $\kappa \sim 10^{-4}$. Eq. (11)

corresponds to $|\delta_k|^2 \propto k^n$ with $n = 6\alpha + 1$. The case $\alpha = 0$ ($n = 1$) is commonly referred to as the Zeldovich spectrum.

Inflationary models predict adiabatic fluctuations with the Zeldovich spectrum (18). In the simplest models κ is several orders of magnitude too large, but it is hoped that this will be remedied in more realistic - possible supersymmetric - models (45).

The important difference between the fluctuation spectra δ_{DM} at late times in the hot and warm DM cases is that $\delta_{DM,warm}$ has power over an increased range of masses, roughly from 10^{11} to $10^{15} M_\Theta$. As for the hot case, the lower limit, $M_X \sim M_{P\ell}^3 m_X^{-2}$, arises from the damping of smaller-scale fluctuations by free streaming. In the hot case, the DM particles become nonrelativistic at essentially the same time as they become gravitationally dominant, because their number density is nearly the same as that of the photons. But in the warm case, the X particles become nonrelativistic and thus essentially stop free streaming at $T \sim m_X$, well before they begin to dominate gravitationally at $T_{eq} \approx 6\Omega h^2$ eV. The subscript "eq" refers to the epoch when the energy density of massless particles equals that of massive ones:

$$z_{eq} = \frac{\Omega \rho_c c}{4\sigma T_0^4 (1+\gamma)} = 2.47 \times 10^4 \, \Omega h^2 \, (\frac{1.681}{1+\gamma}) \, \theta^{-4}. \tag{12}$$

We assume here that there are n_ν species of very light or massless neutrinos, and $\gamma \equiv \rho_\nu^0/\rho_\gamma^0 = (7/8)(4/11)^{4/3} n_\nu$ ($= 0.681$ for $n_\nu = 3$), $\theta \equiv T_0/2.7K$, and σ is the Stefan-Boltzmann constant. During the interval between $T \sim m_X$ and $T \sim T_{eq}$, growth of δ_{DM} is inhibited by the "stagspansion"[3] phenomenon (also known as the Meszaros (46) effect), which we will discuss in detail in the section on cold DM. Thus the spectrum δ_{DM} is relatively flat between M_X and

$$M_{eq} = \frac{4\pi}{3} (\frac{ct_{eq}}{1+z_{eq}})^3 \rho_c \Omega = 2.2 \times 10^{15} (\Omega h^2)^{-2} M_\Theta. \tag{13}$$

Fluctuations with masses larger than M_{eq} enter the horizon at $z < z_{eq}$, and thereafter δ_{DM} grows linearly with $a = (1+z)^{-1}$ until nonlinear gravitational effects become important when $\delta_{DM} \sim 1$. Since for $\alpha = 0$ all fluctuations enter the horizon with the same magnitude, and those with larger M enter the horizon later in the matter-dominated era and subsequently have less time to grow, the fluctuation spectrum falls with M for $M > M_{eq}$: $\delta_{DM} \propto M^{-2/3}$. For a power-law primordial spectrum of arbitrary index,

$$\delta_{DM} \propto M^{-\alpha - 2/3} = M^{-(n+3)/6}, \quad M > M_{eq}. \tag{14}$$

This is true for hot, warm, or cold DM. In each case, after re-combination at $z_r \approx 1300$ the baryons "fall in" to the dominating DM fluctuations on all scales larger than the baryon Jeans mass, and by $z \approx 100$, $\delta_b \approx \delta_{DM}$ (47).

In the simplest approximation, neglecting all growth during the "stagspansion" era, the fluctuation spectrum for $M_X < M < M_{eq}$ is just $\delta_{DM} \propto M^{-\alpha} = M^{-(n-1)/6} = M^{-(n_{eff}+3)/6}$, where $n_{eff} = n - 4$; i.e., the spectrum is flattened by a factor of $M^{2/3}$ compared to the primordial spectrum. The small amount of growth that does occur during the "stagspansion" era slightly increases the fluctuation strength on smaller mass scales: $n_{eff} \approx n - 3$. Detailed calcula-tions of these spectra are now available (19,37).

B. Which Formed First, Galaxies or Superclusters?

For $\alpha \gtrsim 0$, $\delta_X(M)$ has a fairly broad peak at $M \sim M_X$. Conse-quently, objects of this mass - galaxies and small groups - are the first to form, and larger-scale structures - clusters and superclusters - form later as $\delta_X(M)$ grows toward unity on suc-cessively larger mass scales. For a particular primordial spectral index α, one can follow Peebles (48,49) and use the fact that the galaxy autocovariance function $\xi(R) \approx 1$ for $R = 5h^{-1}$, together with the (uncertain) assumption that the DM is distributed on such scales roughly like the galaxies, to estimate when galaxies form in this scenario. For $\alpha = 0$, $z_{galaxies} \sim 4$, which is consistent with the observed existence of quasars at such redshifts. But superclusters do not begin to collapse until $z < 2$, so one would not expect to find similar Lyman α absorption line redshifts for quasars separated by $\sim 1h^{-1}$ Mpc perpendicular to the line of sight (50). Indeed, Sargent et al. (51) found no such correlations. This is additional evidence against hot DM.

C. Potential Problems with Warm DM

The warm DM hypothesis is probably consistent with the ob-served features of typical large galaxies, whose formation would probably follow roughly the "core condensation in heavy halos" scenario (52,29,53). The potentially serious problems with warm DM are on scales both larger and smaller than M_X. On large scales, the question is whether the model can account for the observed net-work of filamentary superclusters enclosing large voids (23,24). A productive approach to this question may require sophisticated N-body simulations with $N \sim 10^6$ in order to model the large mass range that is relevant (54). We will discuss this further in the next section in connection with cold DM, for which the same ques-tion arises.

On small scales, the preliminary indications that dwarf

spheroidal galaxies have large DM halos (31-33) pose problems
nearly as serious for warm as for hot DM. Unlike hot DM, warm DM
is (barely) consistent with the phase space constraint (32-34).
But since free streaming of warm DM washes out fluctuations δ_X
for $M \lesssim M_X \sim 10^{11} M_\odot$, dwarf galaxies with $M \sim 10^7 M_\odot$ can form in this
picture only via fragmentation following the collapse of struc-
tures of mass $\sim M_X$, much as ordinary galaxies form from superslusters
fragmentation in the hot DM picture. The problem here is that
dwarf galaxies, with their small escape velocities $\sim 10 km\,s^{-1}$, would
not be expected to bind more than a small fraction of the X particles, whose typical velocity must be $\sim 10^2\,km\,s^{-1}$ ($\sim$ rotation
velocity of spirals). Thus we expect M_{tot}/M_{lum} for dwarf galaxies
to be much smaller than for large galaxies - but the indications
are that they are comparable (31-33). Understanding dwarf galaxies
may well be crucial for unravelling the mystery of the identity
of the DM. Fortunately, data on Carina, another dwarf spheroidal
companion of the Milky Way, is presently being analyzed (55).

VI. GALAXY FORMATION WITH COLD DM

Damping of fluctuations by free streaming occurs only on
scales too small to be cosmologically relevant for DM which either
is not characterized by a thermal spectrum, or is much more massive
than 1 keV. We refer to this as cold DM.

A. Cold DM Candidates

Quantum chromodynamics (QCD) with quarks of nonzero mass violates CP and T due to instantons. This leads to a neutron electric
dipole moment that is many orders of magnitude larger than the
experimental upper limit, unless an otherwise undetermined complex phase θ_{QCD} is arbitrarily chosen to be extremely small. Peccei
and Quinn (56) have proposed the simplest and probably the most appealing way to avoid this problem, by postulating an otherwise
unsuspected symmetry that is spontaneously broken when an associated
pseudoscalar field - the axion (57) - gets a nonzero vacuum expecta-
tion value $<\phi_a> \sim f_a e^{i\theta}$. This occurs when $T \sim f_a$. Later, when the
QCD interactions become strong at $T \sim \Lambda_{QCD} \sim 10^2\,MeV$, instanton effects generate a mass for the axion $m_a = m_\pi f_\pi / f_a \approx 10^{-5}\,eV(10^{12}\,GeV/f_a)$.
Thereafter, the axion contribution to the energy density is (58)
$\rho_a = 3 m_a T^3 f_a^2 (M_{P\ell} \Lambda_{QCD})^{-1}$. The requirement $\rho_a^o < \rho_c \Omega$ implies that
$f_a \lesssim 10^{12}\,GeV$, and $m_a \gtrsim 10^{-5}\,eV.^4$ The longevity of helium-burning
stars implies (59) that $m_a < 10^{-2}\,eV$, $f_a > 10^9\,GeV$. Thus if the hypothetical axion exists, it is probably important cosmologically,
and for $m_a \sim 10^{-5}\,eV$ gravitationally dominant. (The mass range
$10^{9-12}\,GeV$, in which f_a must lie, is also currently popular with
particle theorists as the scale of supersymmetry (39) or family
symmetry breaking, the later possibility connected with the axion
(60).)

Two quite different sorts of cold DM particles are also possible. One is a heavy stable "ino", such as a photino (41) of mass $m_{\tilde{\gamma}} > 2\,\text{GeV}$ as discussed above. By a delicate adjustment of the theoretical parameters controlling the $\tilde{\gamma}$ mass and interactions, the $\tilde{\gamma}$s can be made to almost all annihilate at high temperatures, leaving behind a small remnant that, because $m_{\tilde{\gamma}}$ is large, can contribute a critical density today (14).

The second possibility may seem even more contrived: a particle, such as a ν_R, that decouples while still relativistic but whose number density relative to the photons is subsequently diluted by entropy generated in a first-order phase transition such as the Weinberg-Salam symmetry breaking (36). (Recall that the m_X bound in Fig. 1 assumes no generation of entropy.) More than a factor $\lesssim 10^3$ entropy increase would over dilute $\eta = n_b/n_\gamma$, if we assume η was initially generated by GUT baryosynthesis; correspondingly, $m_X \lesssim 1\,\text{MeV}$, and $M_X \gtrsim 10^6\,M_\Theta$.

Actually, it is not clear that we have a good basis to judge the plausibility of any of these DM candidates, since in no case is there a fundamental explanation - or, even better, a prediction - for the ratio $\omega \equiv \rho^O_{DM}/\rho^O_{lum}$, which is known to lie in the range $10 \lesssim \omega \lesssim 10^2$. Two fundamental questions about the universe which the fruitful marriage of particle physics and cosmology has yet to address are the value of ω and of the cosmological constant Λ. (We have here assumed $\Lambda = 0$, as usual.)

B. "Stagspansion"

Peebles (49) has calculated the fluctuation spectrum for cold DM, with results that are well approximated by the expression

$$|\delta_k|^2 = k^n(1 + \alpha k + \beta k^2)^{-2},$$

$$\alpha = 6\,\theta^2\,h^{-2}\,\text{Mpc}, \quad \beta = 2.65\,\theta^4\,h^{-4}\,\text{Mpc}^2, \quad \theta = T_o/2.7\text{K}. \qquad (15)$$

This calculation neglects the massless neutrinos; we find qualitatively similar results with their inclusion (61). For an adiabatic Zeldovich ($n = 1$) primordial fluctuation spectrum, the spectrum of rms fluctuations in the mass found within a randomly placed sphere,[5] $\delta M/M$, is relatively flat for $M < 10^9\,M_\Theta$, $\propto M^{-1/6}$ ($n_{eff} \approx -2$) for $10^9\,M_\Theta \leq M \leq 10^{12}\,M_\Theta$, $\propto M^{-1/3}$ ($n_{eff} \approx -1$) for $10^{12}\,M_\Theta \leq M \leq M_{eq}$, and $\propto M^{-2/3}$ ($n = 1$, reflecting the primordial spectrum) for $M \geq M_{eq}$.

The flattening of the spectrum for $M < M_{eq}$ is a consequence of "stagspansion",[3] the inihibition of the growth of δ_{DM} for fluctuations which enter the horizon when $z > z_{eq}$, before the era of matter domination. In the conventional formalism (12,48,62) - synchronous

gauge, time-orthogonal coordinates – the fastest growing adiabatic
fluctuations grow $\propto a^2$ when they are larger than the horizon. When
they enter the horizon, however, the radiation and charged particles
begin to oscillate as an acoustic wave with constant amplitude
(later damped by photon diffusion for $M < M_{Silk}$), and the neutrinos
free stream away. As a result, the main source term for the growth
of δ_{DM} disappears, and once the fluctuation is well inside the horizon δ_{DM} grows only as (46), (48, pp. 56–59)

$$\delta_{DM} \propto 1 + \frac{3a}{2a_{eq}} \tag{16}$$

until matter dominance ($a = a_{eq}$); thereafter, $\delta_{DM} \propto a$. Based on
Eq. (16), it has sometimes been erroneously remarked [also by the
present authors (38), alas] that there is only a factor of 2.5
growth in δ_{DM} during the entire stagspansion regime, from horizon
crossing until matter dominance. There is actually a considerable
amount of growth in δ_{DM} just after the fluctuation enters the
horizon, since $d\delta_{DM}/da$ is initially large and the photon and neutrino source terms for the growth of dark matter fluctuations do
not disappear instantaneously. (See reference 61 for details.)
This explains how $(\delta M/M)_{DM}$ can grow by a factor ~30 between M_{eq}
and $10^9 M_\odot$.

C. Galaxy Formation

When δ reaches unity, nonlinear gravitational effects become
important. The fluctuation separates from the Hubble expansion,
reaches a maximum radius, and then contracts to about half that
radius (for spherically symmetric fluctuations), at which point
the rapidly changing gravitational field has converted enough
energy from potential to kinetic for the virial relation <PE> =
– 2<KE> to be satisfied. (For reviews see (63) and (48).)

Although small-mass fluctuations will be the first to go nonlinear in the cold DM picture, baryons will be inhibited by pressure from falling into them if $M < M_{J,b}$. What is more important
is that even for $M > M_{J,b}$, the baryons will not be able to contract
further unless they can lose kinetic energy by radiation. Without such mass segregation between baryons and DM, the resulting
structures will be disrupted by virialization as fluctuations
that contain them go nonlinear (52). Moreover, successively
larger fluctuations will collapse relatively soon after one another if they have masses in the flattest part of the $\delta M/M$ spectrum,
i.e., (total) mass $\lesssim 10^9 M_\odot$.

Gas of primordial composition (about 75% atomic hydrogen and
25% helium, by mass) cannot cool significantly unless it is first
heated to ~10^4K, when it begins to ionize (65). Assuming a pri-

mordial Zeldovich spectrum and normalizing (49) so that

$$\frac{\delta M}{M} \ (R = 8h^{-1}) = 1, \tag{17}$$

the smallest protogalaxies for which the gas is sufficiently
heated by virialization to radiate rapidly and contract have
$M_{tot} \sim 10^9 \, M_\odot$ (65). One can also deduce an <u>upper</u> bound on galaxy
masses from the requirement that the cooling time be shorter than
the dynamical time (64); with the same assumptions as before, this
upper bound is $M_{tot} \lesssim 10^{12} \, M_\odot$ (65). It may be significant that this
is indeed the range of masses of ordinary galaxies. The collapse
of fluctuations of larger mass is expected in this picture to lead
to clusters of galaxies. Only the outer parts of the member galaxy
halos are stripped off; and the inner baryon cores continue to
contract, presumably until star formation halts dissipation (29).[6]

D. Potential Problems with Cold DM

Dwarf galaxies with heavy DM halos are less of a problem in
the cold than in the hot or warm DM pictures. There is certainly
plenty of power in the cold DM fluctuation spectrum at small masses;
the problem is to get sufficient baryon cooling and avoid disruption. Perhaps dwarf spheroidals are relatively rare because most
suffered disruption.

The potentially serious difficulties for the cold and warm
DM pictures arise on very large scales, where galaxies are observed
to form filamentary superclusters with large voids between them
(23,24). These features have seemed to some authors to favor the
hot DM model, apparently for two main reasons: (1) it is thought
that formation of caustics of supercluster size by gravitational
collapse requires a fluctuation power spectrum sharply peaked at
the corresponding wavelength, and (2) the relatively low peculiar
velocities of galaxies in superclusters are seen as evidence for
the sort of dissipation expected in the baryonic shock in the
"pancake" model. Recent work by Dekel (67) suggests, however,
that <u>nondissipative</u> collapse fits the observed features of superclusters. Results from N-body simulations with $N \sim 10^6$ (54) will
soon show whether broad fluctuation spectra lead to filaments.

VII. SUMMARY AND REFLECTIONS

The hot, warm, and cold DM pictures are compared schematically
in Fig. 2. Although only very tentative conclusions can be drawn
on the basis of present information, it is our impression that the
hot DM model is in fairly serious trouble. Maybe that is mainly
because it has been the most intensively studied of the three possibilities considered here.

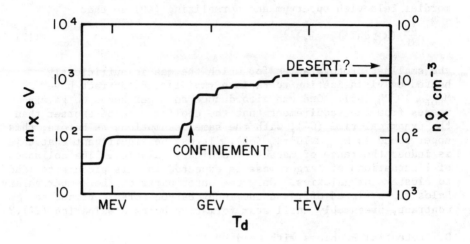

Figure 1. The mass m_X and present number density n_X^o of warm dark matter particles X, calculated assuming the standard particle physics model and no entropy generation. The mass scale should be multiplied by the factor $h^2\Omega$.

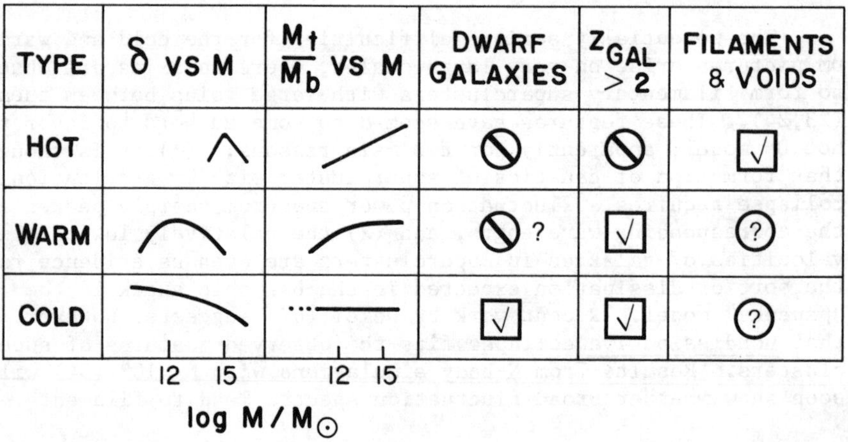

TYPE	δ VS M	$\frac{M_t}{M_b}$ VS M	DWARF GALAXIES	$Z_{GAL} > 2$	FILAMENTS & VOIDS
HOT	∧	╱	⊘	⊘	☑
WARM	⌢	⌣	⊘ ?	☑	?
COLD	⌍	····	☑	☑	?

$$\text{log } M/M_\odot$$

(12 15   12 15)

Figure 2. Consumers' Guide to dark matter. The circle with a bar means trouble and the box with a check means consistency.

Probably the greatest theoretical uncertainly in all three
DM pictures concerns the relative roles of heredity vs. environment. For example, are elliptical galaxies found primarily in
regions of high galaxy density, and disk galaxies in lower density
regions, because such galaxies form after the regions have undergone a large-scale dissipative collapse which provides the appropriate initial conditions, as in the hot DM picture? Or is it
because disks form relatively late from infall of baryons in an
extended DM halo, which is disrupted or stripped in regions of
high galaxy density? An exciting aspect of the study of large
scale structure and DM is the remarkable recent increase in the
quality and quantity of relevant observational data, and the
promise of much more to come.

Perhaps even more remarkable is the fact that this data may
shed important light on the interactions of elementary particles
on very small scales. Fig. 3 is redrawn from a sketch by Shelley
Glashow which recently was reproduced in *The New York Times
Magazine* (68). Glashow uses the snake eating its tail - the uroboros, an ancient symbol associated with creation myths (69) -
to represent the idea that gravity may determine the structure of
the universe on both the largest and smallest scales. But there
is another fascinating aspect to this picture. There are left-
right connections across it: medium-small-to-medium-large, very-
small-to-very-large, etc. Not only does electromagnetism determine
structure from atoms to mountains (70), and the strong and weak
interactions control properties and compositions of stars and
solar systems. The dark matter, which is gravitationally dominant on all scales larger than galaxy cores, may reflect fundamental
physics on still smaller scales. And if cosmic inflation is to
be believed, cosmological structure on scales even larger than
the present horizon arose from interactions on the seemingly
infinitesimal grand unification scale.

ACKNOWLEDGMENTS Our interest in the subject of dark matter grew
out of collaboration with Heinz Pagels (7,38). In preparing this
paper we benefited from conversations with N. Abrams, J.R. Bond,
A. Dekel, M. Davis, G. Efstatiou, C. Frenk, D. Lin, J. Silk,
A. Szalay, M. Turner. S. White, and especially from extensive
discussions with S. Faber at Santa Cruz and M. Rees at the Moriond
Conference and subsequently. We received partial support from
NSF grants and from the Santa Cruz Institute for Particle Physics.

FOOTNOTES

1. This discussion is approximate. Since neutrino decoupling and
    e^+e^- annihilation so nearly coincide, there is actually a little
    heating of the neutrinos too (13).

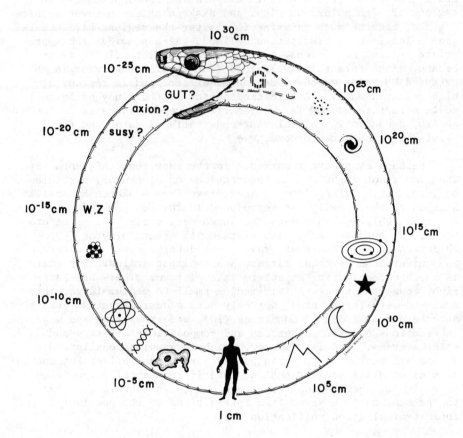

Figure 3. Physics Uroboros (after Glashow (68)).

2. It is also possible that the DM is heavy stable neutrinos with mass $\gtrsim 2$ GeV, almost all of which would have annihilated (14). This is a possible form of cold DM, discussed below.

3. In economic "stagflation", the economy stagnates but the economic yardstick inflates. The behavior of δ_{DM} during the "stagspansion" era is analogous: $\delta_{DM} \approx$ constant but a expanding. We suggest here the term stagspansion rather than stagflation for this phenomenon since it occurs during the ordinary expansion (Friedmann) era rather than during a possible very early "inflationary" (de Sitter) era.

4. One might worry that such a light particle could give rise to a force that at short distances $(10^{-5} eV)^{-1} \sim 2$ cm would be much stronger than gravity. But because the axion is pseudoscalar, its nonrelativistic couplings to fermions are $\sim \vec{\sigma} \cdot \vec{p}$.

5. One calculates δ_k initially. In order to discuss mass fluctuations it is more convenient to use $\delta M/M$ than $\delta\rho/\rho$, the Fourier transform of δ_k (49). Note that there is a simple relationship between $|\delta\rho/\rho|^2$ and $|\delta_k|^2$ only for a power law fluctuation spectrum $|\delta_k|^2 \propto k^n$.

6. The model presented by Peebles at the Moriond conference differs from that sketched here mainly in Peebles' assumption that there is sufficient cooling from molecular hydrogen for baryon condensation to occur rapidly even on globular cluster mass scales.

REFERENCES

1. Faber,S.M. and Gallagher,J.S.1979 Ann.Rev.Astron. and Astrophys.17,pp 135-187.

2. Hegyi,D.J. and Olive,K.A.1982 U.Mich.preprint; Hegyi,D.J.1983 Rencontre de Moriond,Astrophysics.

3. Steigman,G.1981 in ν81 and in Heusch,C.ed.,*Particles and Fields 1981: Testing the Standard Model* (A.I.P.Conf.Proc.,No.81) pp 548-571; Schramm,D.N. and Steigman,G.1981 Astrophys.J.,241, pp 1-7.

4. Bond,J.R.,Efstatiou,G. and Silk,J.1980 Phys.Rev.Lett.45,pp 1980-1984.

5. Doroshkevich,A.G.,Khlopov,M.Yu.,Sunyaev,R.A.,Szalay,A.S. and Zeldovich,Ya.B.1981 *Proc. Xth Texas Symposium on Relativistic Astrophysics* Ann.N.Y.Acad.Sci.375,pp 32-42; Sato,H.,ibid,pp 43-53; and ref. therein.

6. Partridge,R.B.1980 Astrophys.J.,235,pp 681-687.

7. Pagels,H.R. and Primack,J.R.1982 Phys.Rev.Lett.48,pp 223-226.

8. Carr,B.1978 Comments on Astrophys.7,pp 161-173.

9. Canizares,R.1982 Astrophys.J.,263,pp 508-517.

10. Lacey,C.1983 Rencontre de Moriond Astrophysics.

11. Gelmini,G.B.,Nussinov,S. and Roncadelli,M.1982 Nucl. Phys.B209, pp 157.

12. Weinberg,S.,*Gravitation and Cosmology*(Wiley,1972),pp 534.

13. Dicus, D.A.,Kolb,E.W.,Gleeson,A.M.,Sudarshan,E.C.G.,Teplitz,
     V.L.and Turner,M.S.1982 Phys.Rev.D26,pp 2694-2706.
14. Lee,B.W. and Weinberg,S.1977 Phys.Rev.Lett.39,pp 165-168.
15. Lyubimov,V.A.,et al.1980 Phys.Lett.94B,pp 266-268.
16. Boehm,F. 1983 *Proc.4th Workshop on Grand Unification*,in press.
17. Bisnovatyi-Kogan,G.S. and Novikov,I.D.1980 Sov.Astron.24,pp
     516-517.
18. Gibbons,G.,Hawking,S., and Siklos,S.,eds.1983 *The Very Early
     Universe* (Cambridge Univ. Press) and ref. therein.
19. Bond,J.R. and Szalay,A.S.1981 Proc. ν81,1,pp 59; Astrophys.J.,
     in press.
20. Melott,A.1983 Mon.Not.R.astr.Soc.202,pp 595-604, and ref.
     therein.
21. Klypin,A.A. and Shandarin,S.F.1982 preprint.
22. Frenk,C.,White,S.D.M. and Davis,M.1983 Astrophys.J., in press;
     White,S.D.M.,Frenk,C. and Davis,M.1983 Rencontre de Moriond,
     Astrophysics.
23. Zeldovich,Ya.B.,Einasto,J. and Shandarin,S.F.1982 Nature,300,
     407-413, and refs. therein.
24. Oort,J.1983 Ann.Rev.Astron. and Astrophys. in press and refs.
     therein.
25. Kaiser,N. Berkeley preprint.
25a. Bond,J.R.1983 Rencontre de Moriond,Astrophysics; Shapiro,P.R.,
     Struck-Marcell,C. and Melott,A.L.1983 preprint.
26. Bond,J.R.,Szalay,A.S. and White,S.D.M.1983 Nature,301,pp 584-
     585.
27. Press,W.H. and Davis,M.1982 Astrophys.J.,259,pp 449-473.
28. Forman,W. and Jones,C.1982 Ann.Rev.Astron.Astrophys.20,pp
     547-585.
29. Faber,S.M.1982 *Astrophysical Cosmology* (eds. Bruck,H.,Coyne,G.
     and Longair,M.),pp 191-234.
30. Gunn,J.E.1982 ibid.
31. Faber,S.M. and Lin,D.N.C.1983 Astrophys.J.(Lett.)266,pp L17-
     L20.
32. Aronson,M.1983 Astrophys.J.(Lett.)266,pp L11-L15.
33. Lin,D.N.C. and Faber,S.M.1983 Astrophys.J.(Lett.)266,pp L21-
     L25.
34. Tremaine,S.D. and Gunn,J.E.1979 Phys.Rev.Lett.42,pp 407-410.
35. Steigman,G.1979 Ann.Rev.Astron.Astrophys.17,pp 135-
36. Primack,J.R.1981 *Proc.1981 Banff Summer Institute on Particles
     and Fields*.
37. Bond,J.R., Szalay,A.S. and Turner,M.S.1982 Phys.Rev.Lett.48,
     pp 1636-1639.
38. Blumenthal,G.R.,Pagels,H. and Primack,J.R.1982 Nature,299,pp
     37-38.
39. Savoy,C.A.1983 Rencontre de Moriond, Elementary Particles.
40. Ellis,J.1983 private communication.
41. Goldberg,H.1983 Phys.Rev.Lett.50,pp 1419-1422.
42. Olive,K.A. and Turner,M.S.1982 Phys.Rev.D25,pp 213-216.

43.   Harrison,E.R.1970 Phys.Rev.D1,pp 2726-2730; Peebles,P.J.E.
      and Yu,J.T.1970 Astrophys.J.,162,pp 815-836; Zeldovich,Ya.B.
      1972 Mon.Not.R.astr.Soc.160,pp 1P-3P.
44.   Lubin,P.M.,Epstein,G.L. and Smoot,G.F.1983 Phys.Rev.Lett.50,
      pp.616-619; Fixin,D.J.,Cheng,E.S. and Wilkinson,D.T.1983
      Phys.Rev.Lett.50,pp 620-622.
45.   Olive,K.A.1983 Rencontre de Moriond, Astrophysics.
46.   Guyot,M. and Zeldovich,Ya.B.1970 Astron.Astrophys.9,pp 227-
      231; Meszaros,P.1974 Astron.Astrophys.37,pp 225-228.
47.   Doroshkevich,A.G.,Zeldovich,Ya.B.,Sunyaev,R.A. and Khlopov,
      M.Yu.1980 Sov.Astron.Lett.6,pp 252; Chernin,A.D.1981 Sov.
      Astron.25,pp 14-16.
48.   Peebles,P.J.E.1980 *The Large Scale Structure of the Universe*,
      Princeton Univ. Press.
49.   Peebles,P.J.E.1982 Astrophys.J.(Lett.)263,pp L1-L5; Astrophys.
      J.,258,pp 415-424.
50.   Dekel,A.1982 Astrophys.J.(Lett.)261,pp. L13-L17.
51.   Sargent,W.L.W.,Young,P. and Schneider,D.P.1981 Astrophys.J.,
      256,pp 374-385.
52.   White,S.D.M. and Rees,M.1977 Mon.Not.R.astr.Soc.
53.   Silk,J.1983 Nature,301,pp 574-578.
54.   Davis,M.,Efstatiou,G.,Frenk,C. and White,S.D.M.1983 private
      communication.
55.   Schechter,P.1983 private communication.
56.   Peccei,R. and Quinn,H.1977 Phys.Rev.Lett.38,pp 140.
57.   Weinberg,S.1978 Phys.Rev.Lett.40,pp 223; Wilczek,F.1978 Phys.
      Rev.Lett.40,pp 279.
58.   Abbott,L. and Sikivie,P.1983 Phys.Lett.120B,pp 133; Dine,M.
      and Fischler,W.1983 Phys.Lett.120B,pp 137; Preskill,J.,
      Wise,M. and Wilszek,F.1983 Phys.Lett.120B,pp. 127; Ipser,J.
      and Sikivie,P.1983 Phys.Rev.Lett.50,pp 925.
59.   Dicus,D.,Kolb,E.,Teplitz,V. and Wagoner,R.1978 Phys.Rev.D18,
      pp 1829; Fukugita,M.,Watamura,S. and Yoshimura,M.1982 Phys.
      Rev.Lett.48,pp.1522.
60.   Wilczek,F.1982 preprint NSF-ITP-82-100.
61.   Blumenthal,G.R. and Primack,J.R., in preparation.
62.   Press,W.H. and Vishniac,E.T.1980 Astrophys.J.,239,pp 1-11.
63.   Gott,J.R.1977 Ann.Rev.Astron.Astrophys.15,pp 235-266.
64.   Rees,M.J. and Ostriker,J.P.1977 Mon.Not.R.astr.Soc.179,pp
      541-559.
65.   Blumenthal,G.R.,Faber,S.M.,Primack,J.R. and Rees,M., in pre-
      paration.
66.   Peebles,P.J.E.1983 Rencontre de Moriond, Astrophysics.
67.   Dekel,A.1983 Astrophys.J.,264,pp 373-391.
68.   Ferris,T.1982 *The New York Times Magazine*, Sept.26,1982,pp 38.
69.   Neumann,E.1954 *Origins and History of Consciousness*, Prince-
      ton Univ. Press.
70.   Weisskopf,V.F.1962 *Knowledge and Wonder*, Hinemann.

DARK MATTER, GALAXIES AND GLOBULAR CLUSTERS

P. J. E. Peebles

Dominion Astrophysical Observatory, Victoria,
Joseph Henry Laboratories, Princeton University

Recent discussions of how galaxies might have formed have tended to focus on two scenarios: in the pancake model proto-clusters form first and fragment to produce galaxies; in the hierarchy picture the first generation is no larger than galaxies which later collect to form the large-scale clustering hierarchy. (These often are called primeval adiabatic and isothermal models but the connection is not required and in fact the hierarchy scenario described below assumes initially adiabatic density fluctuations.) I explain here why I think there are serious problems with the pancake scenario in an expanding world model, and then outline a hierarchy scenario based on the assumption that the universe is dominated by dark matter with negligible primeval pressure, like axions or massive photinos. This scenario produces two clustering scales, one of which might be identified with galaxies, the other with globular star clusters. Depending on how the observations turn out, it may be a decided problem or a dramatic triumph of the scenario that it predicts that globular clusters are born with massive dark halos.

1. THE PROBLEM WITH THE PANCAKE SCENARIO

The epoch of formation of large galaxies seems to be pretty well over. The only candidates for young galaxies at the present epoch are the low mass "intergalactic H II regions" (1). At redshifts $z \sim 0.5$ to 1 galaxies tend to be somewhat bluer than at low z but the main point is that the colors and luminosities have changed so little, indicating that at $z \sim 1$ bright galaxies typically have already settled down to a fairly stable

J. Audouze and J. Tran Thanh Van (eds.),
Formation and Evolution of Galaxies and Large Structures in the Universe, 185–195.
© 1984 by D. Reidel Publishing Company.

state (2). It is generally thought that the spheroid components
of galaxies formed rapidly, in a few collapse times, because it
is difficult to see what could prolong the lifetime of a protogalactic gas cloud at the density of a galaxy. If so, young
galaxies ought to be bright at $\lambda \sim 1000$ to 2000A, and so easily
visible at redshifts $z \lesssim 3.5$. (As quasars are not observed much
beyond this redshift we can assume young galaxies at higher z
are obscured by intergalactic dust.) Since young galaxies at
$z \lesssim 3.5$ are not seen it appears that the majority of massive
galaxies formed at higher redshifts. If cosmological redshifts
arise from the expansion of the universe, as will be assumed
throughout this paper, it follows that galaxies formed when the
mean density of the universe was at least 100 times the present
value.

In the pancake scenario galaxies form in pancakes or associations roughly as massive as a rich cluster of galaxies (3), and,
we have just seen, at densities $\gtrsim 100$ times the present mean.
But if that were so I would have expected to find that the
majority of galaxies now are in clusters at densities contrast
$\gtrsim 100$, which is not observed. Only 2 percent of bright galaxies
are within one Abell radius ($r = 1.5\ h^{-1}$ Mpc, $H = 100\ h$ km s^{-1}
Mpc^{-1}) of an Abell cluster, 5 percent within $r = 5\ h^{-1}$ Mpc of an
Abell cluster (4). More typical is the Local Group, which is a
part of the Local Supercluster, but at a density (averaged over a
sphere centered on the Virgo Cluster) only two or three times the
large-scale mean (5).

To account for the fact that most galaxies are not in rich
dense clusters we would have to postulate that the pancakes that
formed at $z \gtrsim 3.5$ tended to expand thereafter more or less with
the general expansion. That might be arranged by stipulating
that the velocity field that produced the pancakes was
divergenceless, but then the problem is to account for the origin
of this velocity field. If primeval we have the familiar problem
of diverging fluctuations at high redshifts. If the field is
produced by gravity we know the field is not divergenceless: the
matter tends to pile up in regions of negative energy, and there
is no obvious source of energy to pull the pancakes apart to
produce things like the Local Group.

A simple model for what I have in mind might be useful.
Suppose we start with a homogeneous mass distribution and
collapse all of the mass within a sphere of radius r perpendicu-
larly onto a plane that runs through the center of the sphere.
This produces a thin homogeneous ellipsoid with radius equal to
the radius of the sphere. Real pancakes would have expanded or
contracted relative to the Hubble flow, but as gravity prefers
contraction this is a conservative case for estimating the acceleration of the radius of the disc. If the mass of the disc is M

the acceleration of the rim is

$$g = \frac{3\pi}{4} \frac{GM}{r^2} \quad , \qquad\qquad\qquad\qquad [1]$$

the cosmological acceleration is GM/r^2, so the disc acts as if it were a density enhancement

$$\frac{\delta\rho}{\rho} = \frac{\delta g}{g} = \frac{3\pi}{4} - 1 = 1.4 \quad . \qquad\qquad [2]$$

The disc therefore tends to expand a factor of two or so in radius and then collapse, ending up at a density comparable to the initial density of the disc averaged over spherical shells.

This conclusion would be avoided if the radius of the disc tended to expand faster than the Hubble flow so as to balance the excess gravity. We might therefore consider a model for the peculiar motion of the disc. If the mass density contrast is

$$\frac{\delta\rho}{\rho} = \sum_{\vec{k}} \delta_{\vec{k}} \, e^{i\vec{k}\cdot\vec{r}} \quad , \qquad\qquad [3]$$

then in linear perturbation theory in an Einstein-deSitter cosmological model the peculiar velocity field produced by $\delta\rho/\rho$ is
(6)

$$\vec{v} = \frac{2ia}{3t} \sum \frac{\vec{k}}{k^2} \delta_{\vec{k}} \, e^{i\vec{k}\cdot\vec{r}} \quad . \qquad\qquad [4]$$

In Zel'dovich's (7) approximation the first pancakes appear where an eigenvalue of $\partial v^\alpha/\partial r^\beta$ is negative and large. Let us consider therefore the joint distribution in the three variables $D_x = \partial v_x/\partial x$, $D_y = \partial v_y/\partial y$ and $D_z = \partial v_z/\partial z$. One finds from the above equation for the velocity field

$$\langle D_x D_z \rangle = \langle D_y D_z \rangle = \langle D_z^2 \rangle /3 \quad . \qquad\qquad [5]$$

It will be noted that this follows from the assumption that v is produced by gravity and that $\delta\rho/\rho$ is statistically homogeneous; we need not specify whether $\delta\rho/\rho$ is a gaussian process. Because D_x and D_y are positively correlated with D_z we see that if D_z is negative and large, so that a pancake is forming in the xy plane, then D_x and D_y tend to be negative so that the pancake shrinks as it forms, hastening the collapse of the protocluster.

The N-body model pancake simulations described by White at this conference, and by Melott (8), do show a "frothy" structure

that looks rather like the galaxy distribution; but that is not
an acceptable solution because in these models the dilute "frothy"
part of the structure forms at low redshifts. As noted above,
the problem is that young galaxies at low redshifts are rare.
For further aspects of this discussion see reference (9).

It is possible that while galaxies formed at high redshift
the large-scale features in the galaxy distribution developed
only recently as pancakes, as discussed by Dekel (10). In that
case the N-body pancake models described by White and Melott (8)
could be realistic representations of the way galaxies, which
formed by some other process, arranged themselves in the present
clustering pattern. Two points should be noted here. The galaxy
two-point correlation function $\xi(r)$ is such a good approximation
to a power law that I would be surprised to learn that the power
law is a transient effect. I would expect that one Hubble time
from now $\xi(r)$ has very nearly the same shape. The present N-body
pancake models can produce a $\xi(r)$ that roughly approximates a
power law at one epoch, but the shape varies with time. It will
be interesting to see whether future work can yield a more
stable ξ. Second, if there are physical large-scale sheets or
chains or bridges of galaxies that cannot be ascribed to the
tidal tails that would develop out of an initially random distribution it will be an argument for a recent pancake development.
My impression is that a convincing test of this important point
awaits better dynamical analysis and better data.

Another problem specific to the massive neutrino scenario
(neutrino masses a few tens of eV) is that it is difficult to
arrange parameters so that the coherence length r_0 of the galaxy
distribution and the redshift z_f of galaxy formation both are
reasonable. If z_f is increased to make young galaxies invisible
it increases r_0 because it allows more time for the growth of
clustering by gravitational instability after the first pancakes
form. Analytic arguments suggested that if r_0 is fixed to the
observed value for the galaxy distribution then z_f is uncomfortably small (11). This has been confirmed in the very careful
numerical studies by Frenk, White and Davis (12) described at
this conference by White.

One could imagine that galaxy formation in a pancake was
inefficient, producing only a single galaxy or a tight group, or
that a pancake is no more massive than a tight group of galaxies.
Yet another possibility is that the first generation is much less
massive than a single galaxy. Such a model is described next.

2. A HIERARCHY SCENARIO

The discussion of how galaxies might have formed is

perplexed by the uncertainty in what galaxies are made of. An idea now much discussed is that the mass is dominated by weakly interacting particles. The behavior of this dark matter depends on its initial pressure. I consider here the possibility that nature has simplified the problem by making the primeval pressure negligibly small. There are candidates: massive photinos (13) and axions (14), and it is interesting to consider the astrophysical consequences of this limiting case (15, 16).

A complete cosmogony depends on many additional assumptions; the following are my current favorites for the simplest of all possible worlds.

1) The background cosmological model is the usual Friedman-Lemaitre general relativity model with Hubble's constant $H = 75$ km s^{-1} Mpc^{-1} and background temperature $T_o = 2.7$K.

2) The cosmological constant is $\Lambda = 0$. If Λ were appreciable it would mean $|\Lambda| \sim G\rho$ now, which seems to be an unlikely coincidence because $|\Lambda|$ would be very much less than $G\rho$ at high redshift when these parameters of classical cosmology would have been set.

3) By the same argument the density parameter is $\Omega = 1$.

4) The contribution to Ω by baryons is $\Omega_b \sim 0.03$, there is a small contribution by relativistic quanta and the remainder, $\Omega_x \sim 0.97$, is provided by the dark weakly interacting matter with negligible primeval pressure.

5) The primeval departure from homogeneity is the growing mode of the adiabatic mass density fluctuation $\delta\rho$.

6) $\delta\rho$ is homogeneous and isotropic random process with power spectrum $P \propto k$. The power law form $P \propto k^n$ usually is considered to be the simplest possibility and so the one to try first. Several authors noted at about the same time that the choice $n = 1$ is particularly interesting because in this case the space curvature fluctuations produced by $\delta\rho$ diverge only as $\log \lambda$ at large and small scales λ, so we do not need to specify the cutoff of the power law (17, 18, 19).

7) The spectrum P is normalized to agree with the observed large-scale rms fluctuations in galaxy counts. This assumes that on large scales mass clusters like galaxies.

Under these assumptions the universe at redshift $z \sim 100$ is dominated by the dark matter and the density fluctuations are still close to linear. The power spectrum of the dark matter is (11, 15)

$$P_x \propto k/(1 + \alpha k + \beta k^2)^2 \quad , \tag{6}$$

$$\alpha = 10.7 \text{ Mpc}, \qquad \beta = 8.4 \text{ Mpc}^2,$$

where k is the comoving wave number expressed in units of radians Mpc^{-1} at the present epoch. At small k, $P_x \propto k$, which is the primeval spectrum. At short wavelengths the spectrum was suppressed by radiation pressure, approaching $P_x \sim k^{-3}$ at large k. At $z \sim 100$ the hydrogen has decoupled from the radiation and has relaxed to the same distribution as the dark matter longward of the hydrogen Jeans length. The gas pressure suppresses density fluctuations on smaller scales. The resulting spectrum for the hydrogen is (16)

$$P = P_x/(1 + \gamma k^2)^2, \qquad \gamma = 2.6 \times 10^{-6} \text{ Mpc}^2. \tag{7}$$

At comoving wavelengths ~ 10 kpc to ~ 1 Mpc, $P \propto k^{-3}$, so the contribution to the variance of $\delta\rho/\rho$ per octave of wavelength is about constant. An example of the resulting mass distribution is shown in Figure 1 (20). This example assumes random phases ($\delta\rho$ is a random Gaussian process). The sizes of the smallest structures seen in the graph are fixed by the short wavelength cutoff of the $P \propto k^{-3}$ spectrum. For the hydrogen distribution this cutoff is fixed by the Jeans length, and one finds that the hydrogen mass in one of these gas clouds is $M_b \sim 3 \times 10^6 \, M_\odot$ (16). As noted by Peebles and Dicke (21), it is a suggestive coincidence that this is close to the mass of a globular star cluster. The figure also shows that the clouds tend to appear in associations, an unusually dense cloud tending to be surrounded by other dense clouds. The characteristic mass of an association is fixed by the break in the spectrum $P_x(k)$, and as is discussed next this mass is interestingly close to that of a galaxy (11, 15).

For a more quantitative analysis of the mass distribution implied by the spectrum $P(k)$ in the present model it is useful to consider the fractional mass excess Δ within distance r of a peak of the baryon mass density $\delta\rho$,

$$\Delta = \int_r dV \delta\rho/(<\rho>V) . \tag{8}$$

If $\delta\rho$ is a random Gaussian process we can derive the probability distribution of Δ from the spectrum $P(k)$. If $\delta\rho$ is not Gaussian this distribution may still illustrate the main features of the mass clustering. Figure 2 shows the behavior of Δ around a two standard deviation extremum of $\delta\rho$ (16). The central curve is the mean value of Δ. The top and bottom curves are shifted from the mean by the rms fluctuation of Δ around the mean. The horizontal axis is the comoving distance r from the extremum of $\delta\rho$; the top scale is the mean baryon mass M_b within a sphere of radius r.

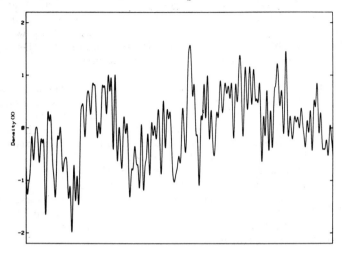

Fig. 1. Realization of a gaussian process with fixed power per logarithmic interval of wavelength (20).

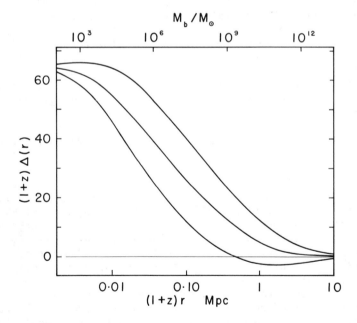

Fig. 2. The fractional mass excess around a 2σ peak in the hydrogen distribution (16).

In linear perturbation theory $\Delta \propto (1 + z)^{-1}$ at fixed M_b. Analytic and N-body model results show that this is a good approximation when the rms value of $\Delta \lesssim 1$ even if the mass fluctuations are highly nonlinear on smaller scales (22). Thus we can normalize the graph by setting $<\Delta^2>$ at large r equal to the observed rms fluctuation $\delta N/N$ in the galaxy counts.

We see from Figure 2 that the denser gas clouds reach density contrast $\Delta \sim 1$ and start to break away from the general expansion at redshift $z \sim 60$. This is the first generation of hydrogen objects. Around such a dense spot there is a net mass excess ($\Delta > 0$) that tends to extend to $M_b \sim 10^{10}M_\odot$, but with considerable scatter: a 1 σ downward fluctuation would limit the zone of substantial positive mass excess, where $(1 + z) \Delta \gtrsim 10$, to $M_b \sim 10^8 M_\odot$, and a 1 σ upward fluctuation could produce $M_b \sim 10^{11} M_\odot$ at the same contrast and a tail of positive mass excess extending to considerably larger scales. Thus at redshift $z = 10$ the objects just forming would have baryon masses about in the range observed for galaxies.

It will be noted that we did not obtain this result by adjusting parameters to produce a reasonable candidate for protogalaxies. The shape of the spectrum of the mass fluctuations follows from the initially scale-free spectrum $P \propto k$, the characteristic scale that might be identified with protogalaxies following from the size of the horizon when the densities of radiation and dark matter are equal. The time scales for formation of our candidates for protogalaxies and protoglobular clusters are set by the normalization of the spectrum, which is tied to the observed large-scale fluctuations in the galaxy distribution.

Let us consider now some details of the candidates for globular star clusters. At redshifts $z \sim 50$ the most prominent (3σ) peaks of the hydrogen distribution would have developed into the first generation of gas clouds with radii about 1 kpc and density comparable to the mean, which amounts to about 10^{-24} g cm^{-3} for the dark matter and about 3×10^{-26} for the hydrogen. The hydrogen temperature is roughly 100K. This is close to the background radiation temperature because the residual ionization is high enough to allow appreciable heat transfer from radiation to matter (23). This also means that once the gas cloud stops expanding and is supported by gas pressure it can lose energy, shrink and heat up. When the temperature reaches 10^4K the gas is collisionally ionized and the rate of loss of energy considerably increased. It seems reasonable to assume that this signals free collapse to stars. If so, the star cluster would have a radius on the order of 20 pc, which is not unreasonable for a globular cluster.

Since the initial density fluctuations were supposed to be adiabatic the cloud would be formed with comparable concentrations of hydrogen and dark matter, the mass density of the latter being some 30 times larger. The dark matter is weakly interacting and so is left behind when the hydrogen cloud contracts. This makes the final central density of the star cluster substantially larger than the dark matter density, which is desirable because it measns the mass-to-light ratio in the core can be close to unity, as observed. But this picture suggests that a globular cluster, like some galaxies, may be born with a dark halo that dominates the low density outer parts of the system.

If globular clusters are born with massive halos it may play an interesting role in the origin of the heavy elements in the cluster stars. In the present picture protoglobular clusters are the first generation so the heavy elements must originate by processing in stars within the cloud or by injection from outside (21, 24). The dark halo would increase the binding energy of the gas cloud and so would enhance the possibilities that a gas cloud resists disruption by the first supernovae within it and accumulates debris from external supernovae. The binding energy of a gas cloud is highly variable because that depends on the density in the dark halo, which in turn depends on the initial density contrast, which Figure 1 shows is highly variable. The figure also shows that the most tightly bound clouds tend to appear toward the centers of the largest protogalaxies. It seems reasonable to conjecture that the most tightly bound clouds have the best chance of surviving collisions and supernovae and ending up with relatively high heavy element abundances. If so, this would account for the correlation of heavy element abundance with galactocentric distance.

Another interesting feature is the relatively high abundance S of globular clusters per unit of spheroid luminosity in M87 and other first-ranked galaxies (25, 26). Van den Bergh (27) has shown that it is exceedingly difficult to see how this could be the result of mergers: galaxies with large S must have been born that way. In the present model that could come about because the most massive protogalaxies would tend to have the greatest initial density contrast and so the most durable protoglobular clusters.

A direct test for the presence of a dark halo can be based on the runs of the star velocity dispersions in the plane of the sky (28) and along the line-of-sight (29, 30). These fix the effective "pressure gradient" near the surface of the cluster, which measures the mass. Because of the uncertainty in velocity anisotropy the motions in the plane of the sky are particularly important. It appears that an interesting test may be feasible (28).

If dark mass is found around nearby globular clusters it certainly will have to be counted as strong encouragement for the
model. If dark mass is not found there is unfortunately a way
out: the expected density of dark matter is comparable to the
mean density in the galaxy at our position, so if a cluster had
passed much closer to the center of the galaxy the halo would
have been stripped by tides. For further discussion see
reference (16).

The model presented here is only one version of the hierarchy
picture. It illustrates just how speculative the subject is: we
are assuming, among other things, that the universe is dominated
by a form of matter that we have no very strong reason to believe
even exists. It is an attractive feature of the model that a set
of particularly simple assumptions yields characteristic mass and
time scales that seem reasonable for two important classes of
astronomical objects, galaxies and globular clusters. If globular clusters were found to have massive halos it would be an
additional point in favor of the model. But in any case the
model may be considered a trial balloon to be examined for virtues
and flaws that may serve as guides to where to turn next.

REFERENCES

1. Sargent, W.L.W. and Searle, L. 1971, Comments on Astrophysics
   3, p. 111.
2. Spinrad, H. 1980, in "Objects of High Redshift," eds. G.O.
   Abell and P.J.E. Peebles (Dordrecht: Reidel Publ.) p. 39.
3. Zel'dovich, Ya. B., Einasto, J. and Shandarin, S.F. 1982,
   Nature 300, p. 407.
4. Seldner, M. and Peebles, P.J.E. 1977, Ap. J. 215, p. 703.
5. Davis, M. and Huchra, J. 1982, Ap. J. 254, p. 437.
6. Peebles, P.J.E. 1980, "The Large-Scale Structure of the
   Universe" (Princeton: Princeton University Press).
7. Zel'dovich, Ya. B. 1970, Astron. Astrophys. 5, p. 84.
8. Melott, A.L. 1983, M.N.R.A.S. 202, p. 595.
9. Peebles, P.J.E. 1983, Ap. J. 274.
10. Dekel, A. 1983, Ap. J. 264, p. 373.
11. Peebles, P.J.E. 1982, Ap. J. 258, p. 415.
12. Frenk, C.S., White, S.D.M. and Davis, M. 1983, Ap. J. 271.
13. Weinberg, S. 1983, Phys. Rev. Letters 50, p. 387.
14. Preskill, J., Wise, M.B. and Wilczek, F. 1983, Phys. Letters
    120B, p. 127.
15. Peebles, P.J.E. 1982, Ap. J. Letters 263, p. L1.
16. Peebles, P.J.E. 1983, Ap. J. in press.
17. Peebles, P.J.E. and Yu, J.T. 1970, Ap. J. 162, p. 815.
18. Harrison, E. R. 1970, Phys. Rev. D1, p. 2726.
19. Zel'dovich, Ya. B. 1972, M.N.R.A.S. 160, p. 1P.
20. Reference 6, p. 385.

21. Peebles, P.J.E. and Dicke, R.H. 1968, Ap. J. 154, p. 891.
22. Peebles, P.J.E. and Groth, E.J. 1976, Astron. Astrophys. 53,
    p. 131.
23. Peebles, P.J.E. 1968, Ap. J. 153, p. 1.
24. Iben, I. 1980, in "Globular Clusters," eds. D. Hanes and
    B. Madore (Cambridge: Cambridge University Press) p. 125.
25. Harris, W.E. and Van den Bergh, S. 1981, A. J. 86, p. 1627.
26. Hanes, D.A. 1980, in "Globular Clusters," eds. D. Hanes and
    B. Madore (Cambridge: Cambridge University Press) p. 231.
27. Van den Bergh, S. 1982, P.A.S.P. 94, p. 459.
28. Cudworth, K.M. 1979, A. J. 84, p. 1312.
29. Gunn, J.E. and Griffin, R.F. 1979, A. J. 84, p. 752.
30. Seitzer, P.O. 1983, Doctoral Dissertation, The University
    of Virginia.

This research was supported in part by the National Science
Foundation.

COSMOLOGICAL CONSTRAINTS ON NEUTRINOS AND OTHER "INOS" AND THE "MISSING LIGHT" PROBLEM

David N. Schramm and Katherine Freese

The University of Chicago

ABSTRACT

   The arguments favoring non-baryonic dark matter are summarized. Cosmological constraints are presented on the masses of neutrinos and other "inos", where "ino" represents any candidate particle for the dark matter. A neutrino mass $10eV \lesssim m_\nu \lesssim 25eV$ is favored for the most massive eigenstate, where the upper limit may be extended to 100 eV if one assumes weaker constraints on the age of the universe.

I.   INTRODUCTION

   Over a decade ago massive neutrinos were suggested as possible candidates for the "missing mass" in the haloes of galaxies[1]. On larger and larger scales, less and less of the dynamically inferred mass can be accounted for by luminous matter[2]. Some form of dark matter must reside in the haloes of galaxies and clusters of galaxies. The density of "ordinary" nucleonic matter (faint stars, gas clouds, etc.) that can contribute to the missing mass is restricted by primordial nucleosynthetic arguments to $0.01 \leq \Omega_b \leq 0.14$ [3]. [Here Ω_b is the ratio of baryonic matter density ρ_b to critical density

$$\rho_c = \frac{3H_o^2}{8\pi G} = 1.88 \times 10^{-29} h_o^2 \ gm/cm^3 = 8.1 \times 10^{-11} h_o^2 \ eV^4$$ where the

Hubble parameter $H_o = 100 \ h_o \ kms^{-1}Mpc^{-1}$. We will also use the notation Ω_ν to represent the ratio of neutrino density ρ_ν to critical density and simply Ω for the ratio of total energy density ρ of the universe to critical density.] Schramm and Steigman[4]

197

J. Audouze and J. Tran Thanh Van (eds.),
Formation and Evolution of Galaxies and Large Structures in the Universe, 197–214.
© *1984 by D. Reidel Publishing Company.*

stressed massive neutrinos as the least "ad hoc" possibility for
the dark matter; other candidates include primordial black holes,
axions, and supersymmetric particles like photinos.

Experimental evidence for neutrino masses[5] and neutrino
oscillations[6] as well as theoretical work in grand unified
theories (for a review, see P. Langacker[7]) further stimulated
interest in massive neutrinos. The original Lubimov et al.[5]
result of 14 eV $\lesssim m_{\nu_e} \lesssim$ 42 eV has been revised to m_{ν_e} = 35 ± 5 eV,
although because of uncertainties in the molecular effects, the
mass may be as low as m_{ν_e} = 0. In any case, these results have
not been verified in other laboratories, and several experiments
are in progress. The negative results from oscillation experiments place strict limits on the squared mass difference of neutrino species and the mixing angle.

We will examine constraints on neutrino masses due to generalized cosmological arguments as well as specific models and
look for masses consistent with all the arguments. After re-
viewing the necessary background material, a discussion of mass-
to-light ratios and density constraints, we discuss fermion densities appropriate to neutrinos and to supersymmetric "inos",
and find mass constraints from phase space arguments[8] and restrictions on the age of the universe[9,10,11]. We examine the
evolution in a neutrino-dominated universe of the adiabatic density perturbations that may be responsible for galaxy formation,
in particular with large-scale structures forming first. Using
the concept of the Jeans mass, the minimum mass that can collapse
under its own self-gravity, we find constraints on masses of
neutrinos and other "inos". The alternative hierarchical view
with "cold" dark matter where small scales form first is also
discussed. Among the alternative to massive neutrinos that have
been proposed for the nonbaryonic dark matter are supersymmetric
particles[12], axions[13], and primordial black holes[14]. We will
argue that if the role of the large-scale structure is dominant
as implied by the large voids and large clusters, dark matter
should not differ significantly from massive neutrinos, either
in mass or decoupling temperature. Although we will argue that
low mass neutrinos are preferred for the large scales we will
also point out that they encounter problems in understanding the
smaller-scale structures as indicated by the correlation function[15] and the relative dynamical equilibration timescales of
different size systems. Possible scenarios[16] which avoid these
problems are discussed. Finally we discuss experimental implications and summarize the results. This talk summarizes the
detailed results presented in the paper by Freese and Schramm[32].

II. COSMOLOGICAL MASS DENSITIES

In this section we will summarize the cosmological density arguments. Since the deceleration parameter q_o in the standard hot big bang model with zero cosmological constant ($\lambda = 0$) is estimated to range from 0 ± 0.5 to 1.5 ± 0.5[17], an extreme upper limit to Ω with $\Lambda = 0$ is $\Omega \lesssim 4$. Dividing the mass of a bound system (obtained by application of the virial theorem) by its luminosity, one can obtain mass-to-light ratios (M/L) and estimates of matter contributions on different scales. Many authors[18] find evidence for M/L increasing linearly with scale from M/L $\sim$ (1-2) for stars to M/L $\sim$ (300-800)h_o for rich clusters (see Table I, drawn largely from Faber and Gallagher[2]).

Table I

Mass-to-Light Ratios

Object	$\frac{M}{L} / (\frac{M}{L})_\odot$	Ω
stars	1-4	$(0.7-2.9) \times 10^{-3} h_o^{-1}$
spiral galaxies	(8-12)h_o	$(5.7-8.6) \times 10^{-3}$
elliptical and S0 galaxies	(10-20)h_o	$(0.7-1.4) \times 10^{-2}$
binaries and small groups	(60-180)h_o	$(0.4-1.3) \times 10^{-1}$
clusters of galaxies	(280-840)h_o	0.2-0.6

Multiplying M/L on a given scale by an average luminosity density (uncertain by a factor of 2) for the universe[19],

$$\mathcal{L} \simeq 2 \times 10^8 h_o (L_\odot/Mpc^{-3}), \tag{2.1}$$

one obtains a mass density (also listed in Table I) implied by assuming M/L on that scale applies to the average light of the universe. Davis et al.[20] have suggested that the M/L curve may be approaching an asymptotic limit (perhaps $\Omega = 1$) on the scales of superclusters, while other authors[21] believe that the curve flattens already on smaller scales. In any case the con-

sensus is that some form of dark matter dominates the dynamics
of objects on scales larger than 100 kpc and, as shown by flat
rotation curves, may be important on scales larger than 10 kpc.

The above arguments are independent of whether or not the
matter is baryonic. Big Bang Nucleosynthesis[3] provides density
constraints on the baryonic components. A lower limit on the a-
mount of baryonic matter in the universe can be derived from combined D and ^{3}He abundances, $\Omega_b h_o^2 \gtrsim 0.01$, and the observed
abundances of ^{4}He (mass fraction Y $\lesssim$ 0.25), D, and ^{7}Li result in
an upper limit to baryonic matter density, $\Omega_b h_o^2 \lesssim 0.034$. These
arguments restrict baryonic matter to the range

$$0.01 \lesssim \Omega_b \lesssim 0.14. \tag{2.2}$$

Helium abundances from Big Bang Nucleosynthesis also constrain
the number of neutrino species; at most four low mass ($\lesssim$ 1 MeV),
long-lived neutrino species are compatible with Y $\lesssim$ 0.26, and
only three with the best observational limit of Y $\lesssim$ 0.25[22]. We
know two of these experimentally, namely the ν_e where
$m_{\nu_e} \lesssim$ 60 eV and the ν_μ where $m_{\nu_\mu} \lesssim$ 570 keV. The experimental
mass limit on the ν_τ is $m_{\nu_\tau} \lesssim$ 250 MeV[23]. In this case we may
already know all the neutrinos and other low mass "inos" which
interact with the strength of neutrinos. Note, however, that
the limit of three increases if the particle couples more weakly
than the neutrino[24] and thus decouples in the early universe at
a temperature $\gtrsim$ 100 MeV.

If M/L really keeps increasing on scales larger than binar-
ies and small groups, then Ω exceeds the upper limit on Ω_b and
we are forced to say that the bulk of the matter in the universe
is non-baryonic. While the general trend towards larger Ω exists,
and in this paper we will assume that Ω is $\gtrsim$ 0.15, the situation
is by no means settled; there is still a possibility that
$\Omega \leq 0.15$ and everything is baryonic, as emphasized by Gott et al.[25]
One should note that inflation[26] explicitly predicts $\Omega = 1.\overline{000}...$
and thus requires non-baryonic matter. On the basis of "sim-
plicity" we believe that a nonbaryonic universe with $\Omega \gtrsim$ 0.15
should satisfy $\Omega = 1$.

III. NEUTRINO (AND OTHER INO) DENSITIES AND MASSES

The equilibrium number density of a relativistic fermionic
species (subscript f) is given by

$$n_f = \frac{1}{2\pi^2} \int dp \, p^2 / [\exp(p - \mu_f)/T_f) + 1] \tag{3.1}$$

(throughout we take $\hbar = c = k_B = 1$). The fermions fall out of

chemical equilibrium at temperature T_D when the reaction rates
for their production (e.g. $e^+e^- \to f\bar{f}$) can no longer keep up
with the expansion of the universe. By entropy conservation it
can be shown[27] that neutrino (T_ν) and photon (T_γ) temperatures
after e^+e^- annihilation are related by $T_\nu = (4/11)^{1/3}T_\gamma$, whereas the temperature of other fermions is given by
$T_f = T(3.9/g_{*f})^{1/3}$ for $T_\gamma \ll m_e$, where g_{*f} is the number of relativistic species at decoupling.[24]

The value of the number density in the present epoch for a
species of neutrinos which are relativistic at decoupling is
given by $n_{\nu_i} = 109 \text{ cm}^{-3}(\frac{T_{\gamma o}}{2.7K})^3$ (i = e, μ, τ and $T_{\gamma o}$ is photon

temperature today), and the energy density in units of the
closure density by

$$\Omega_{\nu_i} = \frac{\rho_{\nu_i}}{\rho_c} = \frac{m_{\nu_i}}{97 \text{ eV}} h_o^{-2}(\frac{T_{\gamma o}}{2.7K})^3 \qquad (3.2)$$

where m_{ν_i} is the mass of a neutrino species. If the sum of the
masses of different neutrino species exceeds $\sim 100 \ h_o^2$ eV
the universe is closed. Requiring $\Omega \lesssim 4$ and $h_o \lesssim 1$ gives only
the weak limit, $\sum m_\nu \lesssim 400$ eV; we will see that age of the universe constraints can strengthen this limit. The ratio of neutrino to baryonic matter is given by

$$\frac{\Omega_\nu}{\Omega_b} \gtrsim \frac{\sum_i m_{\nu_i}}{2.4 eV} \qquad (3.3)$$

where the equality sign corresponds to the largest value of
baryonic matter density consistent with element abundances from
primordial nucleosynthesis, $\Omega_b h_o^2 \lesssim 0.034$. Hence if the sum of
the neutrino masses exceeds a few eV, neutrinos are the dominant
matter in the universe and must play an important role in galaxy
formation.

The above discussion of fermion number densities assumes
the fermions are relativistic at decoupling. For neutrinos more
massive than a few MeV or for other fermions whose mass exceeds
their decoupling temperature, the mass density falls roughly as
$m_f^{-1.85}$ [28]. Thus total density limits $\Omega \lesssim 4$ can be satisfied
for sufficiently massive $(m_\nu \gtrsim 1 \text{ GeV})$ particles, while if
$\sum_i m_{\nu_i} \gtrsim 20$ GeV the density has fallen so low that neutrinos cannot be the dominant matter. Krauss[29] and Goldberg[30]
have recently shown that for certain supersymmetric particles
the annihilation rates can be slower than those for neutrinos,
and the mass limits are pushed to even higher values. Following
Gunn et al.[31] for these very massive particles the appropriate
mass density one has to worry about exceeding is the density of

matter in groups of galaxies ($\Omega \leq 0.13$) not the total density
of the universe (since as we will see these massive particles
must cluster on small scales, in ways such that baryons are good
tracers of their presence. This yields $m_\nu \gtrsim 6$ GeV for neutrinos
and corresponding higher limits for photinos.

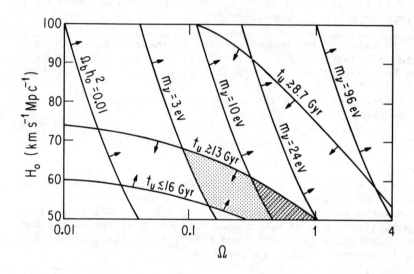

Figure 1: On a plot of Hubble parameter H_o vs. energy density
Ω we have drawn curves for several values of the age of the
universe ($t_u = f(\Omega)H_o^{-1}$), for the energy density in baryons
($\Omega_b h_o^2 \geq 0.01$), and for the total energy density of the universe
with several values of the neutrino mass $\Omega_{total} =$

$$\Omega_b + \Omega_\nu \geq (0.01 + \frac{\sum m_\nu}{97 \text{ eV}})h_o^{-2} \text{ (for } T_{\gamma o} = 2.7\text{K)}. \text{ The firm upper}$$

limit to the age of the universe $t_u < 8.7$ Gyr restricts $\sum m_\nu \lesssim 100$ eV,
while an age range consistent with dynamics and globular clusters
13 Gyr $< t_u <$ 19 Gyr requires $\sum m_\nu \lesssim 25$ eV. The dotted region
indicates the range of neutrinos massive enough to serve as the
dark matter in clusters ($m_\nu > 3$ eV) yet consistent with all the
age arguments. The smaller hatched region indicates the range
of neutrinos which may be responsible for the formation of large-
scale structure in the adiabatic picture ($m_\nu \gtrsim 10$ eV, cf. § 4).

     Figure 1 is a plot of total energy density in the universe
($\Omega \lesssim 4$) vs. Hubble parameter ($0.5 \lesssim h_o \lesssim 1$). The total energy
density is the sum of baryon density ($\Omega_b h_o^2 \gtrsim 0.01$) and neutrino
density ($\Omega_{\nu_i} h_o^2 \simeq m_{\nu_i}/97$ eV); we have plotted this sum for several values of neutrino mass. We have also plotted curves for

several values of the age of the universe, which can be para-
meterized (for $\lambda = 0$) as $t_u = f(\Omega)H_o^{-1}$ (where $f(\Omega)$ is a monotonically decreasing function of Ω with values between 1 and $\frac{1}{2}$
in the range of interest). Several arguments[10] have been used
to restrict the age of the universe: certainly it must exceed
the age of the solar system $t_u > 4.6$ Gyr (1 Gyr = 10^9 yr),
dynamical arguments ($h_o \gtrsim 0.5$) restrict $t_u \lesssim 20$ Gyr, the age of
the globular clusters combined with an upper limit on ^{4}He frac-
tion $Y \lesssim 0.26$ restricts 13 Gyr $\lesssim t_u \lesssim 19$ Gyr[14], and nucleo-
cosmochronology requires 8.7 Gyr $\lesssim t_u \lesssim 19$ Gyr. The range (13-
16) Gyr is simultaneously consistent with all arguments, while
the widest range allowed by the most stringent limits is (8.7 -
19) Gyr. Consistency with the widest range allowed as well as
the restrictions $\Omega \lesssim 4$ and $h_o > \frac{1}{2}$ requires

$$\sum_i m_{\nu_i} \lesssim 100 \text{ eV} \quad (8.7 \text{ Gyr} < t_u). \tag{3.4}$$

Consistency with the "best fit" range of ages restricts

$$\sum_i m_{\nu_i} \lesssim 25 \text{ eV} \quad (13 \text{ Gyr} < t_u) \tag{3.5}$$

(see also 9)), where $\Omega = 1$ is achieved only for $\sum_i m_{\nu_i} \simeq 25$ eV.
As Schramm[11] noted, this best fit age also limits $h_o^i < 0.7$ to
have concordance. By the inversion of this age argument, an
actual neutrino mass gives an upper limit to the age of the universe. For example, if Lubimov et al.[5] are correct and
$m_{\nu_e} \gtrsim 30$ eV, then the universe must be younger than 12 Gyr.

Similar constraints can also be found on the masses of other
fermions which decouple while still relativistic and are candidates
for the dark matter; for this analysis see Freese and Schramm[32]
as well as Olive and Turner[33]. For all standard unified models
with $g_{*f} \lesssim 161$, the limit from our best fit age argument does not
allow m_f to exceed 400 eV, contrary to the limits in previous
papers(12).

Tremaine and Gunn[8] have used phase space arguments to obtain
a restriction on neutrino masses. The smaller the scale on which
neutrinos are confined, the larger the velocity dispersion, and
the easier it is for neutrinos to escape from the region. A
necessary (but not sufficient) condition for trapping neutrinos
on the scale of clusters is the requirement $m_\nu \gtrsim 5h_o^{-2}$ eV, on the
scale of binaries and small groups $m_\nu \gtrsim 14h_o^{-2}$ eV, and in galaxies $m_\nu \gtrsim 20$ eV. If massive neutrinos are to solve the missing
mass problem they must be trapped at least on scales of clusters
of galaxies, i.e., $m_\nu \gtrsim 3$ eV. Of course to actually trap them
requires some cluster formation scenarios. Possibilities will
be addressed in the next two sections, where it will be shown

that this lower limit can probably be strengthened for any
realistic scenario.

IV. ADIABATIC PERTURBATIONS

The formation of galaxies requires the clumping of baryons;
i.e., enhancements $\delta_b = \dfrac{\delta\rho_b}{\rho_b}$ in the baryon density over the background value must grow from small values in the early universe
to nonlinearity ($\delta_b > 1$) by the present-day to achieve the formation of bound structure. In the adiabatic mode the baryon perturbations δ_b are accompanied by radiation perturbations δ_γ,
whereas in the isothermal mode initially $\delta_\gamma << \delta_b$. In general
any primordial fluctuation scheme for galaxy formation can be
treated as a superposition of these two independent modes. Thus
in the adiabatic theory of galaxy formation, initially
$\delta_\gamma = \delta_\nu = \delta_{\overline{\nu}} = \frac{4}{3}\delta_b$ (where $\delta_i = \delta\rho_i/\rho_i$ describes the density
enhancement of particle species i in a perturbation over the
background value). These fluctuations grow together outside the
horizon, and once inside the horizon their evolution depends on
the value of the Jeans mass.

The Jeans mass is the smallest mass unstable to gravitational
collapse. It is given by the rest mass of particles in a sphere
of radius equal to the Jeans length λ_J, the scale on which radiation pressure forces just balance gravitational forces. In
Figure 2 we have plotted the evolution of neutrino (M_J) and
baryon (M_{Jb}) Jeans masses in a neutrino-dominated universe[34].
The neutrino Jeans mass reaches its peak value[35],

$$M_{\nu M} \simeq 1.8\ m_{pl}^{3}/m_\nu^{2} \simeq 3 \times 10^{18}\ M_\Theta/(m_\nu/eV)^2 \qquad (4.1)$$

at $z_M \simeq 1900\ m_\nu(eV)$, where $m_{pl} = G^{-\frac{1}{2}} = 1.2 \times 10^{19}$ GeV is the Planck
mass. Neutrino perturbations on scales $\gtrsim M_{\nu M}$ can grow once the
neutrinos become the dominant matter. However, Bond et al.[34]
have shown that neutrino perturbations on scales smaller than
$M_{\nu M}$ are strongly damped by free-streaming of the neutrinos out
of dense regions (Landau damping). Only perturbations on scales
larger than M_M can survive and grow to nonlinearity[36]. To
enable the formation of large-scale structure, we require this
damping scale to be smaller than the largest structure observed,
superclusters of mass $\simeq 10^{16}\ M_\Theta$, i.e., $M_{\nu M} \lesssim M_{sc} \simeq 10^{16}\ M_\Theta$.
Eq. (4.1) is only approximate; folding an initial power spectrum
$|\delta_k|^2 \propto k^n$ with a transfer function to describe damping by neutrino diffusion, Bond, Szalay, and Turner[36] obtain an n-dependent
power spectrum. Although the peak of the power structure is the
scale on which perturbations first go nonlinear, significant power may exist on somewhat smaller or larger scales. We take the

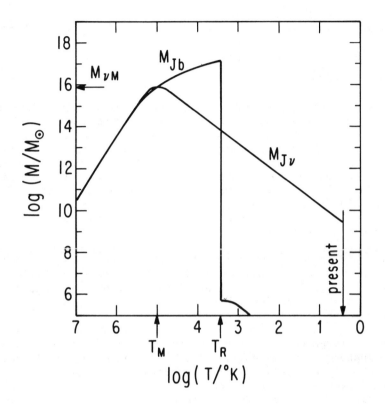

Figure 2: Neutrino and baryon Jeans masses as a function of temperature for $m_\nu \simeq 20$ eV. During the radiation-dominated era, the Jeans mass is approximately the comoving mass inside the horizon and grows as $(1 + z)^{-3}$. Once neutrinos become the dominant matter at $T_M \simeq 10^5$ K, the neutrino Jean mass peaks at $M_{\nu M} \simeq 1.8 \ m_{p1}^3/m_\nu^2 \simeq 7.5 \times 10^{15} \ M_\odot$ for $m_\nu = 20$ eV and thereafter falls as $(1 + z)^{3/2}$. The exact shape of $M_{J\nu}$ near its peak value has been calculated by Bond, Efstathiou, and Silk (1980) and is merely approximate here. M_{Jb} drops at recombination ($T_R \simeq 2700$K) to $\simeq 5 \times 10^5 M_\odot$.

least restrictive limit, the smallest mass for physically plausible values of n that has significant power, and find that

$$M_{\nu_M} \simeq \frac{9 \times 10^{17} \, M_\odot}{(m_\nu/eV)^2} \lesssim 10^{16} \, M_\odot \, . \tag{4.2}$$

In a universe with one massive neutrino species this requires
$m_\nu \gtrsim 10$ eV.

If there are three species of neutrinos with equal mass, the mass
of each species must satisfy $m_{\nu_i} \gtrsim 16$ eV, giving a sum of masses
$\sum_i m_{\nu_i} \gtrsim 48$ eV. This is not compatible with the requirement
$\sum_i m_{\nu_i} \lesssim 25$ eV from consistency of all the arguments restricting
the age of the universe. A "best fit" model does not allow all
the neutrino masses to be equal. Of course if we relax our age
constraint to $t_u > 8.7$ Gyr then equal masses are allowed. If
larger scales than M_{ν_M} in Eq. (4.3) reach nonlinearity first and
tidally strip the smaller scales, the limit on the masses only
becomes more restrictive.

In this adiabatic picture with massive neutrinos, the smallest scales to form initially are large clusters, and smaller
scales come from later cooling and fragmentation. The alternative
model of galaxy formation, where small scales form first and
cluster hierarchically onto larger and larger scales, has not
been shown to give rise to the observed structure on large scales,
namely the large voids seen by Kirshner et al.[38] as well as
large clusters and filaments. Also, dark matter which clusters
first on small scales will yield a constant M/L; however $\Omega \simeq 1$
for the universe is possible only if M/L continues to increase
with scale, or if there are regions with M/L larger than any
measured value. If, indeed, small scale damping is required for
the formation of large-scale structure, we also get a lower limit
on the Jeans mass. Note that the necessity for small-scale damping has not been rigorously proven. The failure of the attempts
of Frenk, White, and Davis[15] to have hierarchical pictures work
and the ease with which their small-scale damped models succeed
in giving rise to observed large-scale structure is certainly
suggestive if not compelling. Damping those scales smaller than
cluster sizes,

$$M_{\nu_M} \simeq (\frac{g_f}{2}) \, (\frac{10.75}{g_{*f}}) \, \frac{1.8 \, m_{p1}^3}{m_f^2} \gtrsim 10^{14} \, M_\odot \tag{4.3}$$

requires

$$m_f \lesssim 200 \text{ eV } (g_f/2)^{\frac{1}{2}} \tag{4.4}$$

for the mass of any non-interacting particle proposed for the dominant matter (for an alternative approach to similar results see (39)). For particles decoupling earlier than neutrinos the number density is lower, and hence to conserve the inequality in Eq. (4.4), the restriction on the mass only becomes tighter. If valid, this argument rules out the high mass branch, $m_f \gtrsim 6$ GeV for the dominant matter (if $\Omega \simeq 1$), since such high mass particles would cluster first on very small scales and would not explain the large voids or large M/L (this does not mean such particles cannot exist; it merely means that their contribution to Ω must be small, $\lesssim 0.2$). This argument would also rule out gravitinos or other supersymmetric particles in the keV mass range.

Since our primary motivation for non-baryonic matter is to get a large Ω, this upper limit on the mass of our candidate particle becomes quite constraining. As we mentioned in Section III, the number density of any species, n_f, decreases roughly stepwise with increasing decoupling temperature T_D for that particle[24]. Given low particle masses m_f, in order to keep a high Ω, where $\Omega \simeq m_f n_f / \rho_c$, the decoupling temperature T_f must also be low. Quantitatively, the constraint $m_f \lesssim 200$ eV and $\Omega \sim 1$ argues that

$$n_f \gtrsim 61 \, h_o^2 \, (g_f/2)^{-\frac{1}{2}} \, cm^{-3}.$$

Comparing this with Fermi-Dirac number density Eq. (3.2), we find that the number of relativistic species at decoupling must satisfy $g_{*f} \leq 16 h_o^{-2} (g_f/2)^{3/2} (\frac{T_{\gamma o}}{2.7K})^3$, i.e. for $g_f \leq 2$ and $T_{\gamma o} = 2.7$K T_D must be $\lesssim O(100$ MeV). In other words, in the favored model where Ω is large ($\Omega \simeq 1$) and large-scale structure forms first, the universe is probably dominated by particles which behave in all ways like massive neutrinos.

Any candidate particle for the dominant matter decoupling at $T_D \lesssim 100$ MeV is constrained by Big Bang Nucleosynthesis arguments in exactly the same way as neutrinos. We have seen in Section II that it is very difficult to add to the three known neutrino species a full additional neutrino type (or equivalently $g_* = 7/4$ additional degrees of freedom, where one bosonic spin state contributes $g_* = 1$ and one fermionic spin state contributes $g_* = 7/8$) without violating limits on the observed [4]He abundance. Thus it is probable that the dominant matter is a neutrino and not some other "ino". Axions have been proposed for the dark matter,[13] but since they have a low Jeans mass they cannot be the dominant matter if the large voids require large scales forming first. These arguments would also prevent small (planetary mass) black holes from being the dominant matter unless they were able to stimulate Ostriker-Cowie[40] explosions[14].

The formation of structure in the universe requires the growth of perturbations from small amplitudes in the early universe to nonlinearity ($\delta\rho/\rho > 1$) by the present day. In the standard adiabatic picture of galaxy formation without massive neutrinos, the coupling of baryons to photons before recombination damps the perturbations on scales smaller than the Silk mass, $M_S \simeq 3 \times 10^{13}\Omega_B^{-\frac{1}{2}}\Omega_v^{-3/4}h_o^{-5/2}M_\odot$ [34,41], and prevents perturbations from growing on larger scales. After recombination at $1 + z_R \simeq 1000$ the surviving perturbations in an Einstein-de-Sitter universe grow as $(1 + z)^{-1}$, allowing a growth factor of only about 1000 between recombination and the present day. To reach nonlinearity by today, the perturbations must have been $(\delta\rho/\rho)_b \simeq 10^{-3}$ at recombination. The resulting microwave anisotropy $\delta T/T \simeq 1/3(\delta\rho/\rho)_b \simeq 1/3 \times 10^{-3}$ would be larger than the observed large-scale anisotropy, $\lesssim 2 \times 10^{-4}$. This dilemma is resolved in a universe dominated by massive neutrinos or other "inos". [41] After the neutrinos become the dominant matter at $1 + z_M \simeq 1900\ m_v$ (time t_M), neutrino perturbations grow as $(1 + z)^{-1}$ on scales larger than the maximum neutrino Jeans mass, $M \geq M_{vM}$. Baryon perturbations are tied to the photons until after the time of recombination (t_R), when they can rapidly catch up to the neutrino perturbations. This extra growth period for neutrino perturbations allows nonlinearity by the present epoch with $\delta_v(t_R) \simeq 10^{-3}$ and

$$\delta_b(t_R) \simeq \delta_v(t_R)(\frac{1 + z_R}{1 + z_M}) \simeq \frac{10^{-3}}{1.9m_v} \simeq 5 \times 10^{-4}m_v^{-1}. \qquad (4.5)$$

The microwave anisotropy, now only $\delta T/T \simeq 2 \times 10^{-4}m_v^{-1}$, no longer violates the observed limit as long as $m_v \gtrsim 1$ eV.

Before leaving the adiabatic mode, it is worthwhile to remember that GUTs naturally allow adiabatic perturbations but not isothermal ones[42]. All attempts at producing isothermals in the context of GUTs have been very ad hoc and unnatural[43]. In fact inflationary scenarios seem naturally to give rise to adiabatic perturbations[44] with a Zel'dovich spectrum, i.e. fluctuations come within the horizon at constant although model-dependent amplitude.

HIERARCHICAL CLUSTERING

The alternative to condensing large scales first is to build up from small scales. We discuss this because the adiabatic pancake scheme of the previous section does not by itself lead to the small scale structure. Such structure requires subsequent cooling and fragmentation of the baryons while leaving the non-interacting neutrinos on the large scale where they produce the large M/L. These cooling and fragmentation processes are not well understood,

so many ad hoc assumptions are required. On the other hand, isothermal perturbations naturally produce these small scales down
all the way to the scale of globular clusters. The hierarchical
picture, whether produced by isothermals or by adiabatics with a
small "ino" Jeans mass so that small scales form first, does
correctly fit the 2 and 3 point galaxy correlation functions[15]
up to cluster scales of $\sim$ 5 Mpc. To fit this in the simple pancake picture (with no cooling or fragmentation assumptions) requires a fine tuning of the parameters or at least a very small
Jeans mass that is not needed in a hierarchical scheme. If neu-
trionos make the large scales first then galaxies might not form
until redshifts z < 1, but quasars are seen with z $\sim$ 3.5. In
addition, the equilibration time scale can be determined from the
dynamics of galaxies on various scales and it is found that the
largest scales are not in dynamical equilibrium yet, whereas small
scales are. For example, the core of the Virgo Cluster as well as
the Coma Cluster are well virialized, whereas the Virgo Supercluster is not. This might argue in favor of isothermals or at
least a hierarchical picture with small ($\lesssim 10^8$ M$_\odot$) Jeans mass as
given by axions, GeV mass photinos, or planetary mass black holes.
But as we've already seen GUTs argues against isothermals and the
large scale voids, superclusters, and $\Omega \sim 1$ argue against any
hierarchical picture on the largest scales (hierarchical models
produce constant M/L).

One might suggest a solution where the adiabatic picture holds
but there are two significant nonbaryonic particles, the dominant
one a 10 to 25 eV neutrino and the second a particle such as the
axion or massive gravitino which has a small Jeans mass and yields
the small scale structure. Such a model is certainly possible but
seems very ad hoc in that we require new particles to solve each
problem. Alternatively, some new hierarchical model may be developed with tidal stripping and a power spectrum which, contrary
to previous attempts, does produce the large scale structure and
enables $\Omega \sim 1$. We feel the most likely model is one in which the
adiabatic picture holds and cooling and fragmentation of the baryons is enhanced at the intersection of caustics[16]. In this case
galaxy formation occurs in filaments. Fry[46] has shown that the
galaxy correlation functions for filaments agrees with the observed
2 and 3 point galaxy correlation functions as well as the hierarchical case. To meet the time scale constraints one merely has to
argue that within the filaments distant regions will take longer
to equilibrate than central ones. Of course, a low density universe with Ω < 0.15 and only baryons is also a possibility.
Scenarios of mixed isothermal and adiabatic components are also
possible[47]; since the adiabatic mode grows more quickly, the
resulting large-scale structure is essentially that of the adiabatic picture.

VI. DISCUSSION AND SUMMARY

We have tried to be very explicit about the assumptions required for each cosmological "ino" mass argument, and then we synthesized the arguments to obtain some very powerful simultaneous constraints.

Table II - Cosmological Matter

	Light emitting matter (glowing baryons)	Dark Matter (no dissipation into galactic cores)			
		COLD[1]		WARM[2]	HOT[3]
		Baryons	Others		
Examples	stars	Jupiters or stellar mass black holes[4]	Axions, (10-100)GeV mass "inos", planetary mass black holes	10eV-400 eV[7] "inos"	10 eV - 25 eV[7] "inos" (if N_ν=3 then ν_e, ν_μ or ν_τ)
Maximum Jeans Mass	Irrelevant	Depends on fluctuation spectrum	$\lesssim M_\Theta$	$\sim 10^{12} M_\Theta$ - $10^{16} M_\Theta$	$\sim 10^{15} - 10^{16} M_\Theta$
Ω_i	~ 0.01	$\sim 0.01 - 0.14$	$\lesssim 0.6$	$\gtrsim 1$ [6]	$\gtrsim 1$ [6]
M/L	$1 - 10 h_o$	$\lesssim 150 h_o$	$\lesssim 800 h_o$[5]	increases with scale	
dark halos of dwarf spheroidals	--	O K		marginal on phase-space, unlikely from Jeans mass	NO
galaxy correlation function	observed	O K		requires special heating and cooling of baryons so mass anti-correlates with light	
large-scale filaments and voids	NO	no existing hierarchical model yields this structure		marginally natural	natural
$\frac{\delta T}{T} \lesssim 10^{-4}$ with adiabatic fluctuations	NO	NO	YES	YES	
galaxy formation epoch	Z $\sim$ 100	Z $\sim$ 100	Z $\sim (100 - 1)$	Z $\sim$ 10	Z $\sim$ 1

[1] "cold": decouples while non-relativistic.
[2] "warm": decouples while relativistic, present temperature $T_f < T_\nu \approx 2K$.
[3] "hot": decouples while relativistic, present temperature $T_f \approx T_\nu \approx 2K$.
[4] stellar black holes were baryons at Big Bang Nucleosynthesis.
[5] it is assumed that low Jeans mass will result in non-dissipative clustering with baryons; no self-consistent calculation shows otherwise.
[6] since these don't fit small scales well, prime motivation is large scales and high Ω's.
[7] assumes $t_u > 13$ Gyr or equivalently $\Omega h_o^2 < 0.25$, i.e. can have $\Omega = 1$ only for $h_o = 1/2$.

Table II gives a summary of the arguments for different possible constituents of the universe. A universe with $\Omega \gtrsim 0.15$ must probably be dominated by non-baryonic matter, although the observational evidence pointing in this direction is not without loopholes. The galaxy formation mode compatible with GUT s, namely the adiabatic mode, produces disagreement with the 3K background unless the universe is dominated by non-baryonic matter. We also mentioned that Big Bang Nucleosynthesis constrains the number of neutrinos to at most 4, probably only 3.

Merely from limits on the total mass density of the universe, neutrino masses are restricted to two ranges, $\sum_i m_{\nu_i} \lesssim 400$ eV or $m_{\nu_i} \gtrsim 6$ GeV (corresponding arguments for other "inos" depend on their decoupling temperatures). However, applying an additional age constraint at $t_u > 8.7$ Gyr, we find a stricter limit $\sum_i m_{\nu_i} \leq 100$ eV ($\sum_{ino} m_{ino} \lesssim 2$ keV) while the range $m_{\nu_i} \gtrsim 6$ GeV remains unchanged. If we further restrict the age by requiring consistency between globular cluster ages, observed helium abundances and Big Bang Nucleosynthesis, then $t_u > 13$ Gyr and $\sum_i m_{\nu_i} \leq 25$ eV ($\sum_{ino} m_{ino} \lesssim 400$ eV) (remember age arguments assume $\Lambda = 0$).

We mentioned that the Tremaine and Gunn[8] phase space argument gives a necessary (but not sufficient) limit of $m_\nu \gtrsim 3$ eV if neutrinos are to cluster on cosmologically significant scales. In the GUT s favored adiabatic scenario, we showed that a maximum neutrino Jeans mass small enough to allow the formation of superclusters requires $m_\nu > 10$ eV. For three neutrino species of equal mass, the limit becomes more restrictive, $m_{\nu_i} > 16$ eV ($\sum_i m_{\nu_i} > 48$ eV).

Hence three equal neutrino masses are allowed only for a universe younger than 12 Byr, an age excluded by the best fit age arguments. Based on the conclusions of Frenk, White, and Davis[15] that the formation of large-scale structure required small-scale damping, a potentially far-reaching argument was presented that the mass of the dominant particle be less than ~ 200 eV. In a high density universe of $\Omega \simeq 1$, with $T_{\gamma o} = 2.7$K, a particle with this mass and $g_f \leq 2$ decouples at $T_D \lesssim 100$ MeV; in other words the dominant matter is similar to a massive neutrino. The nucleosynthetic constraint that there are probably only 3 neutrinos leads to the conclusion that not only does the dominant particle act like a massive neutrino but it probably is one. The mass for this best fit neutrino is $10 \lesssim m_\nu \lesssim 25$ eV. While the adiabatic picture with massive neutrinos successfully gives the large scale structure, some problems exist on smaller scales. Various speculative models are given which might solve the small scale problems while retaining a solution to the large scale. We feel the most promising is the scheme of Bond, Centrella, Szalay and Wilson[16], which investigates the cooling and fragmentation of pancakes into galaxies. It will be interesting to see whether subsequent work verifies this model.

In the most likely scenario the most massive neutrino would have a mass 10 eV $\lesssim m_\nu \lesssim$ 25 eV while the other neutrinos would have negligible masses. If the Lubimov et al.[5] result holds then we may have found the answer, massive ν_e with m_{ν_μ} and $m_{\nu_\tau} \lesssim$ 3 eV. Although one might naïvely expect the $\nu_\tau\mu$ to be the most massive, because these are weak rather than mass eigenstates, depending on the mixing matrix any combination of leptons might be involved in the most massive eigenstate. Ushida et al.[64] in an experiment looking for $\nu_\mu - \nu_\tau$ oscillations sensitive to mixing angles $\sin^2 2\alpha \gtrsim 0.013$ found a limit $|m_{\nu_\mu}{}^2 - m_{\nu_\tau}{}^2| < 3.0$ eV^2 (90% confidence level). This suggests that either the mixing angle is very small, that limits on the age of the universe from globular clusters must be reevaluated (although all current possible corrections go towards longer, not shorter ages), or that there is some form of dark matter other than the three known neutrino species. Unfortunately, if the mass eigenstates is close to ν_τ and the mixing angle is small, the experimental detection of the dominant matter in the universe may take a long time.

We acknowledge useful discussions with Dick Bond, Marc Davis, Jim Fry, Joel Primack, Martin Rees, Gary Steigman, Alex Szalay, and Michael Turner, especially during the 1982 Astrophysics Workshop at the Aspen Center for Physics. This work was supported in part by DOE and NSF at the University of Chicago.

REFERENCES

1) R. Cowsik and J. McClelland, Phys. Rev. Lett. 29 (1972) pp. 669; G. Marx and A. S. Szalay, in Proc. Neutrino '72, Technoinform, Budapest 1 (1972) 123; A. S. Szalay and G. Marx, Acta. Phys. Hungary 35 (1974) pp. 113.
2) S. M. Faber and J. S. Gallagher, Ann. Rev. Astron. Astrophys. 17 (1979) pp. 135.
3) D. N. Schramm and R. V. Wagoner, Ann. Rev. Nucl. Sci. 27 (1977) pp. 37; J. Yang, M. S. Turner, G. Steigman, D. N. Schramm and K. A. Olive, in preparation (1983).
4) D. N. Schramm and G. Steigman, Ap. J. 243 (1981) pp. 1.
5) V. A. Lubimov, E. G. Novikov, V. Z. Nozik, E. F. Tretyakov, and V. S. Kosik, Phys. Lett. 94B (1980) pp. 266.
6) F. Reines, H. W. Sobel, and E. Pasierb, Phys. Rev. Lett. 45 (1980) pp. 1307.
7) P. Langacker, Phys. Rep. 72C (1981) pp. 185.
8) S. Tremaine and J. E. Gunn, Phys. Rev. Lett. 42 (1979) pp. 407.

9)   G. Steigman, Proceedings Europhysics Study Conference
     "Unification of the Fundamental Interactions" eds. J. Ellis,
     S. Ferrara, and P. Van Nieuwenhuizen (Plenum, NY, 1981) pp.
     495; G. Steigman, Proc. Neutrino '81, eds. R. J. Cence, E.
     Ma, and A. Roberts (University of Hawaii, 1981); P. S. Joshi
     and S. M. Chitre, Nature 293 (1981 pp. 679; G. Steigman, to
     appear in Proceedings of ICOBAN (international Colloquium on
     Baryon Nonconservation, 1982).

10) E. M. D. Symbalisty and D. N. Schramm, Rep. Prog. Phys. 44
     (1981) pp. 293.

11) D. N. Schramm, Phil. Trans. R. Soc. Lond. A307 (1982) pp. 43.

34) J. R. Bond, G. Efstathiou, and J. Silk, Phys. Rev. Lett 45
     (1980) pp. 1980.

38) R. P. Kirshner, A. Oemler, P. L. Schechter, and S. A. Schectman,
     Ap. J. 248 (1981) pp. 57.

12) G. R. Blumenthal, H. Pagels, and J. R. Primack, Nature 299
     (1982) pp. 37.

35) J. R. Bond, A. S. Szalay, and M. S. Turner, Phys. Rev. Lett
     48 (1982) pp. 1636.

13) M. S. Turner, F. Wilczek, and A. Zee, Univ. of Chicago pre-
     print No. 83-11; J. Preskill, M. B. Wise, and F. Wilczek,
     Phys. Lett. B120 (1983) pp. 127.

14) B. J. Carr, Astron. Astrophys. 56 (1977) pp. 377; M. Crawford
     and D. N. Schramm, Nature 298 (1982) pp. 538; K. Freese,
     R. Price, and D. N. Schramm, University of Chicago preprint
     No. 83-06.

15) P. J. E. Peebles, Ap. J. 258 (1982) pp. 415; C. S. Frenk,
     S. D. M. White, and M. Davis, preprint (1983).

16) J. R. Bond, J. Centrella, A. S. Szalay, and J. R. Wilson,
     "Cooling Pancakes" preprint (1983).

17) G. A. Tammann, A. Sandage, and A. Yahil, in Les Houches 1979,
     ed R. Balian, J. Audouze and D. N. Schramm (North-Holland
     Publishing Co., Amsterdam, 1979) pp. 53.

18) W. Press and M. Davis, Ap. J. 259 (1982) pp. 449.

19) R. P. Kirshner, A. Oemler, and P. L. Schechter, Astron. Jour.
     84 (1979) pp. 951.

20) M. Davis, J. Tonry, J. Huchra, and D. W. Latham, Ap. J. 238
     (1980) pp. 113.

21) J. R. Gott and E. L. Turner, Ap. J. 213 (1977) pp. 309.

22) B. E. J. Pagel, Phil. Trans. R. Soc. Lond. A307 (1982); pp. 19;
     D. Kunth and M. Sargent, preprint (1983); D. Kunth, in Pro-
     ceedings of the First Moriond Astrophysics Meeting: Cosmology
     and Particles, ed. J. Audouze, P. Crane, T. Gaisser, D. Hegyi,
     and J. Tran Thanh Van (1981) pp. 241.

23) W. Bacino et al., Phys. Rev. Lett. 42 (1979) pp. 749.

24) K. A. Olive, D. N. Slchramm, and G. Steigman, Nucl. Phys.
     B180 (1981) pp. 497.

25) J. R. Gott, J. E. Gunn, D. N. Schramm and B. M. Tinsley,
     Ap. J. 194 (1974) pp. 543.

26)   A. H. Guth, Asymptotic Realms of Physics: A Festschrift in
      Honor of Francis Low, ed. A. H. Guth, K. Thrang, and R. L.
      Jaffe (MIT Press, Cambridge, Mass., 1983).

27)   S. Weinberg, Gravitation and Cosmology (Wiley, NY, 1972).

28)   B. W. Lee and S. Weinberg, Phys. Rev. Lett. 39 (1977) pp. 165;
      D. A. Dicus, E. W. Kolb, and V. L. Teplitz, Ap. J. 221 (1978)
      pp. 327.

29)   L. M. Krauss, Harvard University preprint No. 83/A009.

30)   H. Goldberg, Northeastern University preprint No. 2592.

31)   J. E. Gunn, B. W. Lee, I. Lerche, D. N. Schramm, and G.
      Steigman, Ap. J. 223 (1978) pp. 1015.

32)   K. Freese, D. N. Schramm, submitted to Nucl. Phys. B.(1983).

41)   J. Silk, In Confrontation of Cosmological Theories With
      Observational Data, eds. M. S. Longair (Reidel, Holland,
      1974) pp. 175.

36)   J. R. Bond and A. S. Szalay, Proc. Neutrino '81, eds. R. J.
      Cence, E. Ma, and A. Roberts (University of Hawaii, 1981);
      G. Steigman, Inst. Phys. Conf. Ser. 64 (1982) pp. 65.

37)   J. H. Oort, Astron. Astrophys. 94 (1981) pp. 359.

40)   J. P. Ostriker and L. L. Cowie, Ap. J. 243 (1980) pp. 127.

41)   A. G. Doroshkevich, M. Yu. Khlopov, R. A. Sunyaev, A. S.
      Szalay, and Ya. B. Zel'dovich, Tenth Texas Symposium on
      Relativistic Astrophysics (1981) pp. 32.

42)   M. S. Turner and D. N. Schramm, Nature 279 (1979) pp. 303.

43)   J. D. Barrow and M. S. Turner, Nature 291 (1981) pp. 469;
      J. R. Bond, E. W. Kolb, and J. Silk, Ap. J. 255 (1982) pp.
      341.

44)   A. H. Guth and S. Yi. Pi, Phys. Rev. Lett. 49 (1982) pp. 1110.
      J. M. Bardeen, P. J. Steinhardt, and M. S. Turner, University
      of Pennsylvania preprint No. 2020T (1982).

45)   J. Ellis, D. V. Nanopoulos, K. A. Olive, Phys. Lett 120B
      (1983) pp. 331; S. W. Hawking, and I. G. Moss, "Fluctuations
      in the Inflationary Universe", Cambridge University preprint
      (1982).

46)   J. N. Fry, University of Chicago preprint No. 83-01.

47)   H. Sato and F. Takahara, Prog. Teor. Phys. 66 (1981) pp. 508.

64)   N. Ushida et al., Phys. Rev. Lett. 47 (1981) pp. 1694.

39)   G. Steigman, in Proceedings of the Second Moriond Astrophysics
      Meeting, ed. J. Audouze et al. (1982).

33)   K. A. Olive and M. S. Turner, Phys. Rev. D. 25 (1982) pp. 213.

BLACK HALOS AND DWARF GALAXIES

Colin Norman

Sterrewacht, Leiden and Institute of Astronomy,
Cambridge

Joseph Silk

Institute d'Astrophysique, Paris
and Department of Astronomy, University of California,
Berkeley

INTRODUCTION

Dissipation of infalling gas into a preexisting dark halo of non-dissipative matter provides an attractive scenario for the formation of disk galaxies. Tidal torquing of gas against the gravitationally dominant dark halo can account for the observed angular momentum (Fall and Efstathiou, 1980), if halos form by hierarchical clustering in the expanding universe (c.f. White and Rees, 1978). Moreover recent observations demonstrate that galactic bulges and the less luminous ellipticals are rotationally supported (Davies et al. 1983), suggesting that this model may have more widespread application.

Given these and other attractive features of dissipation in a dark halo, two possibilities come to mind for forming such dark halos. White and Rees (1978) proposed that the first bound systems consisted of low luminosity remnants of Population III and represented the dominant mass constituents of the universe. The dark remnants clustered hierarchically as the universe expanded to provide dark potential wells on galactic and subgalactic scales into which dissipative infall of gaseous matter eventually gave rise to the observed galaxies. An attractive feature of this scheme was the natural emergence of the luminous mass-scale of a typical bright galaxy (Rees and Ostriker 1977, Silk 1977).

J. Audouze and J. Tran Thanh Van (eds.),
Formation and Evolution of Galaxies and Large Structures in the Universe, 215–225.

The most implausible aspect of this scheme is the formation of sufficiently dark potential wells from the initial generation of Population III star formation. Current ideas on primordial star formation suggest that stars of mass $\gtrsim 0.1$ M_0 should form in considerable numbers (Kashlinsky and Rees 1983, Silk 1983) and would violate observational constraints if present in sufficient numbers to dominate the mass density of the universe. A further difficulty has arisen with the realisation that somewhat contrived initial conditions may be required to yield a hierarchical clustering scheme for galaxy formation based on primordial fluctuations in the context of baryonic synthesis models (Bond et al. 1982), since these yield a universal value of the specific entropy. Adiabatic fluctuations, on the other hand, may arise spontaneously in certain inflationary models of the very early universe; in any event they are preserved during baryon synthesis. Moreover, the evolution of primordial adiabatic fluctuations which possess a large initial coherence length apparently do give a more satisfactory fit to the large scale irregularities observed in the galaxy distribution than the isothermal fluctuation theory (Klyphin and Shandarin 1983, Frenk et al. 1983), although a late galaxy formation epoch ($z \sim 1$) appears to be mandatory. Finally, non baryonic forms of matter such as massive neutrinos (Marx and Szalay 1972) and gravitinos (Pagels and Primak 1982) have emerged as interesting candidates for the dark matter. Initially motivated by tentative indications of a finite neutrino rest mass (Lyubimov et al. 1980; Reines et al. 1980) this idea has been reinforced by the growing realisation of elementary particle physicists that the existence of exotic particles predicted by theory should have cosmological implications that can often give the most significant bounds on their properties. A massive neutrino (and to a considerable extent, gravitino) dominated universe leads naturally to anisotropic pancake collapse models, with the first bound structures forming on very large scales associated with superclusters or clusters of galaxies, and ensuing fragmentation resulting in galaxy formation (Doroshkevich et al. 1981).

This type of scheme has met with sufficient success in reproducing the observed galaxy distribution that we reexamine here the issue of dark halo formation. One of the most emphasised aspects of the pancake theory of galaxy formation is that the first protogalactic fragments form in a highly compressed baryonic pancake (Sunyaev and Zel'dovich 1972). Dark halos develop by slow infall of massive neutrinos that have not been heated during the initial collapse with numerical estimates suggesting that halo trapping efficiencies of up to 10 percent of the neutrino matter are attainable (Bond et al. 1983, Melott 1983). Collisionless damping and phase mixing erases all pre existing structure in the neutrino distribution with scales below the maximum Jeans length $\sim 10^{15} (m_\nu/30\text{eV})^{-2} M_0$. In the absence of fluctuations on galactic scales, dynamical fragmentation of the infalling neutrinos does not occur prior to the the fragmentation of the gaseous layer. It would seem then that pancake theory does not provide any possibility for dark halos to form in the absence of a pre-existing baryonic core.

We shall show that this conclusion may be premature. In particular, black halos of scale $\gtrsim 10^9$-$10^{10} M_0$, consisting initially only of massive neutrinos can form as an aftermath of pancaking and provide potential wells into which dissipative infall can subsequently occur.

In what follows, we first discuss the efficiency of massive neutrino accretion by baryonic cores (II). We then show how neutrino halos may form prior to baryonic infall (III), and finally review the implications of our results (IV).

II ACCRETION BY BARYONIC CORES

A large number of one-dimensional simulations of the Vlasov equation were performed more than a decade ago in order to study violent relaxation (c.f. Lecar 1972). These simulations demonstrated that the asymptotic energy distribution attained a core-halo structure that was more pronounced than that predicted by Lynden-Bell's (1967) theory of violent relaxation. A spiral structure generally developed in phase space, and only after several crossing times had elapsed, did the phase space density become significantly diluted, resulting in significant phase mixing. A similar effect has been noted in the development of the one-dimensional beam plasma instability; the formal analogy with the gravitational system has been discussed by Berk and Roberts (1970). Of particular interest are the simulations of Janin (1971), who studied Jeans instability in a one-dimensional collapse and who found that an initially smooth system was stable to fragmentation, even though it was formally Jeans unstable. Only when initial fluctuations were introduced did the system in fact, become Jeans unstable.

A natural source for fluctuations in the adiabatic collapse model of galaxy formation is pancake fragmentation. The dissipative component naturally develops structure that can drive instability in the dissipationless components as discussed for example by Ikeuchi et al. (1972). This effect has been found to occur in one-dimensional simulations of neutrino streaming in the presence of discrete fragments (Bond et al. 1983). Similar effects occur in two-dimensional simulations (Melott 1982). However these calculations fail to incorporate the growing anisotropy of the neutrino velocity distribution which provides a strong source of heating in the collapse direction. It is this heating which is largely responsible for limiting accretion and not the multiple streams.

The importance of anisotropic heating for the neutrinos has recently been emphasized by Zel'dovich and Shandarin (1982) who note that it limits the maximum neutrino density. Here we develop a simple analytic model which takes account of anisotropic heating and allow for one dimensional mixing of the neutrino streams. Since phase space density is incompressible and the structure in phase space is generically a two-armed spiral, phase mixing will occur rapidly when adjacent spirals touch. Assuming equally spaced spirals with constant pitch angle and constant width, corresponding to velocity width Δv on the x axis, then after n turns, where $n \sim V_o/\Delta V$ and V_o is the free fall velocity of the pancake, thermalisation will go rapidly to completion. This gives a time scale for thermalisation of order n collapse times $\sim n\, t_o$, where t_o is the collapse time $\sim (G\rho)^{\frac{1}{2}}$ and $\bar{\rho}$ is the mean density.

Parameterizing the heating by

$$\sigma_v = \sigma_o \left(\frac{t}{t_o}\right)^{1/n},$$

so that $\frac{d\ln\sigma}{d\ln t} \sim \frac{1}{n}$ corresponding to the above thermalisation timescale estimate, with σ_v the neutrino velocity dispersion at time t, where subscript 0 denotes initial values. A baryonic fragment initially of mass M_i accretes, after time t, a mass $M(t)$ of neutrinos given by

$$\frac{dM}{dt} = \frac{M^2 \tau^{-3/n}}{M_{J\nu}},$$

where $M_{J\nu} = \rho(\sigma_o t_o)^3 \xi$, $\xi = 4\pi G^2 \rho^2 t_o^4$ and $\tau = t/t_o$. Thus

$$M = \frac{M_i}{1 - \frac{M_i}{M_{J\nu}} \left(\frac{n}{n-3}\right) [\tau^{1-3}-1]},$$

and significant accretion occurs on a timescale

$$\left(1 + \left(\frac{n-3}{n}\right) \frac{M_{J\nu}}{M_i}\right)^{\frac{n}{n-3}} t_o \quad .$$

Evidently, mass scales $M_i > M_{J\nu}$ will dominate the accretion after an initial crossing time t_o has elapsed.

We assume that the infalling neutrinos form an extended isothermal structure with velocity dispersion of order that of the central baryonic core. In this case, accretion will no longer occur once the neutrino velocity dispersion rises above that of the core. This cut-off occurs at time $\sim t_o \left(\frac{\sigma_{core}}{\sigma_o}\right)^n$, and sets a lower limit to the accreting core mass (for $n > 3$) of $M_B > M_{J\nu}\left(\frac{n-3}{n}\right)\left(\frac{\sigma}{\sigma_{core}}\right)^{n-3}$. Therefore $M_{J\nu}$ emerges as a natural baryonic mass scale. Baryonic fragments can grow by coalescence from smaller scales produced in the initial fragmentation, and we infer that $M_{J\nu}$ should represent a characteristic baryonic mass. The amount of dark halo matter accreted by such fragments is determined by the competition for the overall neutrino reservoir. Smaller baryonic fragments are therefore unlikely to acquire dark halos. At a pancake epoch $z_p = 10$, $M_{Jv} \sim 3 \times 10^{10}(1 + Z_p/10)$ M_0, and corresponds to the baryonic matter that constitute the luminous cores of galaxies of characteristic luminosity. This is in reasonable agreement with the characteristic luminosity as determined by the field or cluster luminosity function, $L_* - 3 \times 10^{10} L_0$ provided $1 \lesssim Z_p \lesssim 5$.

III BLACK POTENTIAL WELLS

Baryonic fragments are capable of driving large fluctuations in the neutrino density as we now demonstrate. Consider a pancake after baryonic fragments have formed and multiple cold neutrino streams are still present. The neutrino density enhancement produced by a baryonic fragment due to gravitational focussing of the wake of a cold neutrino stream is considerable. The exact calculation of the density profile on the symmetry axis behind a baryonic fragment is given by (Danby and Camm, 1967)

$$\frac{\rho}{\rho_0} = \left(\frac{\pi}{2} \right)^{\frac{1}{2}} \left(\frac{2GM_B}{\sigma_v^2 a} \right)^{\frac{1}{2}}$$

for $\frac{V}{q} \gg 1$, where M_B is the mass of a baryonic fragment, ρ_q is the initial density in a cold neutrino fluid streaming past a baryonic fragment with velocity V and mean velocity dispersion σ_v. Here a is the distance from M_B along the symmetry axis. Estimates of the initial baryonic fragment scale lie in the range $10^8 - 10^{10}$ M_0. The unstable baryonic layer is ram-pressure compressed to thickness of several kpc. The neutrinos are assumed to be not yet heated by the processes described in section II and their velocity dispersion $\sigma_v = 6(1 + Z_p)$ kms^{-1} where Z_p is the redshift at which pancaking occurs.

Analytic simplification of this problem appears to be very difficult (Danby and Bray, 1967). We crudely estimate the mass in the wake by assuming it has the density profile given above and is distributed in a cone of apex angle $\sim$ (Mach Number)$^{-1}$ $\sim \sigma_v/V$ of characteristic size $a_w \sim GM_B/\sigma^2$ inside of which $\frac{\delta \rho}{\rho} \sim 1$. The mass of the wake is then $M_w \sim \pi (\sigma_0/V_0)^2 \rho_0 a_w^3$. The initial Jeans mass of neutrinos $M_{JV} \sim \rho_0 \lambda_{Jv}^3$ where λ_{Jv} is the initial Jeans length. We write

$$\frac{M_{wake}}{M_{Jv}} \sim (\sigma_0/V_0)^{2/3} \left(\frac{M_B}{M_{Jv}} \right)^3$$

Therefore the wake with over density $\delta \rho \gtrsim \rho$ can be gravitationally unstable with a mass greater than the Jeans mass if

$$\left(\frac{M_B}{M_{Jv}} \right) \gtrsim \left(\frac{V_0}{\sigma_0} \right)^{2/3} = 10 \left(\frac{V_0/\sigma_v}{10^{1.5}} \right)^{2/3}$$

An accurate numerical integration gives a wake mass inside the $\delta\rho/\rho = 10$ contour that is greater than $M_{J\nu}$ if $M_B \gtrsim M_{J\nu}$ for the case $V_o/\sigma_\nu = 0.1$ (Mulder 1983). Since $M_{J\nu} \sim 10^{10} (1+Zp/10)^{2/3} M_0$ the above condition should be satisfied by some of the baryonic fragments that accumulate in the central layer. Incorporation of the detailed velocity and density profile of a pancake can reduce $M_{J\nu}$ by an order of magnitude of two. Black potential wells roughly of order the initial Jeans mass can form in the wake. These bound neutrino clouds can be convected with the flow and thus be distributed throughout the pancake structure.

Note that the condition that the background Hubble flow in the plane of the pancake can be neglected is obtained from using the impulsive approximation to be $\frac{\Delta v}{H_o a} \sim \frac{\rho_B a}{\rho_p R} > 1$ where $\rho_p R$ is the pancake column density and p_B is the mean density of a baryonic fragment. For such an increased baryonic column density there must have been substantial dissipation and three dimensional collapse of the baryonic fragment.

Consider now the likely evolution of baryonic matter that dissipates and is accreted by these black potential wells. We expect that star formation will only occur once the baryonic density exceeds the neutrino density. The apparent core size of the resulting stellar system can be much less than the core of neutrino cloud. The individual stars are effectively test particles in a nearly constant density core and exhibit harmonic oscillator motion. Straightforward application of the equations of spherical, isotropic, stellar hydrodynamics for the non-self-gravitating material with an isothermal equation of state yields the temperature distribution for a given density profile.

Let us suppose that dwarf spheroidal and elliptical galaxies are to be identified with the result of dissipative baryonic infall into these dark potential wells. We do not restrict our attention to massive neutrinos but consider gravitinos as well as other collisionless particles as candidates for the role of dark matter. The peculiar role of massive neutrinos in dwarf galaxies is discussed in the following section. Now, the light profiles of dwarf ellipticals can be fitted by a tidally- truncated Hubble profile. An equally good fit to the light distribution is given by an exponential profile (Faber and Lin 1983) and we have derived a radial temperature dependence in both cases. For the density profile $\rho_B \sim e^{-\alpha r}$ we find $\sigma_b^2 \alpha 1 + \alpha r$ and for the analytic King profile $\rho_B \sim (1+\beta^2 r^2)^{-3/2}$ we find $\sigma_b^2 \alpha 1+\beta^2 r^2$ where β^{-1} is the core-radius. We interpret this temperature rise with increasing radius as an indication of the role of the dissipation that occured when the stars formed in the dark potential well.

The resulting stellar system should resemble dwarf galaxies but embedded in dark potential wells. Our predicted velocity dispersion increase with radius is testable if enough velocity dispersion data for individual stars is available. Current data on one dwarf spheroidal do not appear to be inconsistent with this prediction (Aaronson 1983).

Another prediction of our model comes from consideration of the rotational velocity of gas within a neutrino halo. The rotation curves will always resemble solid body rotation within a uniform density neutrino core. Indeed Tully et al. (1978) find that several dwarf irregular galaxies display solid body rotation in HI emission out to the edge of the light distribution. This is generally the case for all observations of dwarf galaxy rotation curves we are aware of at present.

IV IMPLICATIONS FOR COSMINOS

Can the masses of dwarf galaxies indeed be explained by dark halos of massive cosminos, such as neutrinos or gravitinos? The mass-to-light ratios of seven spheroidal dwarfs were estimated on the basis of tidal radii to be ~ 30, with at least one high velocity dwarf exceeding 100 (Faber and Lin 1983). Measurement of a lower bound to the velocity dispersion in Draco of 6.5 km s^{-1} obtained from three carbon stars led Aaronson to directly infer that $M/L \simeq 30$ for Draco.

However, these authors noted the following serious objection to accounting for the dark halos with massive neutrinos. According to an argument given by Tremaine and Gunn (1979), the coarse-grained phase space density of the neutrino distribution cannot increase during dynamical processes associated with galaxy formation. Knowing the maximum phase space density, one can then infer a lower bound on the neutrino mass for specified core parameters, notably radius and velocity dispersion. Using the observed values, a lower bound on m_υ was found to be ~ 600 eV in the case of Draco (Lin and Faber 1983). Similar conclusions applied to the other dwarfs.

Our model offers a possible means of circumventing this constraint, since the baryons are test-particles within a much larger neutrino halo. If both baryons and neutrino cores have isothermal structure, $\sigma_\upsilon/r_\upsilon = \sigma_b/r_b$, the phase-space constraint can be rewritten

$$m_\upsilon > 500 \text{ eV } (r_b/r_\upsilon)^{3/4} (10\text{km s}^{-1}/\sigma_b)^{1/4} (100 \text{ pc}/r_b)^{1/2} \ .$$

Evidentally choosing $r_\upsilon \simeq 20 \ r_b$ would enable us to lower the neutrino mass constraint to a plausible value, 50 eV (corresponding to $\Omega \sim 0.5$). However such a large neutrino core dramatically increases the inferred total mass of the dwarfs. The mass-to-light ratios inferred by Faber and Lin (1983) are increased by a factor $(r_\nu/r_b)^3 \sim 8000$ in order to allow a neutrino mass of ~ 50 eV.

The total mass of Draco is now $\sim 5 \times 10^{10}$ M_0; its mass-to-light ratio is $\sim 2 \times 10^5$! In studying Draco-type spheroidal dwarfs, we may only be directly seeing the trace of a vastly more massive black cloud. One might expect that dynamical friction considerations would be fatal to this model. In fact, naive application of the Chandrasekhar formula for the motion of Draco in the massive halo $(\sim 10^{12} M_0)$ of our Galaxy yields an orbital decay time of $t_D \sim \frac{M_{galaxy}}{M_{Draco}} t_G$ where t_G is the crossing time for an orbit of Draco giving $t_0 \sim 10^{10}$ $(M_{draco}/5 \times 10^{10} M_0)$ yr. Indeed the apparent longevity of close binary pairs of galaxies when confronted with a similar constraint, suggests that a common dark halo may develop (Sharp and White 1977). Such may also be the case here, of course. More sophisticated treatment of the dynamical interaction including finite size effects is evidently required to establish a more definitive limit on the mass of Draco-type dwarfs incorporating finite galaxy size.

Such a dramatic conclusion depends on our faith in massive neutrinos. Gravitinos of mass $\leqslant 1$ kcV provide a dynamically similar model for large-scale structure which develops at a late epoch by pancake-like collapse. In this case our pancake fragmentation allows the formation of dark potential wells down t$_0$ $\sim 10^9 M_0$. This would still lead to extreme M/L values $(\sim 10^3 - 10^4)$ for dwarfs. In general, the mass-to-light ratio of dwarfs is toosted by a factor$(500$ eV$/m_\nu)^4$, where m_ν can be taken to be one's favourite candidate for massive particles that were once in thermal equilibrium with the background radiation.

Fragmentation of the dark matter may also be possible on much larger scales than those of dwarf galaxies. One possible mechanism is the two-stream instability as in the Jeans instability in the plane of the pancake. For homogeneous systems both these instabilities will occur. The two-stream instability will be generated when the cold neutrinos stream through a background gas with a velocity that is supersonic with respect to the sound speed in the gas (Ikeuchi et al. 1974, Sweet, 1962, Lynden-Bell, 1967). As we have argued here, the velocity dispersions during pancaking will become highly anisotropic (see also Zel'dovich and Shandarin, 1982) and for the homogeneous problem, instability will arise in the plane on a scale of order the Jeans length in the plane. This is also true for the self-gravitating slab (Kulsrud and Mark, 1970). The problem we face here is that because the pancake is a very highly inhomogeneous, time-variable system with large scale flow both in (Hubble flow) and normal to the plane, we have a classical fragmentation problem. The result is intuitively $\delta\rho/\rho \sim \rho^n$ with n positive and of order unity and if fluctuation exist then fragmentation and collapse will occur at $\delta\rho/\rho \sim 1$. It is not certain that these instabilities could grow sufficiently rapidly to generate $\delta\rho/\rho \sim 1$ perturbation from noise. However, if they do, then much more massive black halos are possible and we shall briefly consider this possibility.

One other example of a galaxy with an exceptionally high mass-to-light ratio may be the intergalactic cloud in Leo recently discovered by Schneider et al. (1983). The gravitational mass of this system is inferred to be $\sim 10^{11} M_0$ and failure to detect any underlying galaxy (Kibblewhite, private communication) implies $M/L_B \gtrsim 300$.

In summary, we have found that the Jeans mass for the dominant component of collisionless particles at the onset of pancaking determined the lower limit to baryonic core masses which can subsequently grow halos. For pancaking at redshift $Z_p \sim 5$ this corresponds to the core of a typical luminous galaxy and varies as $\sim (1+Z_p)^{2/3}$. Smaller cores that have accreted halos must have formed from more recently collapsed pancakes. These baryonic cores form prior to halo accretion, and form moreover near the mid plane of the pancake. We have also found that the fragmenting baryonic layer drives the formation of dark potential wells in the neutrinos. These potential wells can subsequently accrete gas, and considerations of dissipative infall led us to identify them with dwarf spheroidal galaxies.

The more massive examples of such dark clouds may also be observable as X-ray sources containing hot gas. Star formation would be inhibited until the accreted gas locally dominates the gravitational potential. We expect this to provide a natural threshold for star formation. These have been reports of X-ray enhancements in the galaxy cluster A1367 that are not associated with underlying galaxies (Bechtold et al. 1983). Dark halos of $\sim 10^{12} M_0$ suffice to locally bind the hot gas and produce X-ray coronae otherwise indistinguishable from those around luminous galaxies. This phenomenon would be a natural implication of our model.

The preexisting baryonic cores that accrete dark halos will be found predominantly in the densest filaments and sheets while the neutrino wells that accrete baryons will be more widely distributed, since the neutrino potential wells have dissipated little bulk kinetic energy. It is tempting to identify the preexisting baryonic cores with luminous ellipticals. Lack of rotational support is expected in the absence of an initial dark halo. The dark potential wells that accrete baryons provide an attractive site for forming disk galaxies, rotationally supported bulges, and low luminosity ellipticals, as well as the dwarf spheroidals that we have emphasised here. There are some intriguing indications of such a segregation on very large scales

(Giovanelli et al 1982, Oort 1983).

Finally we note that the main advantages of our model is that it permits one type of exotic particles (massive neutrinos, gravitinos, etc.) to account for dark matter on all scales, while retaining the pancake theory of large scale structure. Moreover, the observational implications are remarkable, Stellar systems with mass-luminosity ratios in excess of 10^3-10^4 are predicted by our model for dwarf galaxies.

It is a pleasure to thank J. Audouze for his excellent hospitality at Institut d'Astrophysique and La Plagne, where this work was completed. We specially thank W. Mulder for undertaking the numerical wake calculation with his dynamical friction code, and R. Bond, J. Oort, J. Peebles, M. Rees, B. Tully and S. White for perceptive and stimulating discussions.

REFERENCES

Bechtold, J., Forman, W., Giacconi, R., Jones, C., Schwarz, J., Tucker, W. and van Speybroeck, L., 1983, Ap.J. 265, 26.

Berk, H.L. and Roberts, K.V. 1970 in Methods in Computational Physics 9, 87. B. Adler, S. Fernback and M. Rotenbag (eds.) Academic Press, New York.

Bond, J.R., Centrella, J., Szalay, A. and Wilson, J. 1983, in preparation.

Bond, J.R., Kolb, E.W. and Silk, J. 1982, Ap. J. 255, 391.

Bond, J.R., Szalay, A.S. and White, S.D.M. 1983, Nature, 301, 584.

Davies, R.D., Efstathion, G., Fall, S.M., Illingworth, G. and Schechter, P. 1983, Ap. J. in press.

Doreshkevich, A.G., Zel'dovich, Ya.B., Sunyaev R.A. and Khlopov, M.Yu. 1981, Sov. Astron. Lett. 6, 252.

Efstathiou, G. and Jones, B. 1979, MNRAS 186, 133.

Faber, S. and Lin, D.N.C. 1983, Ap. J. 266, L17.

Fall, S.M. and Efstathiou, G. 1980, MNRAS 193, 189.

Forman, W. and Jones, C. 1982, Ann. Rev. Astron. Astrophys. 20, 549.

Frenk, C., White, S. and Davies, M. 1983 preprint.

Ikeuchi, S., Nakamura, T. and Takahara, F. 1974, Progress of Theoretical Physics 52, 1807.

Janin, G. 1971, Astron. Astrophys. 11, 188.

Kashlinsky, A. and Rees, M.I. 1983, MNRAS, (in press)

Klyphin, A.A. and Shandarin, S.F. 1983 preprint.

Kulstud, R.M. and Mark, J.W.K. 1970, AP. J. 160, 471.

Lecar, M. 1972. Gravitational N. Body Problem, AP and Space Sci. Library 31, 262.

Lyubimov, V.A., Novikov, E.G., Nozik, V.Z., Tretyakov, E.F. and Kozik, V.S. 1980 Phys. Lett. B 94, 266.

Lynden-Bell, D. 1967, MNRAS 136, 101.

Lynden-Bell, D., 1967, Relativity Theory and Astrophysics II, Ed.
    J. Ehlers (Am. Math. Soc.) p. 131.
Marx, G. and Szalay, A.S. 1972 in Proc. Neutrino '72,
    Technoinform, Budapest 1, 123.
Melott, A. 1982, Phys. Rev. Lett. 48, 894.
Melott, A.L. 1982, Nature 296, 721.
Mulder, W.A. 1983, Astron. Astrophys., 117, 9.
Pagels, H. and Primak, J. 1982, Phys. Rev. Lett. 18, 223.
Reines, F., Sobel, S.W. and Pasierb, E. 1980, Phys. Rev. lett. 45,
    1307.
Schneider, S.E., Helow, G., Salpeter, E.E. and Terzian, Y. 1983,
    Ap.J. (Letters), in press.
Silk, J. 1977, Ap. J. 211, 638.
Silk, J. 1983, MNRAS (in press)
Sunyaev, R.A. and Zel'dovich, Ya.B. 1972, Astr. Astrophysics 20,
    189.
Sweet, P.A. 1962, MNRAS, 125, 285.
Tremaine, S. and Gunn, J.E. 1979, Phys. Rev. Lett. 42, 407.
Tully, R.P., Bottinelli, L., Fisher, J.R., Gougenheim, L.,
    Sancisi, R. and van Woerden, H. 1978, Astron. Astrophys. 63,
    37.
White, S. and Recs, M.J. 1978 MNRAS, 179, 541.
White, S.D.M. and Sharp, N.A. 1977, Nature 269, 395.
Zel'dovich, Ya.B. 1970, Astr. Astrophys. 5, 84.
Zel'dovich, Ya.B. and Shandarin, S. 1982, Sov. Astron. Lett. 8,
    139.

MASS-TO-LIGHT RATIOS OF SPIRAL GALAXIES

J. Patricia Vader

Yale University

ABSTRACT. Ratios of mass-to-infrared luminosity are presented for 34 spiral galaxies and compared to model predictions. The observed values exceed the theoretical values by increasingly larger amounts from red to blue galaxies, a behaviour similar to that found by Tinsley for the ratio of mass-to-blue luminosity. This result supports the hypothesis that bluer spiral galaxies have relatively more massive dark haloes.

1. INTRODUCTION

By comparing the observed M/L_B ratios of spiral galaxies to those predicted by stellar population models Tinsley (1981) concluded that blue galaxies have a larger mass ratio of dark to ordinary stellar matter by a factor between two and four as compared to red galaxies. Infrared magnitudes in the H-band at 1.6 μm have been obtained for nearby spiral galaxies by Aaronson et al. (1982). Since the H-band samples a stellar population which is different from that dominating the blue band, and is much less sensitive to the star formation history of the galaxy, it is worthwhile to investigate whether the M/L_H ratios confirm the results inferred from the M/L_B ratios.

2. THE DATA

In Table 1 we present data for 31 spiral galaxies from the samples of Faber and Gallagher (1979) and Burstein et al. (1982) for which H-luminosities are available (Aaronson et al., 1982). To this list we have added NGC 253, NGC 3992, and NGC

J. Audouze and J. Tran Thanh Van (eds.),
Formation and Evolution of Galaxies and Large Structures in the Universe, 227–236.
© 1984 by D. Reidel Publishing Company.

7793 for which rotation curves have recently been obtained by
Pence (1981), Gottesman and Hunter (1982), and Davoust and de
Vaucouleurs (1980), respectively. From the data in Table 1 we
derive (i) a color index $B_T-H_{-0.5}$, with B_T the asymptotic blue
magnitude and $H_{-0.5}$ the infrared magnitude within one third of
the blue isophotal radius R_{25} (i.e. at $\log(A/D_{25}) = -0.5$); (ii)
a mass-to-infrared luminosity $M_*/L_{H-0.5}$, with M_* the mass
measured at the Holmberg radius (or, in a few cases, R_{25}) and
corrected for the contribution by neutral hydrogen, and $L_{H-0.5}$
the luminosity corresponding to $H_{-0.5}$. These quantities are
plotted in Fig. 1. The mass-to-light ratio decreases with
color index by a factor of about four over the whole color
range.

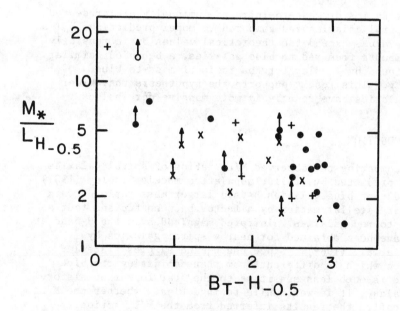

Fig. 1: Mass-to-infrared luminosity ratio versus color.
Symbols indicate the bar parameter of the galaxies: filled
circle = unbarred; cross = intermediate; plus sign = barred;
open circle = unknown. Arrows indicate that the mass has been
measured at R_{25} rather than at R_H; their length corresponds to
an estimated average relation $M(R_H)=1.25\ M(R_{25})$.

3. DATA ADJUSTMENTS

3.1 Extrapolated infrared magnitudes

The comparison of the observed mass-to-light ratio versus color relation to model predictions is complicated by the fact that all three observational quantities involved are measured at different apertures. Since the proportion of dark matter increases with radius in disk galaxies its effect on the mass-to-light ratio is most easily detected at large radii. In order to obtain a more homogeneous data set with respect to aperture we therefore choose to derive H-magnitudes at a larger aperture. To this end we use the mean curves of growth in H for spiral galaxies given by Aaronson, Huchra and Mould (1979). These curves are roughly linear up to the outermost measured points around $\log(A/D_{25}) = -0.2$. We tentatively extrapolate this behaviour to estimate the H-magnitude within R_{25} (i.e. at $\log(A/D_{25}) = 0$). In justification of this procedure we note that $H_{-0.5}$ probably does include a significant contribution from the disk (Wyse, 1982). The curves of growth have slopes of 2.0, 2.1 and 2.4 for galaxies of morphological type in the range 1 to 2, 3 to 4, and 5 to 8, respectively. The mean $B_T-H_{-0.5}$ colors of the corresponding three morphological groups in our sample are 2.84, 2.53 and 1.58, respectively. The above values very closely define the following linear relation between the slope x of the growth curves and the color index

$$x = -0.320(B_T-H_{-0.5}) + 2.908. \tag{1}$$

The infrared magnitude at R_{25}, denoted H, is obtained from eq. (1) and $H-H_{-0.5} = -0.5x$.

3.2 Correction for isophotal radius bias

We consider here the effect of the rate of star formation on mass-to-light ratios measured within a given isophotal radius. As pointed out by Tinsley (1981) a consequence of the observed increase of M/L_B with radius is that even if galaxies were identical except for the rate of star formation we would still measure larger M/L_B values for bluer galaxies since blue galaxies have larger blue isophotal radii than their red counterparts. In the present case both the mass and the infrared magnitude are referred to blue isophotal radii. The correction procedure applied here for the isophotal radius bias is in principle the same as that followed by Tinsley (1981). We construct a theoretical M/L_B versus B-K relation using the theoretical colors and mass-to-light ratios obtained by Larson and Tinsley (1978) and Struck-Marcell and Tinsley (1978) for models with decreasing star formation rates and age 10^{10} years, and models with a constant star formation rate and ages less

Table 1

Galaxy	Type	dist. (Mpc)	$H_{-0.5}$	B_T	$B_T-H_{-0.5}$	B_T-H	R'_H/R_H
(1)	(2)	(3)	(4)	(5)	(6)	(7)	(8)
N3623	1X	11.8	6.76	9.61	2.85	3.85	1.23
N3031	2A	3.6	4.38	7.20	2.82	3.82	1.22
N4698	2A	23.0	8.39	(10.99)	2.60	3.64	1.14
N4826	2A	6.8	6.10	9.21	3.11	4.07	1.32
N224	3A	0.69	0.91	3.58	2.67	3.70	1.17
N891	3A	14.4	6.84	9.33	2.49	3.55	1.11
N2841	3A	12.0	6.90	9.73	2.83	3.83	1.22
N3627	3X	11.8	6.70	9.34	2.64	3.67	1.15
N4501	3A	23.0	7.14	10.05	2.91	3.90	1.25
N4565	3A	18.4	6.70	9.22	2.52	3.57	1.11
N1421	4X	40	9.63	10.61	0.98	2.28	0.73
N3992	4B	22.8	7.98	10.46	2.48	3.54	1.11
N4258	4X	6.8	6.11	8.45	2.34	3.42	1.07
N4536	4X	23.0	8.30	(10.50)	2.20	3.30	1.03
N5055	4A	12.0	6.30	9.12	2.82	3.82	1.22
N7331	4A	21.0	6.44	9.62	3.18	4.13	1.37
N7541	4B	57.5	9.08	(11.85)	2.77	3.78	1.20
N253	5X	5.1	4.74	7.11	2.37	3.44	1.07
N701	5B	37.1	10.06	12.56	2.50	3.55	1.11
N1035	5B	24.7	10.25	12.10	1.85	3.01	0.93
I467	5X	44.6	10.71	11.81	1.10	2.38	0.74
N3198	5B	13.5	8.71	10.47	1.76	2.93	0.89
N4062	5A	15.1	9.30	11.64	2.34	3.41	1.06
N5033	5A	21.0	7.50	10.24	2.74	3.76	1.19
N5907	5A	17.2	7.58	9.90	2.32	3.40	1.06
N598	6A	0.72	4.38	5.84	1.46	2.68	0.81
U3691	6A	41.9	11.38	11.87	0.49	1.87	0.69
N2403	6X	3.6	6.45	8.34	1.89	3.04	0.93
N4244	6A	6.8	8.75	9.42	0.67	2.02	0.71
N4559	6X	18.4	8.34	10.07	1.73	2.91	0.89
N925	7X	14.4	8.74	10.07	1.33	2.57	0.77
N3495	7	19.4	10.11	(10.63)	0.52	1.89	0.69
N4236	8B	3.6	9.08	9.22	0.14	1.57	0.64
N7793	8A	4.7	7.89	9.52	1.63	2.82	0.85

NOTES TO TABLE 1

Column (1):   Galaxy name
Column (2):   Morphological type on the revised Hubble system of
              de Vaucouleurs, de Vaucouleurs and Corwin (1976,
              hereafter RC2).

Table 1

B_T-H'	log $L_{H_{-0.5}}$	M $(10^9 M_\odot)$	M_{HI} $(10^9 M_\odot)$	$M_*/L_{H-0.5}$	M_*'/L_H'	f
(9)	(10)	(11)	(12)	(13)	(14)	(15)
4.03	10.80	94	0.8	1.5	0.6	2.5
4.00	10.73	172	(3.9)	3.1	1.3	2.4
3.76	10.73	270	(3.8)	4.9	2.0	2.5
4.30	10.59	71	0.6	1.8	0.8	2.3
3.84	10.68	196	(3.6)	4.0	1.6	2.5
3.64	10.95	255	11	2.7	1.1	2.5
4.00	10.76	290	6.4	4.9	2.0	2.5
3.80	10.83	176	2.0	2.6	1.0	2.6
4.09	11.23	550	4.6	3.2	1.4	2.3
3.67	11.21	520	19	3.1	1.2	2.6
1.92	10.72	157*	15	2.7	0.8	3.4
3.63	10.89	456	10	5.7	2.2	2.6
3.48	10.59	180	5.5	4.5	1.7	2.6
3.33	10.77	210	11	3.4	1.2	2.8
4.00	11.00	230	13	2.2	0.9	2.4
4.38	11.43	390	22	1.4	0.6	2.3
3.94	11.25	408	37	2.1	0.9	2.3
3.51	10.88	135*	(17)	1.6	0.6	2.7
3.65	10.48	67*	6.8	2.0	0.8	2.5
2.94	10.05	32*	2.3	2.7	0.9	3.0
2.04	10.38	111*	10	4.2	1.3	3.2
2.81	10.14	89	8.9	5.8	1.9	3.1
3.47	10.00	52*	1.5	5.1	1.9	2.7
3.91	11.01	330	18	3.0	1.2	2.5
3.46	10.80	320	17	4.8	1.8	2.7
2.46	9.33	14.6	(1.4)	6.1	2.0	3.1
1.42	10.06	70*	8.3	5.4	1.5	3.6
2.97	9.90	42	4.5	4.7	1.6	2.9
1.61	9.53	30	4.6	7.5	2.2	3.4
2.78	10.56	106	25	2.2	0.7	3.1
2.29	10.18	89	15	4.9	1.6	3.1
0.93	9.90	116*	3.4	14.2	4.1	3.5
1.02	8.84	12.8	1.5	16.3	4.7	3.5
2.74	9.55	12.2*	1.4	3.0	1.0	3.0

Column (3): Distance based on mean group velocity or on radial
             velocity of the galaxy if not a group member.
             Adopted distance scale H_o =50 km s^{-1} Mpc^{-1}.
Column (4): $H_{-0.5}$ magnitude from Aaronson et al. (1982).
Column (5): B_T magnitude from Fisher and Tully (1981) or Tully
             et al. (1982) when available, and from the RC2

otherwise (values in parentheses).

Column (6):   Color index B_T-H-0.5.

Column (7):   Color index B_T-H-0.5 (see text sect. 3.1).

Column (8):   Ratio of the standard to the actual Holmberg radius (see sect. 3.2).

Column (9):   Color index B_T-H' (see text sect. 3.2).

Column (10): Logarithm of infrared luminosity in solar units, adopting $M_{B\odot} = 5.48$ and $(B-H)_\odot = 2.07$.

Column (11): Total mass within R_H or, where followed by an asterisk, within R_{25}, assuming a spherical mass distribution.

Column (12): Neutral hydrogen mass based on HI fluxes from Aaronson et al. (1982) when available, and from the RC2 otherwise (values in parentheses).

Column (13): Mass-to-light ratio $M_*/L_{H-0.5}$ in solar units, with $M_* = M-M_{HI}$.

Column (14): Mass-to-light ratio M_*'/L_H' in solar units, with $M_*' = (R_H'/R_H)M - M_{HI}$ and L_H' the infrared luminosity at $\log(A/D_{25}) = \log(R_H'/R_H)$.

Column (15): Ratio $(M_*/L_{H-0.5})/(M_*'/L_H')$.

than 10^{10} years. Since no theoretical H-K colors are available we use the empirical result that H-K is very nearly constant and about equal to 0.20 for galaxies of all morphological types (Aaronson, 1979) to obtain B-H colors. From the resulting theoretical M/L_B versus B-H curve we calculate for each galaxy the ratio F'/F, with F' = 1.93 the theoretical M/L_B value at B-H = 3.21 which is the mean B_T-H color of our sample of galaxies, and F the theoretical M/L_B value corresponding to the actual B_T-H color of the galaxy. Following Tinsley we define the standard isophotal radius of a galaxy as the radius it would have if its color were equal to the mean value given above, and we calculate the ratio R_H'/R_H of the standard Holmberg radius to its actual value according to her eq. (4). The latter equation is based on an exponential surface brightness profile in B with a fixed central value of 21.65 mag arcsec^{-2}. Since the observed central values vary between 20 and 23 mag arcsec^{-2} (Kormendy 1977, Boroson 1981), a variation which is of the same order as the difference between the surface brightness levels at R_H and R_{25}, a separate correction for the isophotal radius R_{25} is not justified. Therefore we simply take $R_{25}'/R_{25} = R_H'/R_H$ and hereafter drop the indices. Assuming flat rotation curves we obtain the mass within R_H' (or R_{25}') from M' = M R'/R. The infrared magnitude H' at R_{25} follows from the relation $H'-H_{-0.5} = -0.5x$ (1+2log(R'/R)), with x given by eq. (1).

3.3 Results

The mass-to-light ratios M_*'/L_H' resulting after extrapolation of the infrared magnitude and correction for the isophotal radius bias are given in Table 1 and plotted as a function of B_T-H' in Fig. 2. We note that the slope of the mean relation is slightly flatter in Fig. 2 than in Fig. 1 and that the mass-to-light ratios are systematically smaller by a factor of about 3, but otherwise there is little difference between the two figures. Table 1 shows that in fact the ratio f of $M_*/L_{H-0.5}$ to M_*'/L_H' increases slowly but systematically from 2.5 to 3.5 from the reddest to the bluest galaxies. The near constancy of f can be understood as the result of the combined effects of the slope of the infrared curve of growth being larger for bluer galaxies and the ratio of isophotal radii R'/R varying in the opposite sense.

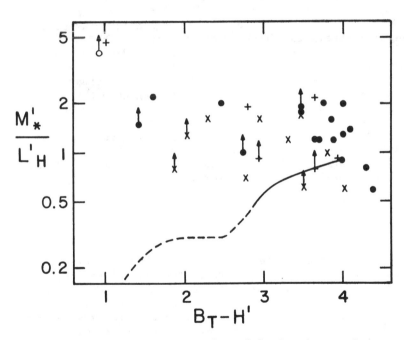

Fig. 2: Corrected mass-to-infrared luminosity ratio versus corrected color (see text). The meaning of the symbols is the same as in Fig. 1. The lines are theoretical relations obtained from the models of Larson and Tinsley (1978) and Struck-Marcell and Tinsley (1978). Solid line: models with monotonically decreasing star formation rates and age 10^{10} yrs; dashed line: models with constant star formation rate and ages in the range $5 \times 10^8 - 10^{10}$ years.

4. COMPARISON WITH MODELS

For each galaxy the values of both M'_*/L'_H and B_T-H' are
expected to differ but little from those corresponding to one
given aperture close to the Holmberg radius. A comparison with
theoretical models can now be made. The empirical
mass-to-infrared luminosity ratio in Fig. 2 is nearly constant
on the average except for the two bluest galaxies which have
larger values, while the theoretical ratio decreases by a
factor of 5 over the observed color range. The systematic
discrepancy between empirical and theoretical M/L_H values is
remarkably similar to that found by Tinsley (1981) in the case
of M/L_B (adjusting the data to the same distance scale and
taking into account a difference of a factor 0.7 between the
M/L_B values at a given color in the models of Larson and
Tinsley (1978) used here and in those of Tinsley (1981)). The
fact that a comparison of observed mass-to-light ratios with
model predictions yields very similar results in the blue and
in the infrared indicates that the observations are little
affected by metallicity for the sample of galaxies considered.
The differentiation of galaxies with respect to the
presence of a bar in Figs. 1 and 2 is meant to display a
possible correlation of the bar parameter with color (Wyse,
1981). Although the fraction of barred galaxies increases
towards bluer colors, this bias is less strong here than in
Tinsley's sample due to a slightly different selection of
galaxies. Moreover there is no indication that M/L is
different for barred and unbarred galaxies. We conclude that
the systematic discrepancy between observed and theoretical
mass-to-light ratios in both the blue and the infrared supports
Tinsley's hypothesis that blue galaxies have a relatively more
massive dark halo.

5. IMPLICATIONS FOR GALAXY FORMATION

Tinsley (1981) suggested that the ratio of mass of a dark
component to that of ordinary stars is fundamental to the
origin of the Hubble sequence of spiral galaxies. Tully, Mould
and Aaronson (1982), on the other hand, claim that the Hubble
sequence of spirals is uniquely determined by total mass on the
basis of both the infrared color-magnitude correlation,
discovered independently by them and by Wyse (1982), and the
correlation between B-H color and HI line profile width. The
fact that total mass seems to uniquely determine the color of a
spiral galaxy, apparently independently of Hubble type, is
interpreted by Tully et al. as due to a systematic decrease of
the star formation time scale with total mass of the galaxy.
The relative importance of the two parameters, mass fraction of
dark matter and total mass, is unclear. The fact that the

observed range of only a factor of 5 in the mass ratio of dark
to stellar matter is an order of magnitude smaller than the
range in total mass of spiral galaxies does not necessarily
imply that the former quantity is a relatively unimportant
parameter because both quantities are measured on a linear
scale determined by the luminous mass. If dark haloes are
effectively the dominant mass component both quantities should
be determined within a radius characteristic of the dark rather
than the luminous matter. This might result in a stronger
dependence of the ratio of dark to luminous matter on color.

Both views given above with respect to the origin of the
Hubble sequence can be accounted for in a single picture if we
tentatively identify the mass of the dark halo as the dominant
parameter in the following way. Given that dark haloes form
first, as proposed by White and Rees (1979), the correlation of
both the luminous mass and the ratio of dark to luminous matter
with color can be explained if the mass of gas trapped in the
potential well of the dark halo increases as the mass of the
dark halo to a power x larger than 1 (possibly as the result of
gravitational accretion of gas by the dark mass). For $M_{H1} >
M_{H2}$, where M_H is the mass of the dark halo, we then have

$$\frac{M_{L1}}{M_{L2}} = \left(\frac{M_{H1}}{M_{H2}}\right)^x > 1,$$

and

$$\frac{M_{H1}}{M_{L1}} < \frac{M_{H2}}{M_{L2}},$$

with M_L the luminous mass, assumed to be equal to the initial
mass of gas. Massive dark haloes tend to tidally inhibit star
formation, and thus act against bulge formation and give rise
to a relatively young stellar population. According to this
picture prominent dark haloes, in the sense of a large M_H/M_L,
are responsible for both the morphological structure (absence
of a bulge) and the blue colors of late-type galaxies.

REFERENCES

Aaronson, M. 1979, in "Photometry, Kinematics Dynamics of
    Galaxies", ed. D.S. Evans, pp. 147-150.
Aaronson, M., Huchra, J. and Mould, J.R. 1979, Astrophys. J.
    229, pp. 1-13.
Aaronson, M., Huchra, J. Mould, J.R., Tully, R.B., Fisher,
    J.R., van Woerden, H., Goss, W.M., Chamauraux, P., Mebold,
    U., Siegman, B., Berriman, G. and Persson, S.E. 1982,
    Astrophys. J. Suppl. Ser. 50, pp. 241-262.
Boroson, T. 1981, Astrophys. J. Suppl. Ser. 46, pp. 177-209.
Burstein, D., Rubin, V.C., Thonnard, N. and Ford, W.K. 1982,
    Astrophys. J. 253, pp.70-85.
Davoust, E. and de Vaucouleurs, G. 1980, Astrophys. J.
    242, pp. 30-52.
Faber, S.M. and Gallagher, J.S. 1979, Ann. Rev. Astron.
    Astrophys. 17, pp. 136-187.
Fisher, J.R. and Tully, R.B. 1981, Astrophys. J. Suppl.
    Ser. 47, pp. 139-200.
Gottesman, S.T. and Hunter, J.H. 1982, Astrophys. J. 260,
    pp. 65-69.
Kormendy, J. 1977. Astrophys. J. 217, pp. 406-419.
Larson, R.B. and Tinsley, B.M. 1978, Astrophys. J. 219,
    pp. 46-59.
Pence, W.D. 1981, Astrophys. J. 247, pp. 473-483.
Struck-Marcell, C. and Tinsley, B.M. 1978, Astrophys. J.
    221, pp. 562-566.
Tinsley, B.M. 1981, Mon. Not. R. Astr. S. 194, pp. 63-75.
Tully, R.B., Mould, J.R. and Aaronson, M. 1982, Astrophys.
    J. 257, pp. 527-537.
de Vaucouleurs, G., de Vaucouleurs, A. and Corwin, H.G.
    1976, "Second Reference Catalogue of Bright Galaxies",
    Austin,University of Texas Press (RC2).
White, S.D.M. and Rees, M.J. 1978, Mon. Not. R. Astr. S. 183,
    pp. 341-358.
Wyse, R.F.G. 1981, Mon. Not. R. Astr. S. 196,
    pp. 911-914.
Wyse, R.F.G. 1982, Mon. Not. R. Astr. S. 199, pp. 1P-8P.

IV

PREGALACTIC STARS AND EARLY NUCLEOSYNTHESIS

PREGALACTIC ACTIVITY, GALAXY FORMATION AND HIDDEN MASS

Martin J. Rees
Institute of Astronomy
Madingley Road
Cambridge CB3 OHA, England

ABSTRACT

The first bound objects are expected to be of sub-galactic scale if <u>either</u> (a) the early universe has isothermal fluctuations <u>or</u> (b) it is dominated by cold non-baryonic matter. The properties of these objects and their implications for galaxy formation and clustering are briefly discussed.

1. INTRODUCTION

Progress in understanding galaxy formation is bedevilled by our ignorance about the $10^6 \lesssim t \lesssim 10^9$ yrs era in our Universe's history (corresponding to $10^3 \gtrsim z \gtrsim 10$). There are still essentially no observations which can tell us directly about this era (see Hogan, Kaiser and Rees 1982 for some discussion); it therefore seems worthwhile to explore the consequences of all the different schemes for galaxy formation. In the purely adiabatic neutrino-dominated picture, nothing of interest happens before $z \cong 5$. However there are various other schemes which would entail pregalactic activity at $z \gtrsim 10$. These schemes have important consequences for:(a) the nature of the hidden mass and Population III;and (b) the possibility that galaxies and clusters may have evolved from fluctuations which are not primordial, but have been causally generated by pregalactic energy input. The work summarised here is mainly collaborative with A. Kashlinsky and B. Carr; further details can be found in three recent papers (Carr and Rees 1983 ab; Kashlinsky and Rees 1983).

J. Audouze and J. Tran Thanh Van (eds.),
Formation and Evolution of Galaxies and Large Structures in the Universe, 239–252.
© *1984 by D. Reidel Publishing Company.*

2.   THE SIZE OF THE FIRST OBJECTS IN A BARYON-DOMINATED UNIVERSE

In the conventional hot big bang Universe, with <u>adiabatic</u>
fluctuations, there are two important mass scales. The first of
these is the maximum radiation Jeans mass, $M_J^{max} \simeq 10^{17} \, \Omega^{-2} \, M_\odot$:
the growth on scales larger than this is never impeded by pressure
gradients, whereas perturbations on scales below M_J^{max} do not
grow between entering the horizon at t_H and decoupling at t_{dec}.
The second important scale is the photon diffusion mass
$3 \times 10^{13} \, \Omega^{-5/4} \, M_\odot$ (Silk 1968): adiabatic fluctuations on scales
below this mass will be damped out before t_{dec}. This gives rise
to the "pancake" picture (Zeldovich 1970) in which clusters bind
first, with galaxies forming later via fragmentation. Adiabatic
perturbations of a hot universe could yield a value of M_{min} far
below the photon diffusion mass if primordial black holes form.
This requires either initial curvature fluctuations of amplitude
$\delta_H \gtrsim 0.1$ on some scale (Carr 1975, Nadejin <u>et al</u>. 1978) or a
special kind of phase transition at some epoch (Carr 1975, Canuto
1978, Kodama <u>et al</u>. 1981, Hawking, Moss and Stewart 1982, Crawford
and Schramm 1982). The mass M_{min} is then rather ill-determined,
but it would usually be of order the horizon mass at the phase
transition, corresponding to some mass in the range 10^{-5}g to 1 $M_\odot$.
Although such holes are very small, they could cluster to form
larger bound objects (Meszaros 1975, Carr 1977, Freese <u>et al</u>. 1983),
thereby generating pregalactic activity indirectly.

The situation is more interesting if one has <u>isothermal</u>
fluctuations in a hot Universe. (This requires initial non-
uniformities in the photon-baryon ratio, or else the hypothesis
that a nonlinear damping process or phase transition deposits
extra entropy in a non-uniform fashion.) Compton drag then
inhibits motion of baryons relative to the radiation before t_{dec},
provided that the baryons are in a form which interacts with the
radiation. (Otherwise the fluctuations can begin to grow once
the baryons dominate the density [Mészáros 1974].) Inhomogeneities
in the baryon-to-photon ratio are therefore preserved until t_{dec},
except on scales $\lesssim 10 \; M_\odot$, where radiative diffusion may have been
significant (Hogan 1978). Isothermal fluctuations on scales
$\lesssim M_J^{max}$ would thus retain the same amplitude at t_{dec} as they had
on entering the horizon. Therefore, provided the amplitude falls
off towards large scales, masses corresponding to the matter Jeans
mass at decoupling, $M_{Jm} \sim 10^6 \Omega^{-\frac{1}{2}} M_\odot$, should condense out before any
larger scale does. However, the fate of isothermal fluctuations
on scales far <u>below</u> M_{Jm}, whose amplitude at t_{dec} may be bigger,
is less obvious. Such fluctuations could not begin to grow until
they became larger than M_J, which itself falls off as $z^{3/2}$ if
matter and radiation are completely decoupled (and more slowly
when the residual coupling is allowed for) at a redshift
$z_J(M) \lesssim z_{dec}(M/M_{Jm})^{3/2}$. If the lower-mass perturbations were
merely "frozen in" for $z_d \gtrsim z \gtrsim z_J(M)$, they might nevertheless

be the first scales to bind if their amplitude at t_{dec} were
large enough: a simple argument along these lines suggests that
if $\delta \propto M^{-\beta}$ at recombination, then $\beta > 2/3$ would be a sufficient
condition if $M_J \propto z^{3/2}$. However, there are three effects which
render it less likely that scales $< M_{Jm}$ develop into important
pregalactic objects:

(i) Possible attenuation of isothermal fluctuations during
     recombination

If recombination were an instantaneous process, dynamical
and thermal coupling between baryons and radiation would cease
suddenly at z_{dec}. The previously "frozen" modes on scales $\lesssim M_{Jm}$
would then become oscillatory, driven by gradients in the gas
pressure ($\propto nT$); but, so long as the amplitudes remained linear
(cf. (ii) below), the initially overdense regions would remain on
a lower adiabat and so would persist, in a time-averaged sense,
despite the oscillations.

The fractional ionization level, $x = n_e/(n_e + n_H)$, plunges
from > 0.99 at $5000°K$ to $< 10^{-3}$ at $2000°K$ (Peebles 1968; Sunyaev
and Zeldovich 1970; Wyse 1982). Though this is rapid compared
with the cosmological expansion timescale t_{exp} (in the sense
that $|\dot{x}/x| > 10 \ t_{exp}^{-1}$ when $T \cong 4000°K$), it is not necessarily
sudden compared with the time taken for an overdense region of
$M \ll M_{Jm}$ to expand owing to its excess internal pressure. The
latter timescale is

$$t_{osc} \cong (M/M_{Jm})^{1/3} \ t_{exp}. \tag{1}$$

Furthermore, as the Universe continues to cool below $\sim 2000°K$,
the ionization fraction x levels off at a value $\sim 3 \times 10^{-5}$ for
$\Omega = 1$ and $\sim 3 \times 10^{-4}$ for $\Omega = 0.01$, when the timescale for the last
few electrons to recombine ($\propto x^{-1}$) becomes as long as the expansion
timescale. Even this small residual fraction is sufficient,
particularly if Ω is low, to preserve some thermal contact
between the matter and radiation.

Dissipative effects during recombination have two distinct
causes: the non-sudden elimination of dynamical coupling with
the radiation, and the gradual loss of thermal contact with it.
The Compton drag timescale is

$$t_{drag} = 10^{13} \ T_3^{-4} \ x^{-1} \ s, \tag{2}$$

where T_3 is the radiation temperature in units of $1000°K$. The
timescale t_{equil} for exchange of thermal energy between gas and

radiation (i.e. the timescale on which the gas temperature would adjust to T) is x^{-1} times the ordinary Compton cooling time for a non-relativistic electron. So, irrespective of x,

$$\left(\frac{t_{equil}}{t_{drag}}\right) = \left(\frac{m_e}{m_p}\right) . \tag{3}$$

This means that there is a range of $\sim 10^3$ in x when radiation drag is not able to inhibit motions driven by gas pressure gradients, whereas the gas temperature is still held close to the radiation temperature (in the sense that $\delta T/T \ll \delta\rho/\rho$). Damping processes - due to Compton drag or to hysteresis effects arising from imperfect thermal coupling of gas and radiation - persist until x and T have fallen so low that $t_{exp} < t_{equil} \ll t_{drag}$.

The cumulative amount of suppression of fluctuations with $M < M_{Jm}$ during the period of decoupling depends on the detailed form of $x(T_3)$. Because recombination is more gradual (i.e. $|x/\dot{x}|$ is larger) during its later stages, when $\dot{x}$ approaches its frozen-in value, the hysteresis damping associated with the gradual loss of thermal contact between matter and radiation is more important than the Compton drag. The net effect (cf. early numerical work by Peebles (1969)) is that fluctuations in the mass range $10^3 - 10^6 M_\odot$ survive if $\Omega_b = 1$, but not if $\Omega_b \ll 1$.

(ii)   Non-linear damping during the oscillatory period between z_{dec} and $z_J(M)$

Any initial isothermal fluctuations below M_{Jm} that do survive damping will be transformed into sound waves, and (on linear theory) would thereafter oscillate until M_J falls sufficiently that they become gravitationally unstable. During the oscillatory stage, the adiabatic invariant is $\delta^2 c_s^2 P$, where P is the period. If the gas temperature varies as z^2, then $c_s \propto z$ and $P \propto z^{-2}$, so the amplitude δ is constant. The number of oscillations for a mode of mass-scale M is $\sim (M/M_{Jm})^{-1/3}$. Thus we have an upper limit $\delta < (M/M_{Jm})^{1/3}$ to the amplitudes that can survive to the stage where these small-scale modes become gravitationally unstable. Any initial spectrum with amplitude rising towards small masses will, owing to this effect, turn over to $\delta \propto M^{1/3}$ for scales when non-linear attenuation has occurred.

(iii)   Processes after turn-around

Even if objects of $\ll 10^6\ M_\odot$ were the first to condense out after t_{rec} (as is conceivable if the initial spectrum rises very steeply towards low masses) they may merely remain as bound clouds of neutral gas. When $T \ll 10^{4}°K$, atomic cooling is negligible, so the collapse will be adiabatic. If the collapse proceeds without shocks occurring, a cloud's temperature evolution will therefore retrace its T(R) trajectory until it has contracted to the radius at which pressure can support it (i.e. it will just return to the temperature T_J which it had when it first went above the Jeans mass). It will therefore still be at $< 10^4 °K$. If shocks occur, then the cloud will stabilise at a larger radius and be even cooler. When clouds with $M > M_{Jm}$ can heat up to $\sim 10^4 °K$, when hydrogen reionizes, and thereafter contract isothermally and fragment, these lower masses may just persist as thermally-supported clouds. This conclusion would be somewhat modified if H_2 molecules formed, since cooling could then occur at temperatures as low as $\sim 10^3 °K$; only clouds below $\sim 10^4\ M_\odot$ would then be stable against collapse and fragmentation (Hartquist and Cameron, 1977).

Because of the three physical effects we have outlined above, the first bound systems are unlikely to have masses much below M_{Jm}. This conclusion holds particularly strongly in a low Ω universe, where the hysteresis damping effect would be more severe. Peebles (1969) presents computations which include the effects we discuss under (i) and (ii) above for masses $10^4 - 10^6 M_\odot$, showing that their combined effect is to impede the growth of $10^4 M_\odot$ perturbations relative to those of $10^6 M_\odot$ by a factor exceeding 50, even for $\Omega = 1$. (We learn from the present discussion, however, that this attenuation is mainly due to the hysteresis effect when thermal coupling breaks down; Compton drag is less severe in its effects, because it comes into play when x is falling steeply.) If some pregalactic energy input into the gas switched on before x had fallen below the range where decoupling was complete, then the damping would be still more severe. Moreover, masses $\ll M_{Jm}$, even if they were the first to bind, may merely persist as non-radiating neutral clouds. Thus, the important agents of pregalactic activity in a "hot" baryon-dominated Universe are likely to be of mass $M_{Jm} \cong 10^6 \Omega^{-\frac{1}{2}} M_\odot$; lower mass "seeds" or sources of energetic input would be likely only if they involved primordial black holes.

3. VIRIALISATION AND FRAGMENTATION: THE NATURE OF POPULATION III OBJECTS

The next important question concerns the fate of $\sim 10^6 M_\odot$ clouds when they collapse and virialise: do they form single massive objects (VMOs), or do they fragment into stellar mass bodies? It is important to emphasise that the absence of heavy elements does not in itself preclude fragmentation into stellar objects. In 'opacity-limited' schemes (Hoyle 1953) where fragmentation continues until the optical depth through a Jeans length is too large to permit further isothermal collapse, general cooling considerations (Rees 1976) show that the final fragment mass is almost independent of the opacity source and is $M_{Ch} (kT/m_p c^2)^{1/4}$ (here $M_{Ch} = (hc/Gm_p^2)^{3/2} m_p$ is the Chandrasekhar mass), implying that the first stars could have had low masses.

Kashlinsky and Rees (1983) further investigate the problem of Population III star formation. They note that clouds of mass $M(\gtrsim M_{Jm})$ may acquire prior to their collapse some angular momentum due to tidal interactions with their neighbours. Any fragment forming within such a rotating cloud would itself have some spin and would therefore be able to collapse by only a limited factor in radius, thus leaving a large enough geometrical cross-section for coalescence to be important. Once the paternal cloud has collapsed below some radius the coalescence time becomes shorter than the free-fall time and fragmentation then becomes impossible during the collapse. At earlier stages in the collapse frag- mentation could be inhibited by the trapped background radiation (it is more efficient for clouds more massive than M_{Jm}). Also, any small density fluctuation in a collapsing cloud which survived damping during decoupling would lead to fragmentation only after the cloud has collapsed by a considerable factor in radius. Fragmentation is thus unlikely until centrifugal forces halt the collapse and a disk forms. H_2 molecules would cool the disk to 10^3 K; the Jeans mass (minimum fragment's mass) can then be $\lesssim 0.1 M_{\odot}$. After the disc has fragmented, two-body interactions between stars (and perhaps collective processes as well) would redistribute angular momentum and thicken the (now stellar) disk. After the system has 'sphericalised' escape of stars from the system will leave some fraction of the stars in a dense core, where stellar collisions will become important leading to the formation of supermassive objects. Thus we envisage that the evolution of the primordial clouds would lead to a formation of two distinct types of objects: one being a generation of low mass stars and another a generation of supermassive objects.

The general scheme we envisage is summarised in Figure 1. We emphasise, however, that the uncertainties are such that we cannot confidently state whether the bulk of Population III would be of low mass, or VMOs - either seems possible. The conclusions of Kashlinsky and Rees (1983) can be summarised as follows:

(i) Most of the material in these clouds could have been transformed into very low mass stars - low enough to yield an M/L as high as is required for the hidden mass despite the absence of heavy elements and the presence of background radiation which prevented cooling below $\sim$ 1000 K.

(ii) These protostars would have escaped physical collisions while the initial disk-like system redistributed its angular

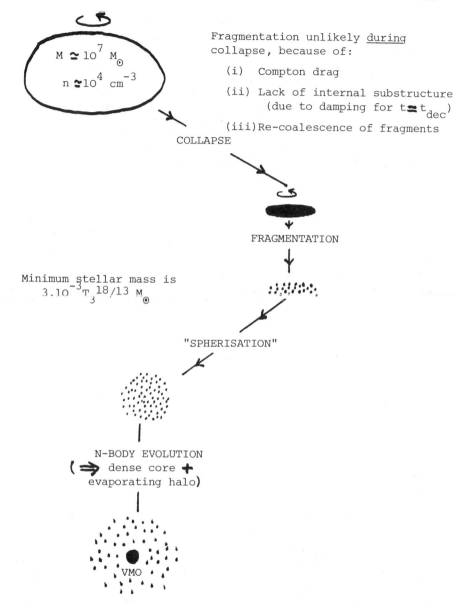

M $\simeq 10^7$ M$_\odot$

n $\simeq 10^4$ cm^{-3}

Fragmentation unlikely <u>during</u> collapse, because of:

(i)   Compton drag

(ii) Lack of internal substructure (due to damping for t $\simeq$ t$_{dec}$)

(iii) Re-coalescence of fragments

COLLAPSE

FRAGMENTATION

Minimum stellar mass is 3.10^{-3} T$_3^{18/13}$ M$_\odot$

"SPHERISATION"

N-BODY EVOLUTION ($\Rightarrow$ dense core **+** evaporating halo)

VMO

<u>Figure 1.</u> Stages in the evolution of a bound cloud which condenses out soon after recombination, according to Kashlinsky and Rees (1983). The cloud is expected to have a small amount of spin, owing to tidal interactions with its neighbours. A rotating disc of H$_2$ forms, and fragments into low mass stars. Dynamical evolution then causes most of these stars to evaporate from the original bound system; those left in the core may coalesce into a VMO.

momentum and 'sphericalised'. Most of them would have evaporated from the original 10^6 - $10^8 M_\odot$ units in which they formed, as a consequence of stellar-dynamical relaxation processes.

(iii) The relaxation process gives rise to a tightly bound core of stars which does not evaporate, possessing the entire gravitational binding energy of the original system. As the number of stars in the core diminishes, conditions eventually become so extreme that physical collisions are important (even though the individual stars would by then have contracted onto the main sequence). The endpoint of stellar disruption and coalescence in this core would be the formation of a very massive object (VMO).

(iv) Although the low mass stars would be the main constituents of Population III, the small fraction evolving into VMOs could generate some pregalactic heavy elements and energy input into the diffuse intergalactic medium. Even though the original low mass stars would have formed at z > 100, the violent VMO phase may occur at a more recent epoch because this happens only after the relaxation within the star clusters has led to a dense core.

4. INFLUENCE OF NON-BARYONIC MATTER

Any initial perturbations in their distribution would have been homogenised by free-streaming on scales up to

$$M_\nu \cong m_{Planck}^3/m_\nu^2 \simeq 5 \times 10^{14} \, \Omega_\nu^{-2} \, M_\odot \qquad (4)$$

(Bond, Efstathiou and Silk 1980, Bisnovatyi-Kogan and Novikov 1981, Doroshkevich et al 1980; Wasserman 1981). If $\Omega_\nu > \Omega_b$, the neutrinos would constitute a homogeneous background impeding the linear growth after decoupling of any small-scale isothermal fluctuations of the kind envisaged in § 2. Growth of baryonic perturbations of mass $M_b(> M_{Jm})$ according to the usual low $\delta \propto (1 + z)^{-1}$ could commence only after the neutrino Jeans mass ($\sim M_\nu(kT/m_\nu c^2)^{3/2}$, where T is the radiation temperature) fell below $M_b(\Omega_\nu/\Omega_b)$. If $\beta < 2/3$, the first objects to bind in a neutrino-dominated universe would have the mass M_ν. If $\beta > 2/3$, the first objects would have $M_{min} \cong 10^6\Omega_\nu^{-5} \, M_\odot$ and bind at $z \cong \Omega_\nu^{-1}$. They could be smaller than galaxies for $\Omega_\nu > 0.1$ but they could not be smaller than $10^6 M_\odot$ (since $\Omega_\nu < 1$) even if isothermal fluctuations on this scale survived radiative damping during decoupling. (Note the $\beta > 2/3$ is perhaps rather unlikely, since it would require curvature fluctuations rising very steeply towards small scales).

On the other hand, other kinds of non-baryonic matter can actually <u>promote</u> rather than inhibit the condensation of small-scale pregalactic objects. If the dominant contribution to Ω comes from particles which are individually much more massive than neutrinos, but correspondingly less numerous, then the damping by free-streaming (cf. eqn. (4)) is restricted to much smaller scales (Bond and Szalay 1983; Peebles 1982). Another conjectured type of non-baryonic matter, axions, would be "cold" and also not subject to free-streaming. These options would permit small-scale inhomogeneities in the non-baryonic distribution to persist and amplify, even if the initial conditions were purely adiabatic; the baryonic component could then respond to their gravitational influence after t_{dec}. Fluctuations in the cold non-baryonic matter do not grow significantly between entering the horizon and the epoch $t = t_{eq}$, corresponding to $z = z_{eq} \cong 4 \times 10^4$, when this matter becomes dynamically dominant. If the fluctuations entered the horizon with amplitudes $\propto M^{-\beta}$, all scales below $M_H(t_{eq})$ would consequently retain this spectrum at t_{dec}, but with amplitudes enhanced by $\sim (z_{eq}/z_{dec})$.

The baryons would accumulate in these gravitational potential fluctuations provided that gas pressure gradients do not prevent this. A sufficient condition for this accumulation to occur in a potential well of total mass M is that M_{Jm} (the Jeans mass of the baryonic component) be less than $M[(\Omega_b + \Omega_\nu)/\Omega_b]^2$. This scheme would permit pregalactic systems of far below the Silk (1968) mass to condense out even if the initial fluctuations were purely adiabatic. They would then evolve as envisaged in § 3. Moreover, the radiative damping processes during decoupling (cf. § 2) are irrelevant, since baryonic fluctuations build up after t_{dec}, induced by the (undamped) fluctuations in the cold non-baryonic matter.

If constant curvature fluctuations are present initially, then the post-recombination fluctuations have an almost constant amplitude. A particular model for galaxy formation based on this scheme is proposed in Peebles' contribution to these proceedings. If however the amplitude of the initial curvature fluctuations increased steeply towards smaller scales, then small clusters of non-baryonic matter might already have virialised by t_{dec}. If the virial velocity within such a cluster were ≥ 10 km s^{-1} (i.e. exceeded the sound speed c_s in gas at $< 10^4$°K), then gas would accrete onto it even if its mass were below M_{Jm}. Such objects could serve as "seeds" (Carr and Rees 1983b) in the same way as primordial black holes, but require slightly less extreme initial amplitudes in order to form.

5. FORMATION OF "SECONDARY" LARGE-SCALE FLUCTUATIONS BY PREGALACTIC OBJECTS

An important aspect of pregalactic objects is that they could themselves generate large-scale density fluctuation and thus give rise to galaxies and clusters of galaxies indirectly. Several scenarios have been suggested (see Carr and Rees 1983b). One important possibility is that bound objects can produce energy in some way - energy which may in principle be used to generate large-scale peculiar velocities. A simple argument shows that the energy input required can be far less than the binding energy of the system which the secondary fluctuations eventually form.

After decoupling, the peculiar velocities in a growing perturbation vary with expansion factor as $R^{1/2}$ (because δ goes as R, whereas the differential Hubble velocity across a comoving sphere goes as $R^{-1/2}$), so the energy required to generate these peculiar velocities at redshift z_i is just (z_B/z_i) times the eventual binding energy, where z_B is the redshift at which the perturbation stops expanding. Equivalently, the peculiar velocities generated at z_i can be smaller than the virial velocity of the final bound system by a factor $(z_B/z_i)^{1/2}$. One consequence of this is that, if the specific binding energy of a virialised system is εc^2, then it entered the horizon at a redshift $\varepsilon^{-1} z_B$. If we regard this as the largest redshift when the velocity perturbations could have been generated, then the specific energy input required is $\varepsilon^2 c^2$. This result assumes that the system entered the horizon after t_{eq}. Otherwise, it is optimal to feed the energy in just after t_{eq} (or just after decoupling in a hot Universe) and the energy input required is $\varepsilon c^2 (z_B/z_{eq})$.

As a crude illustrative example, a "galaxy" of $10^{12} M_\odot$ and velocity dispersion 300 km/sec (whose present binding energy is $\sim 10^6 M_\odot c^2$) could have been produced "causally" by the injection of only 1 $M_\odot c^2$. Of course, this is unrealistic because we have seen that galactic mass perturbations do not necessarily grow uninterruptedly after entering the horizon. On the other hand, dissipative effects may imply that the binding energy of protogalaxies was smaller than that of present-day galaxies. By the same argument, a "globular cluster" of $10^6 M_\odot$ and velocity dispersion 10 km/sec could have been generated from an energy input of only $10^{-12} M_\odot c^2$.

Carr and Rees (1983b) have explored and reviewed various ways in which pregalactic seeds can generate fluctuations on galactic and cluster scales which exceed the amplitude of any fluctuations that might have been present initially on those same scales. The reader is referred to that paper for a full discussion.

One interesting possibility is that an energy input comes from galactic explosions. Any star with initial mass M above about 10 $M_\odot$ but below a critical value $M_c \simeq 200$ $M_\odot$ (Bond et al 1983) should end up exploding. The explosive energy generated can be expressed as $\varepsilon M c^2$, where the efficiency parameter ε is in the range 10^{-4} - 10^{-5} over all mass ranges (Bookbinder et al 1982). Stars larger than M_c probably collapse to black holes, although it is possible that some of the envelope may be ejected even in this case due to hydrogen shell burning (Bond et al 1983). Supermassive stars in the range $M > 10^5$ $M_\odot$ could explode only if they had non-zero matallicity (Fricke 1982), but this would not be a characteristic of the first stars.

As Doroshkevich et al (1967), Ostriker and Cowie (1980) and Ikeuchi (1981) have emphasised, the energy input from exploding stars - or clusters of them - could induce the formation of new bound objects. One might envisage a shock front sweeping up a spherical shell which expands adiabatically until its cooling time becomes comparable to the cosmological expansion time. (For z>5 the dominant cooling process is inverse Compton cooling off the 3K background).

For a runaway process to be initiated, the shell must sweep up more than the seed's initial mass; also, the shell must be able to fragment into second-generation seeds. One can thereby initiate a bootstrap process which progressively amplifies the fraction of the Universe in stars. Carr and Rees (1983b) discuss under what conditions, in terms of seed mass and redshift range, this type of effect might occur.

One general point about the explosion scheme is that, though it may work in principle, it is not "efficient" in using energy to generate fluctuations in the sense discussed at the beginning of this section. A lot of energy is wasted in cooling. Moreover, most of the momentum goes into generating density contrasts > 1. To generate fluctuations early, one needs to produce velocities of only $V_{pec} = V_{virial} (1 + z)/(1 + z_B)^{-1/2}$, where z_B denotes the redshift when the system condenses out and binds. However, the differential Hubble velocity across the relevant scale is $V_{pec} (1 + z)/(1 + z_B) >> V_{pec}$, so the mean propagation speed of the agent generating the prculiar velocity must be much larger than the velocity that is generated. This is not possible for blast waves, except at very late stages after their momentum (and, a fortiori, their energy) has been reduced by the cosmological expansion (cf Schwartz, Ostriker and Yahil 1975; Ikeuchi, Tomisaka and Ostriker 1983).

This suggests that it may be easier to generate large-scale fluctuations from some sort of dissociation front: we need the

"front" to change the sound speed of the gas by a small fraction
of its propagation speed. The resulting pressure gradients then
create peculiar velocities. The efficiency of converting radiation
into kinetic energy can be high provided that cooling is not too
serious. A specific mechanism highly efficient in this respect,
invented by Hogan (1983), involved dissociation of H_2, which
releases about 6 ev per molecule. If the primordial hydrogen
were largely molecular, this could be a highly efficient mechanism.
Ionization of atomic hydrogen by UV can be equally effective if
the UV is released in a pulse lasting less than the recombination
time of $\sim 10^5 \, n_e^{-1}$ yrs (which is itself much less than the Hubble
time at early epochs). When the pulse is over, the gas recombines,
but remains on a higher adiabat than material that was never
ionized. However, if the ionization is caused by UV from an
object at the Eddington limit, rather than by an explosion with
$L \gg L_{Edd}$, then the kinetic energy generated corresponds only to
the luminous output over a recombination time.

A related proposal (Hogan 1983 and these proceedings, Hogan
and Kaiser 1983) is that inhomogeneities are generated by gradients
in radiation pressure arising from randomly-distributed sources.
If the sources were sufficiently intense to ionize the primordial
gas, the resultant fluctuations could extend up to the Jeans mass
of $\sim 10^{17} \, M_\odot$ (this being the largest scale over which a wave driven
by radiation pressure can propagate), with a natural cut-off on
still larger scales.

In principle one could envisage combinations of the various
amplification processes. For example, Freese et al (1983) have
suggested an interesting scheme in which primordial black holes
cluster gravitationally, and then gas accumulates in the potential
wells to form stars.

A related question concerns the rareness of the seeds: what
is the minimum fraction f_1 of the mass that has to go into them,
if they are to initiate a chain reaction leading to the present
large-scale structure? Some scenarios require that the initial
objects involve only a small fraction of the mass. For instance,
in the explosion scenario, we require $f_1 < 10^{-5}$ in order to avoid
over-enrichment (Carr et al 1983). In such a context, one would
like to have a mechanism to prevent f_1 becoming too large. One
way of doing this would be to suppose that the first objects form
from the Gaussian tail of the fluctuation distribution and reheat
the Universe, thereby suppressing further star formation by
increasing the Jeans mass. This results in a value for f_1 in the
range 10^{-4} - 10^{-6} (Hartquist and Cameron 1977, Carr et al 1983).
It is of course possible (cf § 3 and Kashlinsky and Rees 1983) that
only a small fraction of the initial mass goes into VMOs anyway,
the remainder being in inert low-mass Population III stars.

The possibility of generating galaxies via smaller pre-galactic objects alleviates the demands made on any primordial fluctuations in explaining the Universe's present structure. It is however still unclear whether the apparent "voids" on very large scales, and the giant linear structures, can be attributed to secondary perturbations (see Hogan's contribution to these proceedings).

A key question relevant to any interpretation of these large-scale features is the extent to which they delineate genuine high-amplitude density contrasts in the matter distribution. In neutrino-diminated cosmologies, it is generally expected that the neutrino distribution would be less clumped than the baryons. However one must ask further whether the galaxies are good tracers even of the baryons. It has been argued that this may not be so in the neutrino picture (Zeldovich et al 1982): baryons in underdense regions may remain in a diffuse gas. This same possibility may arise in other models. The energy output from the first galaxies to form may heat the remaining gas to $>10^5$K, so that its Jeans mass becomes too high to permit further galaxies to condense (the energy requirements for this, < 100 ev per particle, are far less than would be needed to evacuate a large void). Of great observational value would therefore be any evidence on the baryon density in diffuse regions, or any clues as to whether the velocities of galaxies in large scale filaments were close to the unperturbed Hubble velocities, as would be expected if the overall matter density were much smoother than the distribution of galaxies.

Acknowledgments   I am particularly grateful to Bernard Carr and Sasha Kashlinsky for their collaboration on much of the work which is summarised here. I acknowledge also useful discussions with Dick Bond and Craig Hogan.

REFERENCES

Bisnovatyi-Kogan, G and Novikov, I.D., 1981, Sov. Astron. Lett. 24, 516.
Bond, J.R., Efstathiou, G. and Silk, J.I., 1980, Phys. Rev Lett 45, 1980.
Bond, J.R., Arnett, W.D. and Carr, B.J., 1983, Astrophys. J., in press.
Bond, J.R., and Szalay, A., 1983, Astrophys. J., in press.
Bookbinder, J., Krolik, J., Cowie, L., Ostriker, J.P. and Rees, M.J. 1980, Astrophys. J. 237, 647.
Canuto, V., 1978, M.N.R.A.S., 184, 721.
Carr, B.J., 1975, Astrophys. J. 201, 1.
Carr, B.J., 1977, Astr. Astrophys., 56, 377.
Carr, B.J., Bond, J.R. and Arnett, W.D., Astrophys. J., in press.
Carr, B.J. and Rees, M.J., 1983a, M.N.R.A.S., in press.

Carr, B.J. and Rees, M.J., 1983b, M.N.R.A.S., in press.

Crawford, M. and Schramm, D.N., 1982, Nature, 298, 538.

Doroshkevich, A.G., Zeldovich, Y.B., and Novikov, I.D., 1967,
     Sov. Astron., 11, 233.

Doroshkevich, A.G., Zeldovich, Y.B., Sunyaev, R.A. and Khlopov, M.Y.,
     1980, Sov. Astron. Lett., 6, 459.

Freese, K., Price, R. and Schramm, D.N., 1983, preprint.

Fricke, K.J., 1982, Astr. Astrophys., in press.

Hartquist, T.W. and Cameron, A.G.W., 1977, Astrophys. Sp. Sci.
     48, 145.

Hawking, S.W., Moss, I.G. and Stewart, J.M., 1982, Phys Rev. D26,
     2681.

Hogan, C.J., 1978, M.N.R.A.S., 188, 781.

Hogan, C.J., 1983, M.N.R.A.S., 202, 1101.

Hogan, C.J. and Kaiser, N.,1983, Astrophys. J., in press.

Hogan, C.J., Kaiser, N. and Rees, M.J., 1982, Phil. Trans. Roy.
     Soc. 307, 97.

Hoyle, F., 1953, Astrophys. J.,118, 513.

Ikeuchi, S., 1981, Publ. Astron. Soc. Japan., 33, 211.

Ikeuchi, S., Tomisaka, K. and Ostriker, J.P., 1983, Astrophys. J.
     265, 583.

Kashlinsky, A. and Rees, M.J., 1983, M.N.R.A.S., in press.

Kodoma, H., Sasaki, M., Sato, K., and Mueda, K., 1981, Prog. Theor.
     Phys., 66, 2052.

Meszaros, P., 1974, Astr. Astrophys., 37, 225.

Meszaros, P., 1975 Astr. Astrophys., 38, 5.

Nadejin, D.K., Novikov, I.D. andPolnarev, A.G., 1978. Sov. Astron.
     22, 129.

Ostriker, J.P. and Cowie, L.L., 1980, Astrophys. J.(Lett) 243, L127.

Peebles, P.J.E., 1968, Astrophys. J. 153, 1.

Peebles, P.J.E., 1969, Astrophys. J. 157, 1075.

Peebles, P.J.E., 1982, Astrophys. J. (Lett) 263, L1.

Rees, M.J., 1976, M.N.R.A.S., 176, 483.

Schwartz, J., Ostriker, J.P. and Yahil, A., 1975, Astrophys. J.
     202, 1.

Silk, J.I., 1968, Astrophys. J., 151, 459.

Sunyaev, R.A. and Zeldovich, Y.B., 1970, Astrophys. Sp. Sci., 7, 3.

Wasserman, I., 1981, Astrophys. J., 248, 1.

Wyse, R., 1982, Cambridge University Thesis.

Zeldovich, Y.B., 1970, Astr. Astrophys. 5, 84.

Zeldovich, Y.B., Einasto, J. and Shandarin, S.F., 1982, Nature,
     300, 407.

ON THE NATURE OF THE FIRST STARS

Joseph SILK

Institut d'Astrophysique, 98bis, bd Arago, 75014 PARIS, France.

The first stars formed from matter that was almost comple-tely devoid of heavy elements. The lack of significant numbers of old disk stars with low metallicity signifies either that the ini-tial mass function (IMF) of the first disk stars was weighted towards massive, short-lived stars, or else that the rate of star formation was reduced when the metallicity was low (Audouze and Tinsley, 1976). This problem arises because star formation is an ongoing process in the disk. However in the halo, which contains the oldest stars, gas is not retained. Consequently, any two-component model of our galaxy is consistent with a Salpeter-like IMF for the halo stars (Ostriker and Thuan 1975), as indeed are the observations of metal-poor halo stars (Caloi et al., 1983). This means that even the first stars which formed out of primor-dial gas may have had masses in the range 0.1-100 $M_\odot$. By first stars, I shall mean those stars that generated the heavy elements seen in the most metal-deficient Population II stars. For most as-pects of star formation, a metallicity $\lesssim 10^{-3}$ that of the sun is indistinguishable from zero metallicity (Silk 1977).

Further information on the IMF of the first stars is obtai-ned from studying abundances in extreme Population II stars. These show an enhancement in [0/Fe] $\approx$ 0.7 dex relative to the sun, while [C/Fe] $\approx$ [CNO/Fe] $\approx$ 0 dex and is not enhanced (Sneden et al. 1979). This is suggestive of a precursor population which inclu-ded an excess (relative to the present IMF) of massive stars.

The enrichment of the intergalactic gas detected in rich galaxy clusters was also produced at an early stage of galactic evolution by a generation of stellar precursors to the observed Population II stars in cluster ellipticals. This follows simply

J. Audouze and J. Tran Thanh Van (eds.),
Formation and Evolution of Galaxies and Large Structures in the Universe, 253–258.
© 1984 *by D. Reidel Publishing Company.*

from the low rates of stellar mass loss at the present epoch.
The discovery that [O/Fe] $\approx$ 0.7 dex in the Virgo cluster gas
(Canizares et al. 1982) while [Fe] $\approx$ [S] $\approx$ [Si] $\approx$ -0.5 dex
(Mushotzky et al. 1981) is suggestive of an origin in stars of
mass similar to those responsible for extreme Population II
abundances. The difference in [Fe] between the two sites in
our halo and in the intergalactic medium where [O/Fe] is
enhanced presumably reflects the continuous history of star
formation in our galaxy. On the other hand, primordial enriched
debris accumulates in the intergalactic medium where it is effectively undiluted by the much reduced mass loss from the observed
Population II stars.

In summary, observational evidence suggests that the mass
range of the first stars extended over 10-100 $M_\odot$ and perhaps up
to several hundred $M_\odot$, since only such stars are capable of
accounting for the nucleosynthetic evidence. In addition, the
discovery (Bessel and Norris 1981) of a halo star with [Fe] $\approx$
-4.5 dex indicates that the IMF of the first stars extended to
below 1 $M_\odot$. Perhaps their IMF was not too different from that of
stars forming at the present epoch, apart from the enhanced contribution by massive stars. The goal of the ensuing discussion is
to seek a theoretical understanding of the IMF of the first stars.

The initial conditions prior to formation of the first
stars are reasonably unambiguous in the context of the conventional hot big bang cosmology. One has an initial composition
with helium mass fraction Y $\approx$ 0.25 and heavy element fraction
Z $\approx$ 0. The two alternative models for the growth by gravitational instability of primordial density fluctuations both predict
that the first bound units in which star formation occurs are
uniform clouds of mass 10^6 -10^8 $M_\odot$. The lower value corresponds
to the Jeans mass immediately after the decoupling of matter and
radiation at redshift z $\approx$ 1000. This scale characterizes a primordial isothermal fluctuation spectrum, in which the first nonlinear lumps form above the Jeans mass at z < 1000, and undergo
hierarchical clustering to eventually form galaxies at z $\lesssim$ 100
(Peebles and Dicke 1968). The larger mass for the initial clouds
out of which galaxies may have formed is characteristic of the
minimum fragmentation scale of massive primordial adiabatic
fluctuations. All preexisting primordial substructure was erased
by radiative smoothing of these fluctuations on scales below
~ 10^{14} $M_\odot$ prior to the decoupling epoch. Adiabatic fluctuations
first become large and non-linear, and undergo pancake-like asymmetric collapse at z $\lesssim$ 10_8, the pancakes being unstable to fragmentation on scales $\gtrsim$ 10^8 $M_\odot$ (Sunyaev and Zel'dovich 1972). In
either situation, the precursor clouds where the first stars
formed were initially uniform on stellar mass-scales. What
process then determines the IMF of the first stars ?

One process which has received much attention is opacity-limited gravitational fragmentation. Hoyle (1953) first pointed out that since the Jeans mass is proportional to $T^{3/2}\rho^{-1/2}$, it will continuously decrease during the early isothermal stage of cloud collapse, allowing fragmentation to occur. The principal problem is that the growth time for density fluctuations is $\sim(G\rho)^{-1/2}$, while the density itself is changing on a similar time-scale. Hence one does not find the exponential growth rates that normally apply to instabilities for density fluctuations of scale exceeding the Jeans length, but rather only an algebraic growth rate. This means that the initial fluctuations in density or velocity play a crucial role in determining the final fragmentation scale. For a sufficiently uniform cloud, fragmentation may not occur at all prior to the cloud becoming opaque. In this situation, one might expect to form a supermassive object, with little prospect of nucleosynthesis of heavy elements prior to catastrophic collapse.

The following example illustrates this difficulty. In uniform, spherical collapse of a sufficiently cold cloud, small fluctuations in density grow according to

$$\delta\rho/\rho = \delta_i (1 - t/t_{ff}^i)^{-1},$$

where δ_i is the initial fluctuation amplitude and $t_{ff}^i = (3\pi/32G\rho_i)^{1/2}$ is the initial free-fall time at density ρ_i. The density fluctuations first become large when the mean cloud density has increased to $\rho \approx \delta_i^{-2} \rho_i$. Only if at this stage is the density high enough that any given scale is opaque has one some hope of forming a protostar on this scale. The minimum fragment mass that can form a protostar will then be given by the Jeans mass, which however depends on ρ and therefore on the initial fluctuation level δ_i. Allowance for such complications as finite pressure (Tohline 1980) or aspherical collapse (Silk 1982) does not help avoid this conclusion. If the initial fluctuations are sufficiently great ($\delta_i > 0.1$), there may still be reason to hope that the minimum Jeans mass could be indicative of the minimum masses of protostars. Such large fluctuations are likely to be present in interstellar clouds once substantial star formation has already occurred, since preexisting stars are responsible for driving sizeable inhomogeneities via the action of stellar winds, expansion of HII regions, and supernova explosions. However primordial clouds are quite another matter, and are uniform on stellar scales at the onset of collapse.

The resolution of the puzzle of primordial star formation is attained, as I will now argue, by consideration of the thermochemical stability of clouds of primordial abundance. Such clouds are found to be thermally unstable, with exponential growth of

fluctuations over well-defined mass scales. The thermal instability is associated with the rapid formation of molecular
hydrogen at sufficiently high density. Excitation of rotational
and vibrational levels of H_2 by collisions provides the principal cooling mechanism, and is effective between temperature $\sim$
300 K and $\sim$ 2000 K, above which the H_2 is destroyed. Below a density of $\sim 10^9$ atoms cm^{-3}, H_2 molecules form by attachment of residual electrons to H atoms,

$$H + e^- \longrightarrow H^- + \gamma,$$

followed by $H^- + H \longrightarrow H_2 + e^-$

At higher densities, the 3-body process (Palla et al. 1983)

$$3H \longrightarrow H_2 + H$$

dominates. In the regime where the optical depth τ in the H_2
rotational lines is of order unity and 3-body formation of H_2
prevails, the molecular hydrogen fraction rises rapidly. It is
in this regime that the thermal instability occurs. An infinitesimal density fluctuation produces a significant increase in the
H_2 abundance, which in turn enhances the cooling. Since the
temperature will then slightly decline, the density fluctuation
which remains in pressure equilibrium is further enhanced. Full
details of the linearized stability calculation are given by
Silk (1983), where the following dispersion relation is derived
for the growth (as exp wt) of fluctuations :

$$w^2 + w[t_f^{-1} + 0.3 t_c^{-1}(5-h)] - (3/2 t_c t_f)[(T_o/T)(1-h) - 1 - h] = 0$$

Here t_c is the H_2 cooling time-scale, t_f is the H_2 formation
time-scale, $T \equiv 53000$ K, and h is a function of optical depth
equal to $\tau/2$ for $\tau \ll 1$ and unity for $\tau \gg 1$. A large-scale
velocity gradient model was adopted to solve the H_2 line transfer in the collapsing cloud.

Now t_f may be an order of magnitude less than the instantaneous free-fall time t_{ff} when the molecular hydrogen fraction
is rising. Since $t_c \gtrsim t_{ff}$ ordinarily, with equality if the gravitational binding energy released from cloud collapse is available for heating, one can readily see that the condition $t_f \ll$
t_{ff} is necessary for instability. An additional requirement for
instability is $\tau \sim 1$. These conditions effectively constrain the
cloud temperature $T \approx 1500$ K, the density $n = 10^9 - 10^{10}$ cm^{-3},
and the molecular hydrogen fraction to $10^{-1.5} - 10^{-3}$. The maximum
growth rate for the instability is about $20 t_{ff}$. The maximum
scale of the instability corresponds to a mass of about $M_{th} =$
$0.1\ M_\theta$, at a density and temperature where the Jeans mass $\sim$
$10^3\ M_{th}^\theta \sim 100\ M_\theta$.

This result has the following significance. Clouds of primordial composition, which cool by excitation of H_2 rotational levels (that is to say, with $[Z] \lesssim -3$ dex), are thermally unstable and form clumps of mass $\lesssim 0.1\ M_\odot$. These clumps are not gravitationally bound but are in pressure equilibrium. However the Jeans mass when the clumps form is $\sim 100\ M_\odot$, and this means that the clumps will aggregate together via gravitational instability on this scale. As the Jeans mass decreases during the ensuing collapse, smaller and smaller aggregations become bound. A nearly flat spectrum of bound clumps is produced, extending from $\sim 0.1\ M_\odot$ to $\sim 100\ M_\odot$. These may be identified with the first stars. Thus the IMF of the first stars appears to be enhanced in massive stars of $\lesssim 100\ M_\odot$ relevant to the Salpeter IMF, but extends down to $\sim 0.1\ M_\odot$. Moreover this conclusion is exclusive to clouds with metallicity $\lesssim 10^{-3}$ that of the sun. Once heavy element cooling becomes important, the thermal instability no longer operates. Presumably star formation then operates by processes similar to those occurring in the present interstellar medium.

The thermal instability depends on a local criterion, and should not be grossly modified by such effects as rotation or anisotropic collapse. It seems that without rather contrived initial conditions, supermassive objects are unlikely to be the first objects to condense out of pregalactic clouds. Rather, the first stellar objects are stars in the conventional mass range of $0.1\ M_\odot$ to $100\ M_\odot$. This means that pregalactic and protogalactic nucleosynthesis should produce abundance ratios of heavy elements in concordance with what is observed in old halo stars. The excess of primordial massive stars can result in an oxygen excess, and may also produce a considerable amount of helium.

This research has been supported in part by NASA and by the CNRS. I gratefully acknowledge the hospitality provided by Jean Audouze at the Institut d'Astrophysique.

REFERENCES

Audouze,J. and Tinsley,B.: 1976, Ann. Rev. Astron. Ap.,14,43.
Bessel,M. and Norris,J.: 1981, Astrophysical Parameters for
        Globular Clusters, Ed. A.G. Davis Philip and D.S. Hayes
        (New York: L. Davis Press), p. 137.
Caloi,V., Castellani,V. and Gerola, H.: 1983, preprint.
Canizares,C., Clark,G., Jernigan,J. and Markert,T.: 1982, Ap.
        J., 262,33.
Hoyle,F.: 1953, Ap. J., 118,513.
Mushotzky,R., Holt,S., Smith,B., Boldt,E. and Serlemitsos,P.:
        1981, Ap. J. Letters, 244;L47.
Ostriker,J.P. and Thuan,T.X.: 1975, Ap. J., 203,353.
Palla,F., Stahler,S. and Salpeter,E.: 1983 Ap. J. (in press).
Peebles,P.J.   and Dicke,R.N.: 1968, Ap. J., 154,891.
Silk,J.: 1977, Ap. J., 211,638.
Silk,J.: 1982, Ap. J., 256,514.
Silk,J.: 1983, M.N.R.A.S. (in press).
Sneden,C., Lambert,D. and Whitaker,R.: 1979, Ap. J., 234,964.
Sunyaev,R.A. and Zel'dovich,Ya.B.: 1972, Astron. Ap., 20,189.
Tohline,J.: 1980, Ap. J., 239,417.

ON THE FORMATION OF POP.II STARS VIA THERMAL INSTABILITY

H. Zinnecker and S. Drapatz

Max-Planck-Institut für Physik und Astrophysik
Institut für extraterrestrische Physik
8046 Garching, F.R.G.

ABSTRACT

The fragmentation of a collapsing uniform protogalactic cloud of metal-poor gas is investigated. It is shown that thermal instability is able to induce the fragmentation at an early, diffuse stage of the cloud. The mass scales associated with this non-gravitational instability depend on metallicity and can lead to the formation of a first generation of massive Pop.II stars which seems to be required by the observational data on halo subdwarfs.

1. INTRODUCTION

The problem of the formation of Pop.II stars (halo stars) was originally addressed by Hoyle (1953) in his classic paper on hierarchical fragmentation of a protogalactic gas cloud. While that theory of (Pop.II) star formation is based on gravitational instability on successively smaller scales, this paper is going to investigate another possibility: fragmentation driven by thermal instability.

It is our working hypothesis that star formation in the protogalactic halo first (i.e. at very low metallicities) favored the formation of high mass stars which would rapidly enrich the halo gas with heavy elements; only then (i.e. at moderately low metallicities) the formation of low mass halo stars could proceed. A first generation of high mass Pop.II stars (cf. Twarog & Wheeler 1982, Chiosi & Matteuci 1983) would account for the observational fact that oxygen is relatively overabundant with respect to iron in metal-poor halo subdwarfs (Sneden et al. 1979, Clegg et al. 1981). In addition, a rapid chemical

J. Audouze and J. Tran Thanh Van (eds.),
Formation and Evolution of Galaxies and Large Structures in the Universe, 259–266.
© *1984 by D. Reidel Publishing Company.*

enrichment of the halo gas might explain the observation that extremely metal-poor low mass halo stars, i.e. subdwarfs with
$\log(Z/Zo) < -3$, do not seem to exist (Bond 1981). The situation of the
halo subdwarfs is reminiscent of the famous G-dwarf problem in the
galactic disk for which one way of solving it was to postulate an
Initial Mass Function (IMF) richer in massive stars at early times in
the disk when the metal abundance there was still low (Schmidt
1963). At this point, one is prompted to ask whether there is a
theoretical process for Pop.II star formation in which the Pop.II IMF
is metallicity-dependent. Here, we shall propose that metallicity-
affected-thermal instability could be such a process (we propose to
call it the MATHIN-process). A related concept is the concept of
metal-enhanced star formation (MESF) introduced by Talbot & Arnett
(1973) in which it is the star formation rate rather than the Initial
Mass Function that depends on metallicity. (The star formation is
suggested to be higher in regions with above average metallicity due
to the enhanced heavy element cooling of these regions.)

Literature related to the present investigation includes Hartwick
(1976) and Searle (1977) as for chemical evolution of the halo,
Pikelner (1976) and Gunn (1980) as for the origin of globular clusters,
and Yoshii & Sabano (1981) as for thermal instability.

2. SCENARIO

a) Initial Conditions

We envisage a large uniform gaseous protogalactic cloud which
is gravitationally collapsing. We assume that the cloud contains a
small mass fraction of heavy elements ($Z \gtrsim 10^{-5}$) owing to the
nucleosynthesis of a previous pregalactic generation of Pop.III stars
(e.g. Silk 1983). Moreover, the cloud is assumed to be chemically
homogeneous. The set of parameters adopted for the protogalactic
cloud of gas is summarized in Table 1 (cf. Rees & Ostriker 1977).

Table 1: Initial Conditions for the Protogalactic Cloud

mass M_{cl}	$\sim 10^{11}\ M_o$	temperature T	$\sim 10^4$ K
radius R_{cl}	~ 50 kpc	degree of ionization ξ	$\sim 90\%$
density n	$\sim 10^{-2}$ cm^{-3}	metallicity Z	$\gtrsim 10^{-5}$

The density and the degree of ionization are roughly comparable
to the hot component of the present interstellar medium, while the
temperature is taken to be lower because continuous supernova heating is not considered to occur then; thus, even if the cloud had been
at its virial temperature (3×10^5 K) in the beginning, the temperature

will relatively fast drop to $\sim 10^4$ K, the temperature above which the excitation of atomic hydrogen completely controls the cooling (the cooling time from 3×10^5 K to 10^4 K at $n \sim 10^{-2}$ cm^{-2} is $\sim 3 \times 10^5$ yr including random walk of Ly α photons, i.e. much shorter than the free fall time which is $\sim 3 \times 10^8$ yr). Note that by the time when $T \sim 10^4$ K most of the gas is still ionized, since ionization equilibrium has not yet been reached (the timescale for recombination in the absence of sources of ionization would be $\sim 10^7$ yrs $(T/10^4$ K$)^{1/2}(10^{-2}$ cm$^{-3}/n)$ or about a tenth of the free fall time. However, during the contraction phase below 10^4 K the degree of ionization will decrease substantially.

b) Evolution

The evolution of the protogalactic cloud depends on whether a small amount of heavy elements (C, N, O etc.) is present or not. If there were no heavy elements at all, the cloud would collapse isothermally at $T \sim 10^4$ K until the density becomes high enough that a small amount of molecular hydrogen forms which can act as a cooling agent. In case there is a small mass fraction of heavy elements (as assumed here), it is possible to bypass the "thermal trap" at $T \sim 10^4$ K at an earlier stage of contraction, i.e. at a stage when the cloud is still large (~ 10 kpc) and diffuse (~ 1 cm^{-3}).

The time evolution of the cloud is governed by the coupling of three system parameters: density, temperature, and the degree of ionization. The density is imposed at any instant by the free fall collapse behaviour. The density influences the temperature because it enters the cooling rate as well as the heating rate. The temperature is governed by the thermal balance between the cooling rate C and the heating rate H. Compressional energy input during contraction ("pdV-work") heats the gas, while cooling of the gas is due to hydrogen or electron impact excitation of low lying levels in OI and CI, and probably CII, SiII, FeII if uv background radiation exists. Finally, we assume collisional ionization to be dominating. Then the degree of ionization ξ is a very sensititve function of the temperature and independent of the density as soon as ionization equilibrium is obtained. If $\xi > 10^{-2}$ electron impact excitation dominates, while for $\xi < 10^{-4}$ hydrogen impact becomes the relevant excitation mechanism (see Dalgarno & McCray 1972). The cooling rate is linearly proportional to the metallicity Z, since the protogalactic cloud is optically thin to the radiation of the heavy elements (volume cooling, the OI line with highest opacity starts to become optically thick at line center for the whole cloud at $n \gtrsim 100$ cm^{-3}). Detailed investigations of fine-structure transition excitations have been carried out by Smeding and Pottasch (1979). Cooling by dust grains, if existent, would be irrelevant at the low densities in which we are interested here (no coupling of the dust grains to the gas). Furthermore, we emphasize that collisional deexcitation does not quench radiative cooling (e.g. of OI lines) for hydrogen densities $n \leq 10^5$ cm^{-3} around $T \sim 5000$ K (cross sections from Launay & Roueff 1977), so that photons in higher opacity lines can penetrate the cloud by random walk.

Thermal instability proceeds as follows: Gas in a density enhancement cools faster than its surroundings such as to restore pressure equilibrium. For spherically symmetric perturbations and neglecting the effects of a weak magnetic field, the pressure gradient across the boundary of the perturbation initiates an inward radial flow, so that eventually the core region of the condensation becomes dense (and cool) enough to be gravitationally bound and to collapse to a stellar object (see Stein et al. 1972, Hunter 1969).

3. INVESTIGATION OF THE THERMAL INSTABILITY AND FRAGMENTATION

a) Fragmentation criteria

Two conditions must be fulfilled for thermally driven fragmentation to occur:

i)   The isobaric thermal instability criterion must be met (the condensation mode is due to the isobaric perturbations: see Field 1965).

ii)   The cooling time of the density fluctuation must be shorter than about half the free fall contraction time of the cloud as a whole (otherwise the fragments would not become isolated against the collapsing background).

Let us investigate these two conditions in turn. Firstly, to check the instability criterion, we have to study the thermal properties of the cloud, i.e. we need the heat-loss function L, defined as energy losses minus energy gains per gram of material and per second. Given L = H-C, the instability criterion reads (Hunter 1970, Shchekinov 1978)

$$(\partial L / \partial T)_n - n_0/T_0 (\partial L / \partial n)_T - L/T_0 < 0 \qquad (1)$$

where n_0 and T_0 are the unperturbed density and temperature at the onset of the thermal instability, respectively. For L = 0 (thermal equilibrium) equ. (1) becomes the criterion already derived by Field (1965). Note that for a uniformly collapsing cloud starting from rest the motion has no effect on the stability criterion (Arny 1967).

Using the density and temperature dependence of H and C, i.e.

$$H \approx 3 \, nkT/t_{ff} \qquad (2a)$$

where $t_{ff} \propto n^{-1/2}$, and

$$C = n^2 \Lambda (T) \; Z/Z_0 \qquad\qquad (2b)$$

equ. (1) translates into

$$d\ln \Lambda/d\ln T < 2 - H/2C \qquad\qquad (3)$$

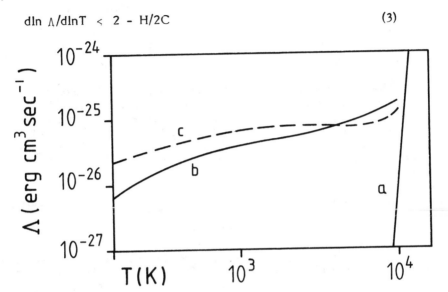

Fig. 1: Cooling function Λ (T) for the following processes: a) Collisions of electrons and H atoms (Dalgarno and McCray 1972), when
ionization is caused by electron collisions and the recombinations are
radiative. b) Excitation of ^3P fine structure levels of O (5 times solar
abundance) and C (solar abundance); the function at $T \lesssim 10^3$ K
(Dalgarno and McCray 1972) is extended to higher temperatures $\propto$
$T^{1/6}$ (Bahcall and Wolf 1968) and excitation of metastable ^1D levels
of O and C (estimated after Bahcall and Wolf 1968) is included. c)
Hydrogen atom and electron excitation of gas particles with solar
abundances and fractional ionization 10^{-2} (Dalgarno and McCray 1972)
is presented for comparison. In b) H-atom collisions are considered only.

The cooling function Λ (T) is plotted in Fig. 1. As can be seen
from this figure, for $T \leq 5000$ K $d\ln \Lambda /d\ln T \sim 1$ when oxygen is
overabundant relative to carbon from Pop.III enrichment (Ober et al.
1983), what has been assumed here. Thus, from equ. (3), we have to
require $H \leq 2C$ in order to meet the thermal instability criterion.
Moreover, since

$$H \approx 10^{-27} \; (T/5000 \; K)(n/1 \; cm^{-3})^{3/2} \; erg/(cm^3 \; sec) \qquad (4a)$$

while at $T = 5000$ K

$$C \approx 10^{-25} \, (Z/Z_o)(n/1 \; cm^{-3})^2 \; erg/(cm^3 \; sec) \tag{4b}$$

we infer that a fractional (oxygen) abundance $Z_{oxygen} \gtrsim 5 \times 10^{-3} Z_o$ is about what we need to satisy $H \leq 2C$. Indeed, $Z_{oxygen} \sim 5 \times 10^{-3} Z_o$ is consistent with the adopted initial metallicity $Z \sim 10^{-3} Z_o \sim 10^{-3}$ (see Table 1), because metallicity is usually defined in terms of heavy elements other than oxygen. We conclude that thermal instability is possible in a metal-poor $(Z > 10^{-5})$, diffuse $(n \sim 1 \; cm^{-3})$, neutral $(T \sim 5000 \; K)$ protogalactic hydrogen cloud. Since oxygen is a major cooling agent young galaxies should be powerful OI emitters.

The next question is: will fragmentation actually take place? As indicated by the second condition above, this question can be answered by comparing the fragment cooling timescale with the overall collapse timescale for the protogalactic cloud. Taking $Z = 5 \times 10^{-3} Z_o$, we find that the cooling timescale $(t_{cool} = nkT/C \sim n^{-1})$ becomes shorter than half the free fall time $t_{ff} = 10^{15} (n/1 \; cm^{-3})^{-1/2} \; sec$ for

$$n \gtrsim 8 \; cm^{-3} \tag{5}$$

largely independent of temperature, because approximately $C(T) \sim T$ for $T \leq 5000 \; K$ as seen from our $\Lambda(T)$-plot. The temperature at $n \sim 8 \; cm^{-3}$ when fragmentation induced by thermal instability becomes first possible should be $T \sim 5000 \; K$, for cloud cooling can begin only as soon as $t_{cool} < t_{heat} \sim t_{ff}/2$, i.e. at $n > 8 \; cm^{-3}$. So we expect isobaric clumping at a pressure $P = nkT = 5.5 \times 10^{-12} \; erg/cm^3$.

Thermal evolution of the cloud is difficult to predict without a detailed calculation (see e.g. Hunter 1969, Miki & Nakano 1975). It also depends very much on the assumptions made about the ionisation process, and the presence of external heat sources (quasars?).

b) Mass Scales

What is the characteristic mass scale of a condensation formed by thermal instability? Since the characteristic lengthscale for (isobaric) thermal instability is given by $\lambda_{th} \approx a_s t_{cool}$ where a_s is the sound speed at the unperturbed temperature and t_{cool} is the cooling time of the gas in the perturbation, the mass of the incipient thermal condensation will be $M_{th} = \rho_o (a_s t_{cool})^3$, a fraction f of which will end up in a stellar object. Thus the mass of the first prospective Pop.II stars will be

$$M_{*II} \sim f \, \rho_o \, (a_s \, t_{cool})^3 \tag{6}$$

where the index II stands for Pop.II, and $f \gtrsim 0.016$ (according to Stein et al. 1972). Inserting numbers ($\rho_o \sim 1.8 \times 10^{-23} \; g/cm^3$, $a_s =$

8×10^5 cm/sec, $t_{cool} \sim 1.25 \times 10^{14}$ sec corresponding to $n \sim 8$ cm^{-3}, $T \sim 5000$ K, and $Z \sim 5 \times 10^{-3}$ Z_0),

$$M_{*II} \gtrsim 100 \ M_0 \tag{7}$$

Note that this mass is only a very rough figure. Notice, however, that the Jeans mass for the above density and temperature by far exceeds the characteristic mass scale (7) derived from the thermal instability picture ($M_{Jeans} \sim 2 \times 10^7$ M_0). Concerning the dependence of the mass scale on metallicity, we find $M_{*II} \sim Z$, since the density at the onset of fragmentation $\rho_0 \sim Z^{-2}$ and $t_{cool} \sim (Z \rho_0)^{-1} \sim Z$. (The temperature at the onset of fragmentation should not depend on Z.) Mind that $M_{*II} \sim Z$ pertains only to first-generation Pop.II stars.

4. OUTLOOK

The whole picture of Pop.II star formation becomes very complicated once the first generation of massive stars has formed. Clearly the gas will become very inhomogeneous with shock dissipation occurring, which enhances the cooling rate locally (Rees & Ostriker 1977, Struck-Marcel 1982). This effect may be more important than the enhancement of the cooling rate due to newly injected heavy elements from those stars. The heat input from supernovae may exceed that from the protogalaxy's gravitational contraction and our thermal instablity criterion does no longer apply in this case. Moreover, in order to enrich 10^{11} M_0 to $Z \sim 0.1$ Z_0 requires $\sim 10^8$ M_0 of injected metals, i.e. $\sim 10^7$ supernovae, each releasing $\sim 10^{51}$ erg into the system, hence the total energy released amounts to $\sim 10^{58}$ erg, which is about the binding energy of the system, if the system size is 5 kpc (corresponding to a mean density $n \sim 8$ cm^{-3}). Given all these problems, there is still a long way to go until we will have a firm understanding of Pop.II star formation, which will eventually enable us to understand the origin of elliptical versus spiral galaxies.

ACKNOWLEDGEMENT

H.Z. would like to thank Prof. R. Wielen who stimulated the present investigation without knowing it in a talk at the Heidelberg Workshop on star formation in October 1982. Discussions at the MORIOND meeting in March 1983 have helped to improve the paper.

REFERENCES

Arny, T., 1967, Astrophys. J. 149, 391.
Bahcall, J.N. and Wolf, R.A., 1968, Astrophys. J. 152, 701.
Bond, H.E., 1981, Astrophys. J. 248, 606.

Chiosi, C. and Matteuci, F., 1983, ESO-Workshop on Primordial
    Helium.
Clegg, R.E.S., Lambert, D.L. and Tomkin, J., 1981, Astrophys. J. 250,
    262.
Dalgarno, A. and McCray, R.A., 1972, Ann. Rev. Astron. Astrophys.
    10, 375.
Field, G.B., 1965, Astrophys. J. 142, 531.
Gunn, J., 1980, in Globular Clusters, p. 301 (Hanes and Madore, eds.;
    Cambridge Univ. Press).
Hartwick, F.D.A., 1976, Astrophys. J. 209, 418.
Hoyle, F., 1953, Astrophys. J. 118, 513.
Hunter, J.H., 1969, M.N.R.A.S. 142, 473.
Hunter, J.H., 1970, Astrophys. J. 161, 451.
Launay, J.M. and Roueff, E., 1977, Astron. Astrophys. 56, 289.
Miki, S. and Nakano, T., 1975, Publ. Astron. Soc. Japan 27, 147.
Ober, W.W., El Eid, M.F. and Fricke, K.J., 1983, Astron. Astrophys.
    119, 61.
Pikelner, S.B., 1976, Sov. Astron. 20, 255.
Rees, M. and Ostriker, J., 1977, M.N.R.A.S. 179, 541.
Schmidt, M., 1963, Astrophys. J. 137, 758.
Searle, L., 1977, in "The Evolution of Galaxies and Stellar
    Populations", p. 219, (Tinsley and Larson, eds.; Yale Univ. Obs.,
    New Haven).
Shchekinov, Yu.A., 1978, Sov. Astron. 22, 182.
Silk, J., 1983, ESO-Workshop on Primordial Helium.
Smeding, A.G. and Pottasch, S.R., 1979, Astr. Astrophys. Suppl. 35,
    257.
Sneden, C., Lambert, D.L. and Whitaker, R.W., 1979, Astrophys. J.
    234, 964.
Stein, R.F., McCray, R. and Schwarz, J., 1972, Astrophys. J. 177,
    L125.
Struck-Marcel, C., 1982, Astrophys. J. 259, 127.
Talbot, R.J.Jr. and Arnett, W.D., 1973, Astrophys. J. 186, 69.
Twarog, B.A. and Wheeler, J.C., 1982, Astrophys. J. 261, 636.
Yoshii, Y. and Sabano, Y., 1981, IAU-Symp. 93, p. 68.

POSSIBLE SYNTHESIS OF DEUTERIUM AND LITHIUM BY PREGALACTIC MASSIVE STARS

Jean Audouze and Joseph Silk

Institut d'Astrophysique du CNRS
98bis Bld Arago
75014 Paris - France

Comparison of the primordial abundances of D, ^{3}He ^{4}He and ^{7}Li (deduced from the available observations) and the theoretical production of these elements in the framework of the best "cosmonical" Big Bang models reveals some discrepancies between these two sets of values. While the discrepancies may turn out to be largely due to various systematic uncertainties it is nevertheless important to consider the possibility of the nucleosynthesis of these elements in pregalactic massive stars. We argue that one cannot rule out a model in which D (and ^{7}Li) are produced by cosmic ray spallation reactions induced by ^{4}He accelerated in the envelopes or in the winds of first generation (metal free and massive) stars on the interstellar primordial hydrogen atoms and where ^{3}He and ^{4}He are produced by similar stars. The cosmological consequences of such models are briefly outlined.

I INTRODUCTION AND MOTIVATIONS

Most recent papers devoted to primordial nucleosynthesis during the Big Bang (see eg Steigman 1983, Beaudet and Reeves 1983) claim that there is a very striking agreement between the observed abundances of the light elements D, ^{3}He, ^{4}He and ^{7}Li and the values deduced from the current models. The following implications which are discussed in these papers are generally accepted :
 1) From D and ^{7}Li the baryonic density should be much lower than the critical density. If neutrinos have a mass lower than 50 eV the Universe should be open.

J. Audouze and J. Tran Thanh Van (eds.),
Formation and Evolution of Galaxies and Large Structures in the Universe, 267–270.

2) From ^{4}He there should be a limitation on the number of different families of leptons (and neutrinos) ; the best determination seems to be consistent with at most three different families of leptons. In view of the recent discovery of the tau lepton, this consequently means that we have now discovered all types of existing leptons.

By contrast, in a few papers (Stecker 1980, Rana 1982, Gautier 1983, Audouze 1983, Vidal-Madjar and Gry 1983) attention is drawn on the fact that the agreement between the theoretical estimates and the observational constraints concerning the abondances of the light elements might not be finally settled. Vidal-Madjar 1983 has displayed convincingly the fact that in the case of the existence of three different neutrinos one does not find a single value of the parameter η (the ratio between the number of baryons and the number of photons which is another way to express the present baryon density) capable to simultaneously account for the abundances of D, ^{4}He and ^{7}Li (see Gry et al in these proceedings for a more complete presentation of this point).

This prompted us to reexamine different hypotheses regarding the origin of the light elements. In a few contributions (Epstein 1977 - Woltjer 1981) it has been previously proposed that might be synthetized by spallation processes during the pregalactic era while ^{4}He would be produced by Big Bang nucleosynthesis. This idea encounters the fatal difficulty of a catastrophic overproduction of ^{6}Li and ^{7}Li due to the contribution of the ^{4}He + ^{4}He reactions. In this presentation we propose that not only D but also ^{3}He, ^{4}He and ^{7}Li are produced during this pregalactic era by first generation massive stars.(This idea has also been presented in Audouze and Silk 1983).

The next section provides the astrophysical framework for the pregalactic formation of the light elements. The implications of this hypothesis are briefly presented in section III.

II INTERACTION BETWEEN PREGALACTIC MASSIVE STARS AND THE INTERSTELLAR MEDIUM

The first hypothesis of the scenario presented here is to assume that the primordial phases of the Big Bang have been such that the resulting material is only made of pure hydrogen. This could be a consequence of the fact that these phases have not been hot enough for nucleosynthesis to take place or that neutrons were already absent when the temperature was suitable ($T \simeq 10^9$ K) for the occurence of these processes. In order to account for the early production of ^{4}He we make the second assumption about the formation of a first generation of very

massive stars   as suggested by Silk (these proceedings). When
they form, these stars are made of pure H. These stars are
stable for a sufficiently long time to allow them to undergo
nucleosynthetic processes producing He and metals in their cores
(Ober private communication). It appears that stars with masses
in the range of 100-1000 M☉ produce ^{4}He copiously during their
evolution (see eg Talbot and Arnett 1971 and Bond et al 1983). A
third assumption supported in particular by Cassé and Paul 1982
and Maeder 1983 is that massive stars are important sources of
cosmic rays. Therefore some of the He nuclei which are
synthesized inside the massive stars and which will eventuallly
contaminate the primitive H gas are accelerated and can induce
spallation reactions on the interstellar H nuclei. In this way
one would avoid the overproduction of Li by He + He spallation
reactions.

Deuterium is produced directly through the ^{4}He (rapid and
stellar) + H (slow and interstellar) spallation reactions
leading to ^{3}He + D at low energy ($E \sim 25 - 50$ MeV) and to D +
nucleons at higher energies. From Meyer (1972) the overall
spallation cross section is ~ 50 mb. In Audouze and Silk (1983)
it is argued that the formation of D through the neutron
production and absorption processes is not significant unless
these processes take place in media dense enough ($n_H > 10^9$
cm^{-3}) for the neutron absorption to take place before their
decay . This would mean in this case that the spallation
processes have to take place in the envelopes of these stars.

Therefore with a D production cross section of about 50 mb,
integrated fluxes of $\gtrsim 210^{20}$ particles cm^{-2} of energy above
25 MeV are sufficient to account for D/H $\gtrsim 10^{-5}$. The occurence
of such fluxes cannot be ruled out by arguments based on energy
requirements : we showed in Audouze and Silk (1983) that if the
time available for this spallation process to take place is
about 10^8 years (corresponding to the cosmic ray life time or
the time scale of mixing of the stellar ejecta) the required
energy density is ~ 700 (E/100) ev cm^{-3} i.e. ~ 1000 times
larger than the present energy density in galactic cosmic rays.
This prescription is not unreasonable in a pregalactic phase
since quasars which are associated with this phase an known to
have luminosities at least 10^4 times larger than that of our
Galaxy.

One should stress that the acceleration of He rich galactic
cosmic rays originating from massive stars should take place
long before a significant contamination occurs of the
interstellar medium with He. Moreover in order to avoid any
overproduction of Li these massive stars cannot release material
in which Z/Y would be higher than 10^{-3}.

III CONCLUSION AND IMPLICATIONS

We acknowledge that in order to explain D, ^{4}He and ^{7}Li production in pregalactic phases we have proposed a somewhat contrived model : these massive pure hydrogen stars must last sufficiently long to accelerate He rich cosmic rays before the phases where they contaminate the interstellar medium into helium. Quantitative models have to be performed to follow the evolution of such stars and verify that in the accelerated material the $(Z/Y \stackrel{<}{\sim} 10^{-3})$ condition is effectively fulfilled. According to an interesting alternative scheme (D.D. Clayton, private communication), during the early pregalactic phases the accelerated particles may originate from the pure hydrogen interstellar medium and bombard the ambient helium rich material still embedded in and ejected by stars. In either situation, quantitative work remains to performed in order to put these ideas on firmer ground. If our models survive all objections they would have the dramatic implication that one could abandon the somewhat classical and universally accepted view of a very hot Big Bang. Such heresies have of course to be investigated with great caution but given the consequences of the current models of primordial nucleosynthesis, very alternative scenarios should be analyzed to see if they are viable.

REFERENCES

(1)   Audouze J., 1983 in ESO Workshop on primordial Helium eds.
        P.A. Shaver, D. Kunth and K. Kjär p.3
(2)   Audouze J., and Silk J., 1983 in ESO Workshop on primordial
        Helium eds. P.A. Shaver, D. Kunth and K. Kjär p.71
(3)   Beaudet G. and Reeves H. 1983 in ESO Workshop on primordial
        Helium eds. P.A. Shaver, D. Kunth and K. Kjär p.53
(4)   Bond R. Carr B. and Arnett W.D. 1983 MNRAS in press.
(5)   Cassé M. and Paul J.A. 1982 ApJ. 258, 860
(6)   Epstein R. 1977 ApJ 212, 595
(7)   Gautier D. 1983 in ESO Workshop on primordial Helium eds.
        P.A. Shaver, D. Kunth and K. Kjär p. 139
(8)   Maeder A. 1983 Astron. Astroph. 120, 130
(9)   Meyer J.P. 1972 Astron. Astroph. suppl. 7, 417
(10) Rana N. 1982 Phys. Rev. Letters 48, 209
(11) Stecker F.W. 1980 Phys. Rev. Letters 44, 1237
(12) Steigman G., 1983, in Eso workshop on primordial Helium
        eds. P.A. Shaver, D. Kunth and K. Kjär p.13
(13) Talbot R.J. and Arnett W.D. ApJ 170, 409
(14) Vidal-Madjar A. 1983 in Diffuse Matter in galaxies eds. J.
        Audouze, J. Lequeux, M. Levy and A. Vidal-Madjar. NATO
        ASI Series C110 Reidel Dordrecht p.57
(15) Vidal-Madjar A. and Gry C. 1983 Astron. Astroph. in press
(16) Woltjer L. 1982 in Astrophysical cosmology eds. H.A. Brück
        et al Pontificia Academia Scientiarium Scriptum Varia p.423

DEUTERIUM AND LITHIUM FROM POPULATION III REMNANTS

Martin J. Rees
Institute of Astronomy
Madingley Road
Cambridge CB3 OHA, England

ABSTRACT: Even if the mean baryon-to-baryon ratio in the early universe were high (corresponding to Ω_b in the range 0.1 - 1), D and Li could still be produced pregalactically. There may have been large-amplitude initial isothermal fluctuations, the material in the high density regions now being incorporated in Population III objects. Alternatively, spallation and neutron captive processes in material at kinetic temperatures 10 - 100 Mev around black holes could, if some of the material was subsequently ejected, produce D, and produce (or even overproduce) ^{7}Li.

1. INTRODUCTION

Primordial nucleosynthesis in the standard hot big bang yields gratifying agreement with the observed abundances of light elements, provided that the baryon density is fairly low, with $\Omega_b h^2 \cong 0.025$ (h being Hubble's constant in units of 100 km s^{-1} Mpc^{-1}). The dynamically inferred hidden mass in clusters and groups contributes $\Omega \gtrsim 0.1$ - marginally consistent with the orthodox nucleosynthetic inference if h $\cong$ 0.5, but apparently requiring some non-baryonic matter if h $\cong$ 1. Advocates of inflationary cosmology favour $\Omega \cong 1$ (which is compatible with the dynamical evidence if the hidden mass is less clumped than the galaxies even on scales as large as 10 h^{-1} Mpc), and this further fuels support for the idea that the hidden mass in non-baryonic.

In this brief contribution, I suggest that the case for non-baryonic matter is not quite so clear-cut, because if one indeed hypothesises a large amount of dark baryonic matter, one has (to be self-consistent) to explore other new effects that could

271

J. Audouze and J. Tran Thanh Van (eds.),
Formation and Evolution of Galaxies and Large Structures in the Universe, 271–277.
© 1984 by D. Reidel Publishing Company.

influence the abundance of light elements. If the hidden mass
is baryonic, it must now be in Population III objects or their
remnants (contributing a high M/L). In a baryon-dominated hot big
bang universe, large amplitude inhomogeneities in the baryon-
photon ratio are prerequisites for the formation of Population
III; and nuclear processes involving black hole remnants at
$z \lesssim 100$ may complicate the picture further.

In what follows, I shall neglect possible thermonuclear
effects due to Population III objects (cf. Arnett, Bond and Carr
1983); I shall also ignore more radical ideas involving cold big
bangs and non-primordial helium (see, for instance, the contribution by Audouze and Silk to these proceedings). Two separate
effects are nevertheless relevant to D and ^{7}Li production.

(a) Pregalactic Population III objects can form in a baryon-
dominated universe only if isothermal fluctuations of substantial amplitude are present before recombination. If the
photon-baryon ratio at nucleosynthesis varies from place to place,
so also will the primordial yield of light elements. In inter-
preting the measured abundances of these elements, we must note
that the observed baryons (i.e. those not incorporated in
Population III) are not a "fair sample", having come preferentially
from underdense regions.

(b) Population III may consist of very massive stars that evolve
into black hole remnants. These remnants will accrete intergalactic gas (especially at earlier epochs when the gas is denser).
Some of this gas may be ejected again, after having been
"processed" through an environment near the hole where ions
attain thermal kinetic energies of 10 - 100 Mev, and where
spallation can occur. The present light element abundances may
be modified as a consequence of this.

2. LARGE AMPLITUDE ISOTHERMAL FLUCTUATIONS

If the scales of structure now observed have formed hierarchically via gravitational clustering of smaller units since
the recombination time t_{rec}, then the masses of these units would
have to be in the range 10^6 - 10^8 $M_\odot$ (the precise value depending
on the fluctuation spectrum). In a baryon-dominated universe,
this scenario requires isothermal fluctuations frozen in before
recombination. Moreover, the amplitude must be $\gtrsim 1$ on the
relevant scales; the population III systems would then form from
overdense regions. In consequence, the gas that escapes incorporation into population III objects and remains available for
condensation into luminous matter would come selectively from
underdense regions. The yields from primordial nucleosynthesis
depend on the local value of the photon-baryon ratio. Therefore,

the material available for present-epoch observations could
have D, He and Li abundances characteristic of low n_b/n_γ, even
if $\langle n_b/n_\gamma \rangle$ were high enough to contribute $\Omega_b = 1$.

Large amplitude isothermal fluctuations have been discussed
by Hogan (1978). If we define $\delta = (n_b - \bar{n}_b)/\bar{n}_b$, then a region
with $\delta > 1$ can still be treated as a small-amplitude perturbation
of the cosmological model - in the sense of being a small perturbation of the metric - provided that, when the background
temperature is T,

$$\delta \times (n_b/n_\gamma) \frac{m_p c^2}{kT} (\ell/\ell_H)^2 \ll 1, \qquad (1)$$

ℓ being the lengthscale of the perturbation and ℓ_H the horizon
scale. Perturbations with δ as large as 10 - 100 satisfy this
requirement right back to the nucleosynthesis stage even on mass
scales as large as $10^6 M_\odot$. Isothermal perturbations of mass
$> 10^6 \delta^{-\frac{1}{2}} M_\odot$ would condense out immediately after recombination,
having maintained constant amplitude ("frozen in") until that
time - they would form Population III objects (either single very
massive objects (VMOs), or clusters of low-mass stars). Only the
underdense material between such perturbations would survive to
form the luminous content of galaxies. It is plainly possible to
envisage an $\Omega_b \cong 1$ universe with isothermal perturbations such
that (for instance) half the volume of the universe has
$n_b/n_\gamma \cong 3 \times 10^{-8}$ and half has $n_b/n_\gamma \cong 3 \times 10^{-10}$. If the high-
density matter were in lumps of $10^5 - 10^6 M_\odot$ then only the
chemical composition of the remainder (characteristic of a low
density universe) might now be measurable.

No very satisfactory model for isothermal fluctuations yet
exists (irregular deposition of entropy in a phase transition?
spatial variations in the CP-violation parameter?); there is
certainly no theoretical basis for specifying the spectrum such
fluctuations might have. The above example is merely illustrative
of the principle that a high initial n_b/n_γ or Ω_b can be compatible
with primordial production of D, ^{3}He and ^{7}Li.

3. EFFECTS OF VMO REMNANTS

The argument summarised in § 2 tells us that the light
element abundances in the gas that escapes incorporation into
Population III objects may be atypical of overall abundances.
These considerations are independent of whether the Population III
objects are "Jupiters", or collapsed remnants of VMOs. However,
if the hidden mass is in VMO remnants - black holes which perhaps
formed as early as z > 100 - then subsequent accretion onto these
objects may have further consequences for the light elements.

Accretion by pregalactic collapsed objects was considered
in some detail by Carr (1979). There are uncertain complications
connected with the motions of the objects through the gas, and the
way the accretion-powered heating of the gas modifies the
accretion rate. But a rough estimate can be made by supposing
that the accretion takes place from a homogeneous gas at $\sim 10^4$ °K,
of particle density 10^{-5} Ω_{gas} h^2 $(1 + z)^3$ cm^{-3}. If motions
relative to the gas are neglected, then

$$\frac{\dot{M}}{\dot{M}_{crit}} = 10^{-11} \left(\frac{M_h}{M_\odot}\right) (1 + z)^3 \, \Omega_{gas} \, h \qquad (2)$$

In this expression, $\dot{M}$ is parameterised in terms of $\dot{M}_{crit} =$
$4\pi GM_h \, m_p/c\sigma_T$, this being the rate that would yield a luminosity
L_{Edd} x (efficiency). Motions at > 10 km s^{-1} would reduce $\dot{M}$
below (2); on the other hand, if the gas were clumped $\dot{M}$ could be
higher.

Another relevant quantity is the fraction $\mathcal{F}$ of the gas that
can be processed onto the holes in the cosmological expansion
timescale $\sim 10^{10}$ $h^{-1}(1 + z)^{-\frac{3}{2}}$ $\Omega^{-\frac{1}{2}}$ yrs. If there are enough holes
to contribute $\Omega \cong 1$, then

$$\mathcal{F} \simeq 10^{-10} \left(\frac{M_h}{M_\odot}\right) h (1 + z)^{3/2} . \qquad (3)$$

We thus see that for VMO remnants in the plausible mass
range 10^3 - 10^6 $M_\odot$, $\mathcal{F}$ can be of order unity if the accretion
starts as early as $z \cong 100$. What, then, could be the consequences
for light element abundances at the present epoch?

Material falling towards a hole generally has too much angular momentum to permit radial infall: it will instead spiral
inwards on a timescale controlled by viscosity. When $\dot{M}$ is well
below $\dot{M}_{crit}$ (as is likely in the present context, from (2)), the
inward drift timescale - though longer than for radial free fall -
may still be shorter than the electron-ion coupling time. When
this is the case, the ions are unable to cool, because their
energy cannot be drained away (and then radiated) by the electrons.
The inflowing material will then form a thick torus, supported by
the pressure of the ions; the ions have thermal kinetic energies
of order the gravitational binding energy $m_p c^2 (r/r_g)^{-1}$, where r_g
is the gravitational radius, and reach > 100 Mev near the hole.
The physics of "ion-supported tori" is discussed in some detail
by Rees et al. (1982). The inward drift timescale for a thick
torus can be written α^{-1} x $t_{free\ fall}$, where α is the usual
viscosity parameter, and the condition for the ions to remain at
the virial temperature of 10 - 100 Mev is

$$\frac{\dot{M}}{\dot{M}_{crit}} < 50 \, \alpha^2 \tag{4}$$

This is equivalent to the requirement that, while stored in the torus, a typical ion moving with thermal velocity $(0.1 - 0.3)c$ cannot pass through more than ~ 10 gm cm^{-2}. (The cross-section for Coulomb energy exchange with ~ 10 Mev electrons being ~ 0.1 barns at these energies). Around a black hole, there will there-fore be a torus of very hot gravitationally-confined plasma. The particle densities would be $\sim 10^{18} (M_h/M_\odot)^{-1} (\dot{M}/\dot{M}_{crit})^{-1}$ $(r/r_g)^{-3/2}\alpha^{-1}$ cm^{-3} - quite low, particularly for large M_h. (The torus is of course not in LTE - it is generally optically thin - the γ-ray density being well below that of the particles.)

Spallation and neutron production can obviously occur under these conditions. But these processes would be completely irrel-evant to present-day observations (except for a possible γ-ray background) if all the material falling into the deep gravitational potential well were swallowed by the hole. There are, however, two reasons why there is a real likelihood that some of the material can be re-expelled:

(i) Accretion flows such as those satisfying (4) where radiative losses are inefficient, maintain an enthalpy close to the gravitational binding energy. Consequently, the internal energy (or P/ρ) of a fluid element need be enhanced only by a small fraction in order for the element to be able to escape. In a somewhat different context, Shakura and Sunyaev (1973) pro-posed a flow pattern in which material spirals inward at equatorial latitudes, but outwards in a cone around the rotation axis.

(ii) Another possibility is that a fraction of the material gets heated (e.g. by power extracted electromagnetically from the hole) and squirts out in jets.

The flow patterns near black holes are very poorly under-stood. We cannot reliably estimate what fraction of the material is expelled, nor how close it gets to the hole before its inward motion is reversed. The viscosity parameter α is very uncertain - its value is important because the density and the residence time of material both scale as α^{-1}, so the "grammage" experienced by each ion, for a given $\dot{M}$, scales as α^{-2} (cf. (4)).

Let us therefore consider in general terms the nuclear processes that could occur in the volume within ~ 100 gravitational radii of a hole, where ion kinetic temperatures would be $\gtrsim 10$ Mev, and their possible implications for light element abundances.

The path length experienced by a typical ion is $\mathcal{G} \sim 0.2 \, \alpha^{-2}$ $(\dot{M}/\dot{M}_{crit})$ gm cm^{-2}. Hence (4) implies an upper limit $\mathcal{G} \lesssim 10$ gm cm^{-2}. Since the spallation reactions have cross-sections of up to ~ 100 mb, the fraction of ^{4}He in the torus which is spalled could be almost of order unity - this fraction would of course be proportional to $\alpha^{-2}(\dot{M}/\dot{M}_{crit})$. Free neutrons released in these reactions would follow individual orbits around the hole until they freely decayed or were captured - in contrast to the ions and electrons whose bulk flow is expected to be fluidlike (cf. Rees et al. 1982). The orbital period at radius r is $\sim 10^{-5}(r/r_g)^{3/2}(M_h/M_\odot)$ sec. Thus, a cloud of neutrons can be maintained around the hole, each neutron surviving for a time which may exceed the inward drift timescale ($\sim \alpha^{-1} t_{free\ fall}$).

The implications for the light elements depend on the value of $\mathcal{F}_{exp}$ (less than, but perhaps comparable with, $\mathcal{F}$ (eqn (3)), the fraction of the intergalactic gas which is accreted into the potential well of a black hole and then ejected. They depend also on the mean value of $\mathcal{G}$, the "grammage" experienced by an ion while in the region where its energy is $\gtrsim 10$ Mev. The compounded uncertainties of the Population III mass spectrum and the physics of accretion prevent us from drawing firm quantitative inferences. However, we can state the following:

(i) To produce all the observed ^{7}Li would require only that $\mathcal{F}_{exp} \cong 10^{-6}$. Since a vastly higher value of $\mathcal{F}_{exp}$ lies within present uncertainties, this consideration places a significant constraint on VMOs, their redshift at formation, and accretion flows around their remnants.

(ii) Deuterium and ^{3}He production by black hole processes is further discussed by Ramadurai and Rees (1983). Although, at $z \gtrsim 100$, $\mathcal{F}_{ext}$ could be high enough ($>10^{-3}$) to yield all the observed D as spallation products, this possibility is apparently ruled out by the concurrent overproduction of ^{7}Li. (Associated γ-rays would then be absorbed (cf. Epstein (1977)) and neutrinos would now be redshifted to $\lesssim 1$ Mev). A more contrived scheme for producing D would be the following. Efficient spallation may occur in the torus, all the spalled material (including the excessive ^{7}Li, etc.) being swallowed by the hole. The free neutrons cloud may extend into a region (e.g. a jet or funnel along the rotation axis) where directed outflow is occurring. This outflow would involve material that either had remained too cool to have undergone spallation or else had been completely broken down into protons. Deuterium could be formed by neutron capture onto outflowing protons.

Acknowledgement   I am grateful to Dr S. Ramadurai for discussions of the ideas summarised in § 3.

REFERENCES

Arnett, W.D., Bond, J.R. and Carr, B.J.   1983.   Astrophys.J.
    (in press).
Carr, B.J.   1979, MNRAS, 189, 123.
Epstein, R.   1977, Astrophys.J., 212, 595.
Hogan, C.J.   1978, MNRAS, 185, 889.
Ramadurai, S. and Rees, M.J.   1983, in preparation.
Rees, M.J., Begelman, M.C., Blandford, R.D. and Phinney, E.S.
    1982, Nature, 295, 17.
Shakura, N.I. and Sunyaev, R.A.   1973, Astr.Astrophys., 23, 337.

CHEMICAL EVOLUTION OF THE LIGHT ELEMENTS
AND THE BIG BANG NUCLEOSYNTHESIS

C. Gry[1*], G. Malinie[2], J. Audouze[2], A. Vidal-Madjar[1]

1. Laboratoire de physique stellaire et planétaire,
   91370 Verrières-le-Buisson, France
2. Institut d'Astrophysique, 75014 Paris, France
* present address: Villafranca del Castillo Satellite
   Tracking Station P.O. Box 54065 Madrid, Spain

New observational results lead us to reconsider current ideas about the formation and evolution of the light elements. The purpose of this study is to see if their observed abundances can be accounted for by a coherent galactic evolution model and how the primordial abundances resulting from this analysis compare with the cosmological predictions of the canonical Big-Bang model.

I PRESENT KNOWLEDGE OF THE LIGHT ELEMENTS ABUNDANCES

a. Deuterium

The evaluations of the deuterium abundance in the nearby interstellar medium present a large scatter in the different lines of sight (see e.g. (1),(2)). This scatter was first explained by a segregating mechanism in the interstellar clouds (3), an interpretation which favoured a quite large value for the actual D/H ratio: 2.25 10^{-5}.

However the recent study of the line of sight towards Per (B0.5 III) (4) showed that the perturbation in the D/H determination could come from the star itself. The deuterium line profiles proved indeed to vary drastically in less than four hours (5) (see Figure 1). This phenomenon can be explained by the presence of a transient component of high velocity neutral hydrogen, which could be due to a shell ejected by the star and moving away with the wind (6). If this hypothesis is valid, the deuterium lines might be blended by high velocity hydrogen in

J. Audouze and J. Tran Thanh Van (eds.),
Formation and Evolution of Galaxies and Large Structures in the Universe, 279–290.

several of the lines of sight studied so far, thus leading to an
overestimation of the D/H ratio. So the net result of this
analysis is to shift downwards the previously admitted value :
Vidal-Madjar et al. (4) deduced the new value D/H = (5 ± 3)
10⁻⁶. One should note that this value is significantly lower
than the Solar System one determined from Voyager experiment on
Jupiter (7): D/H = (3.6 $_{-1.4}^{+1.0}$) 10⁻⁵.

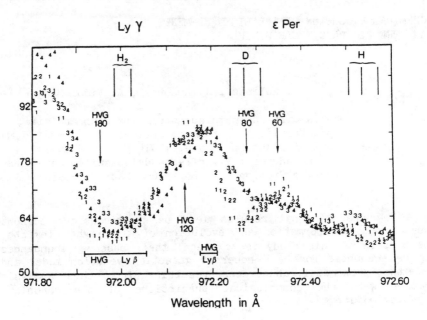

Figure 1: Blue wing of the Lyman γ profile in ε Persei, with
the deuterium line. The data are divided in four
successive spectra (1, 2, 3 and 4). One can note in the
deuterium line the effect of an hydrogen component with
a velocity of -80 km/s, which disappear in the four
hours separating spectrum 2 from spectrum 3.

b. helium

      The helium abundance is derived either in hot stars or,
with the best precision, through optical and radio recombination
lines in galactic or extragalactic HII regions. In the Solar
Neighborhood, this leads to the following value for the helium
mass fraction: Y = 0.28±02 (8). Since a linear relation between
Y and the metallicity Z is expected, the primordial helium
abundance Y_p can be extrapolated from the value of Y obtained
in HII region of different metallicities. The average of all

results presented by different observers and reviewed by Pagel
(8) leads to Y_P = 0.23±0.01. Very recently Kunth and Sargent
(9) derived the value Y_P = 0.245± 0.003 resulting from the
analysis of a sample of 12 metal poor galaxies. However their
result should rather be taken as an upper limit for the
primordial helium abundance, since they have not found any
correlation between Y and Z value in the sample and simply take
Y_P as the average of the Y values. In the Solar System, the
most recent determination carried out by Voyager in Jupiter and
Saturn (7) provides a strict upper limit for the helium abundance
at the birth of the Sun which is $Y_\odot \leqslant 0.24$.

c. Lithium

     Spite and Spite (10) have determined the lithium abundance
at the surface of very old halo stars. By arguing that ^{7}Li has
not been destroyed in these metal poor stars due to the reduction
of the convective zone, they deduce a value of the lithium
abundance as low as Li/H = (1.2 ± 0.3) 10^{-10}. The Solar System
value in meteorites (see e.g. (11)) is 10-20 times higher. This
implies a significant enrichment in lithium during the course of
the galactic evolution. Moreover, Ferlet and Dennefeld (10) have
determined the ^{7}Li/^{6}Li isotopic ratio in one line of sight
in the interstellar medium (ζ Oph line of sight). Their result is
^{7}Li/^{6}Li = 38.5 (from 25 to 180), which is surprisingly higher
than the Solar System value: $(^7$Li/^{6}Li$)_\odot$ = 12.5.

II THE CHEMICAL EVOLUTION OF DEUTERIUM AND LITHIUM

     The equations describing the evolution of the gas density
and of the abundance (by mass) Z_A of an element A can be
written in the following way (11) :

$$\frac{d\mu}{dt} = -\nu\,\mu\,(t) + \int_{m_1}^{m_2} \frac{E(m)}{m}\,\phi(m)\,\nu\,\mu\,(t-\tau_m)\,dm + \delta$$

$$\frac{dZ_A \cdot \mu}{dt} = -\nu\,\mu\cdot Z_A + \int_{m_1}^{m_2} E_A(m)\,\phi(m)\,\nu\,\mu(t-\tau_m)\,dm + \delta\,Z_{A\,\text{infall}}$$

in these equations ν is the rate of astration, (in the solar
neighborhood, ν =0.25 if there is no infall, and ν =0.4 to 0.6
with standard rates of infall, E(m) is the stellar gas fraction
returned by a star of mass m to the interstellar gas, and E_A is
the mass fraction of the element A released by the stars (this
term accounts for the production as well as the destruction of
the considered nuclear species), δ is the rate of infall and
ϕ is the initial mass function.

The production and destruction sites of the elements are represented in figure 2 as a function of the mass of the stars :

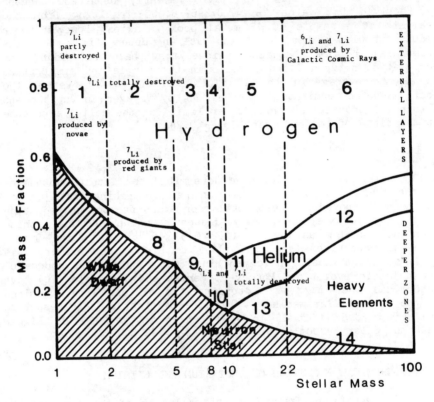

Figure 2: Zones of element production and destruction by stars of different masses.

- in first approximation, we assumed that D and ^{6}Li are in first approximation we assumed that totally destroyed in any layer of any star via the reactions $D(p, \gamma)^3$He and ^{6}Li$(p, \alpha)^3$He.
- ^{7}Li is assumed to be preserved in the external layers of the intermediate and high mass stars and to be partly destroyed in the external layers of stars less massive than 2 M$_\theta$.
- None of the mechanisms of deuterium production during the galactic evolution proposed so far has been proved to play a significant role. So we consider here that D is never produced during this evolution.
- Both ^{6}Li and ^{7}Li are produced by the galactic cosmic rays (GCR). We model this source of lithium by introducing a production factor in the external zones of the very massive stars (M $\geq$ 22M$_\theta$). The thresholds of the spallation reactions

responsible for the ^{7}Li production are smaller than those producing ^{6}Li. Thus the production ratio of ^{7}Li/^{6}Li is 1.6 if there is no low energy component in the GCR (13), and could be 6 to 7 if such a component does exist (14). The two possibilities are considered here.
-An additional source of ^{7}Li production can be found in any stellar event in which the ^{3}He(α, γ)^{7}Be(e$^+$)^{7}Li reaction can take place (novae, red giants) (see e.g. (15) for more details). To account for this possibility we introduced in our model a production term in the external layers of low mass stars ($M \leq 2M_\odot$).

III RESULTS AND INTERPRETATION

In order to reproduce i) the evolution of the metallicity with time (16), and ii) the present fractional gas density observed in the Solar Neighborhood, we considered models with infalling matter with the following parameters: ν = 0.5 and δ = 0.015 which corresponds to an infall rate of 1.5 $M_\odot$ per 10^9 years.

a. Primordial infall

First we considered the case of infall made up with primordial matter. The computed time variation of the isotopic ratio ^{7}Li/^{6}Li is displayed in figure 3 for different assumptions concerning the production sites of ^{7}Li. Figure 4 shows the evolution with time of the abundances of D, ^{4}He, ^{6}Li and ^{7}Li for the model including a stellar production of ^{7}Li. The primordial abundance of deuterium has been adjusted in order to reproduce the present value of D/H = 5 10^{-6}.

From the analysis of these figures, the following points should be stressed :
- The isotopic ratio ^{7}Li/^{6}Li at the Solar System birth cannot be reproduced in any case if no production by novae and/or red giants is taken into account.
- In the case of a stellar ^{7}Li production, one can reproduce the lithium abundances and the solar system ^{7}Li/^{6}Li ratio in both cases of presence or absence of a low energy component in the GCR.
- The stellar production rate of ^{7}Li required in the external zones of stars with mass $1 \leq M/M_\odot \leq 2$ is about ^{7}Li/H = 10^{-8}. This is quite compatible with both the rate of novae outbursts and the enrichment in ^{7}Li by red giants predicted by Scalo (17). Since only ^{7}Li is produced primordially by the Big-Bang nucleosynthesis, the ^{7}Li/^{6}Li ratio decreases during the first 2-3 10^9 years of the galactic life and increases afterwards due to the difference in the destruction and

production rates of ^{6}Li and ^{7}Li.
- Although the ^{7}Li/^{6}Li ratio increases signigicantly after
the birth of the sun, its highest possible present value is
about 20. This is still insufficient to account even for the
lower limit quoted by Ferlet and Dennefeld (12).
- As for the deuterium evolution, it is impossible,with an
infall made up with primordial matter, to reproduce a difference
of abundance between the birth of the solar system and the
present time as big as the one induced by the observational
values (a decrease by a factor 4 to 20). Even if we suppose that
there is no infall, and thus no supply with fresh deuterium, the
decrease of D/H cannot exceed a factor 2 because of the
constraints we have on the rate of astraction due to the present
value of the interstellar gas density. In this case, in order to
lead to a present value in agreement with the new interstellar
results, the primordial value has to be (in mass ; $(X_D)_p$ =
1.5 $(D/H)_p$) : $4 \cdot 10^{-6} \leq (X_D)_p < 4 \cdot 10^{-5}$.

b. Infall of non-primordial composition

One way to explain a steeper decrease of D/H in the last
$4.5 \cdot 10^9$ years, would be to suppose that there is indeed a
significant infall but made up with matter containing no
deuterium at all.

Two remarks led us to consider two kinds of exotic
composition for the infall.
- First, one could assume that the primordial matter was
actually pure hydrogen, as it has been proposed by Audouze and
Silk (18). We thus calculated what infall rate of pure hydrogen
was needed to reproduce the required steep variation of D/H and
what would be the effect on the evolution of the other species.
In particular, Figure 5 displays the evolution curves in this
model of D/H and of Y, the helium abundance in mass. D/H
presents a satisfactory evolution but Y would undergo a dramatic
decrease due to the dilution of the interstellar gas with
helium-free material.
-Secondly, as astrated (and thus deuterium free) matter is
likely to be ejected out of galaxies by supernovae and stellar
winds (e.g.(19)), the infalling gas could be composed of such
material. Again Figure 6 displays the results for the following
composition of the infall: X_H = 0.76, Y=0.235, Z=0.005,
X_D=0.; the rate of infall being again δ =0.015 per 10^9
years, combined with an astration rate ν =0.5. As far as
X_H, X_D, Y, X_{Li}, and Z are concerned, there is no internal
inconsistency in the model. The required primordial deuterium
abundance is then about $(X_D)_p = (2 \pm 1) \cdot 10^{-4}$.
-However, concerning the case of ^{3}He (which is presently not
included in our model) a problem could arise from the fact that
the destruction of important amount of deuterium during the

course of the galactic evolution would enhance significantly the ^{3}He abundance.

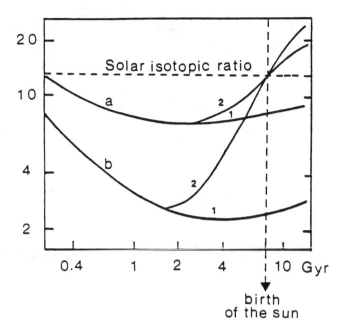

Figure 3: ^{7}Li ratio as function of time for different hypothesis 1: no ^{7}Li stellar production. 2: ^{7}Li production by low mass stars. a: presence of a low energy component in the GCR. b: no low energy component in the GCR.

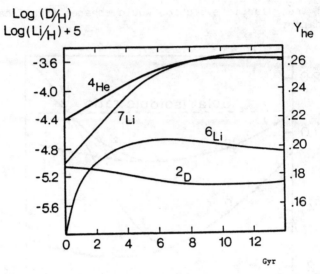

Figure 4: Evolution of the abundances with time (D/H, Y_{He}, $^7Li/H$, $^6Li/H$) in the case 2b with an infall made up with primordial matter.

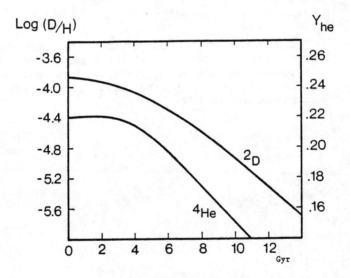

Figure 5: same as Fig. 4 (D/H and Y_{He}) with an infall made up with pure hydrogen.

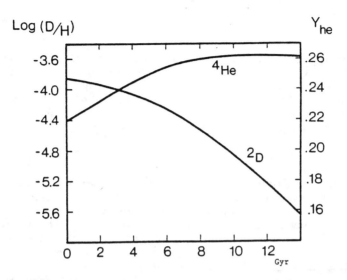

Figure 6: same as Figure 5 with an infall made up with processed
          matter.

IV CONCLUSIONS AND CONSEQUENCES

In this study we have attempted to relate guesses of the
primordial abundances of some light elements with current models
of galactic evolution. The main results are underlined here:

a. Deuterium

It is impossible to match within a standard model of
galactic evolution the abundances observed in the interstellar
medium (now) and inside the Solar System ($4.5 \ 10^9$ years ago).
Nevertheless assuming an infall of processed material
(containing no deuterium) these two observed values could become
compatible. This type of model leads to a primordial deuterium
abundance of $10^{-4} \lesssim X_{D_p} \lesssim 3 \ 10^{-4}$.

b. Helium

In the frame of the current models of chemical evolution
of galaxies it is impossible to account for both a low Y value
in Jupiter $Y \sim 0.24$ and a high value in the young objects of the
Solar Neighborhood $Y \sim 0.28$. An increase of 0.04 of the helium
mass fraction would lead to an increase of at least 0.01 of the

heavy element mass fraction between the birth of the Sun and
now, which is not the case according to Twarog (16).

c. Lithium

Abundance and isotopic ratios could be well predicted if
one assumes a ^{7}Li production by novae and/or red giants.
However the ^{7}Li/^{6}Li ratio as observed in one line of sight
of the interstellar medium seems incompatible by that model. It
is interesting to note that an infall of processed material
(enriched in ^{7}Li) might also solve the discrepancy.

d. Primordial nucleosynthesis

We show on figure 7 the primordial abundances of the three
light elements considered in this study. For the deuterium value
we used both models including a standard infall made of
primordial material (model A) and an infall of processed matter
(model B). They are presented together with the theoretical
predictions of the standard Big-Bang calculations given by
Schramm (20) as a function of η the baryons to photons ratio.
As it can be seen, the two ranges for η inferred by the
estimated deuterium (model B) and the observed Lithium
abundances are quite compatible, but they do not overlap
the η range inferred by the observed primordial helium
fraction. This fact, underlines again that the helium case might
not yet be perfectly understood in the frame of the primordial
nucleosynthesis and of the chemical evolution of the galaxies.

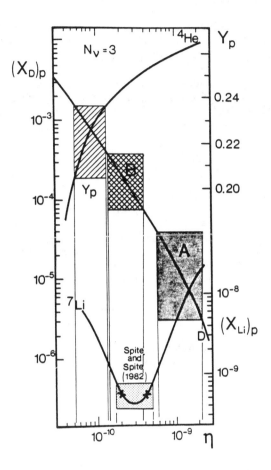

Figure 7: Primordial abundances of the light elements D, He and
^{7}Li as functions of η , the baryons over photons ratio. The
curves show theoretical calculations in the case of 3 neutrinos
and the boxes show the ranges deduced from the observations. In
the deuterium case, we display the two different ranges inferred
by the two different models A and B.

REFERENCES

(1) Laurent C., Vidal-Madjar A. and York D.G., 1979, Astrophys.
      J., 229, 923
(2) Dupree A.K., Baliunas S.L. and Shipman H.L., 1977, Astrophys.
      J., 218, 371
(3) Vidal-Madjar A., Laurent C., Bruston P., and Audouze J., 1978,
      Astrophys.J., 223, 589
(4) Vidal-Madjar A., Laurent C., Gry C., Bruston P., Ferlet R.,
      and York D.G., 1983, Astron. Astrophys., 120, 58
(5) Gry C., Vidal-Madjar A., 1983, Astron. Astrophys., in press
(6) Gry C., Lamers H.J.G.L.M., Vidal-Madjar A., Astron. Astrophys.
      submitted
(7) Gauthier D., Owen T., 1983, Nature, 302, 215
(8) Pagel B.E.J., 1982, Phil. trans. R. Soc. Lond., A307, 19
(9) Kunth D., Sargent W.L.W., 1983, Astrophys. J., in press
(10) Spite F., Spite M., 1982, Astron. Astrophys., 115, 357
(11) Audouze J., Reeves H., 1982, in Essays in Nuclear Astrophy-
      sics, C.A. Barnes et al eds, Cambridge University press,
      p.355
(12) Ferlet R., Dennefeld M., 1983, in "Primordial Helium",
      D. Kunth and P. Shaver eds, to be published
(13) Meneguzzi M., Audouze J., Reeves H., 1971, Astron. Astrophys.
      15, 337
(14) Meneguzzi M., Reeves H., 1975, Astron. Astrophys., 40, 99
(15) Audouze J., Boulade O., Malinie G., and Poilane Y., Astron.
      Astrophys., submitted
(16) Twarog B.A., 1980, Astrophys. J., 242, 242
(17) Scalo J.M., 1976, Astrophys. J., 206, 795
(18) Audouze J., Silk J., these proceedings, p. 267.

V

STRUCTURE AND DYNAMICAL EVOLUTION OF GALAXIES

STRUCTURE AND EVOLUTION OF TRIAXIAL GALAXIES

Tim de Zeeuw

Sterrewacht Leiden

ABSTRACT: Confocal ellipsoidal coordinates are the natural
coordinates for the description of triaxial stellar systems.
Their properties are discussed. An inhomogeneous triaxial mass
model is presented with a gravitational potential that is
separable in these coordinates. All orbits in this model have
three explicitly known isolating integrals of motion, and they
can be classified by analytical means. Three families of general
orbits occur; they are identical to the major orbit families
found by Schwarzschild to be of prime importance for the
structure of elliptical galaxies. Some applications of this
result are mentioned.

I. INTRODUCTION

Models for elliptical galaxies should have at least the
following two properties:
1. The density distribution must be stratified roughly on
concentric ellipsoids. Such a triaxial mass model is the simplest
one that is compatible with the observed brightness contours of
ellipticals.
2. Most individual stellar orbits in it must possess three independent isolating integrals of motion, i.e., two in addition to
the classical energy integral. These integrals are required in
order to maintain the anisotropic velocity dispersions which are
needed to support a triaxial equilibrium shape [1-3].

Few realistic models exist. To date, all are numerical and
have been constructed either by brute force N-body calculations
[4,5], or by Schwarzschild's linear programming method [6,7].

J. Audouze and J. Tran Thanh Van (eds.),
Formation and Evolution of Galaxies and Large Structures in the Universe, 293–313.
© 1984 by D. Reidel Publishing Company.

No simple expressions are known for the two nonclassical integrals
of motion enjoyed by most of the stars in these models. As a
result, the phase space distribution function, which is a function
of all three isolating integrals [8,9], is known only as a set of
occupation numbers of individual orbits.

Although the numerical models have considerably deepened
our understanding of triaxial galaxies [10], still many questions
regarding the structure and evolution of these systems remain.
One would, e.g., like to know what the nonclassical integrals of
motion are. It is also not yet known whether triaxial equilibrium
shapes can exist for all values of the axial ratios, and if so,
how their properties depend on these ratios. In order to answer
such questions it would clearly be helpful to have analytic
models that are easy to use and are realistic in the above sense,
or close to it.

Recently it has become clear that indeed such models exist
[11]. There are triaxial mass models in which all stellar orbits
have three exact isolating integrals which are explicitly known.
Motion in these models is tractable by analytic means. Moreover,
the orbits in them can be divided in three families; these are
just the three major orbit families that — on the basis of the
numerical models — are thought to be important for the equilibrium structure of elliptical galaxies.

The existence of these mass models is based on the existence
of potentials that admit three exact isolating integrals. These
are the special potentials for which the equations of motion are
separable. They have a long history [12-15], were first introduced
in stellar dynamics by Eddington [16], and later repeatedly
classified [17-19]. The most general of these potentials are the
ones for which the equations of motion separate in confocal
ellipsoidal coordinates. We will refer to these as Eddington
potentials.

The main aim of this contribution is to show that the
confocal ellipsoidal coordinates are the *natural coordinates* for
the description of triaxial elliptical galaxies, and that much
insight can be gained in the structure and evolution of these
systems by a study of mass models with potentials of Eddington
form. This we do by first describing some of the properties of
confocal coordinates in two- and three dimensions (§II), and
then (§III) introducing a specific triaxial mass model that has
all the nice properties mentioned in the above. Applications are
discussed briefly in §IV.

II. CONFOCAL COORDINATES AND EDDINGTON POTENTIALS

a. *Confocal Elliptic Coordinates*

Let (x,y) be Cartesian coordinates in a plane. Following the notation of Lynden-Bell [19], we define confocal elliptic coordinates (λ,μ) as the roots τ of

$$\frac{x^2}{\tau+\alpha} + \frac{y^2}{\tau+\beta} = 1, \tag{1}$$

where α and β are constants and we assume $\alpha \leqslant \beta$. The following inequality holds

$$-\beta \leqslant \mu \leqslant -\alpha \leqslant \lambda. \tag{2}$$

Figure 1 shows the coordinate lines. Curves of constant λ are ellipses with the major axis in the y-direction. They all have identical foci at $x=0$, $y=\pm\sqrt{\beta-\alpha}$. The ellipse $\lambda=-\alpha$ is the segment of the y-axis between the foci. Curves of constant μ are hyperbolae, with the same foci. For $\mu=-\beta$ they coincide with the x-axis; $\mu=-\alpha$ corresponds to the parts of the y-axis outside the foci.

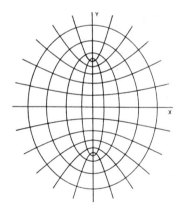

Figure 1. Confocal elliptic coordinates in a plane.

Near the origin the coordinates (λ,μ) are approximately Cartesian. At large distances from the origin the ellipses become progressively rounder $(\lambda\to\infty)$ and the hyperbolae are nearly identical to their asymptotes, which are straight lines through the origin. Thus, very far from the origin the confocal elliptic coordinates behave as polar coordinates.

Through each point (x,y) there is exactly one ellipse $\lambda = c^{st}$ and exactly one hyperbola $\mu = c^{st}$, and these curves are perpendicular to each other in this point. Thus, confocal elliptic coordinates are curvilinear orthogonal coordinates. Accordingly, the lineelement ds^2 is given by

$$ds^2 = P^2 d\lambda^2 + Q^2 d\mu^2, \tag{3}$$

with metrical coefficients P and Q given by

$$P^2 = \frac{\lambda - \mu}{4(\lambda + \alpha)(\lambda + \beta)}, \qquad Q^2 = \frac{\mu - \lambda}{4(\mu + \alpha)(\mu + \beta)}. \tag{4}$$

The most general form of the potential V_E in which the motion is separable in these coordinates is

$$-V_E = \psi_E = \frac{\zeta(\lambda)}{P^2} + \frac{\eta(\mu)}{Q^2} = \frac{\zeta^*(\lambda) - \eta^*(\mu)}{\lambda - \mu}, \tag{5}$$

where $\zeta(\lambda)$ and $\eta(\mu)$ are essentially arbitrary functions of λ and μ, respectively, and

$$\zeta^*(\lambda) = 4(\lambda + \alpha)(\lambda + \beta)\zeta(\lambda), \qquad \eta^*(\mu) = 4(\mu + \alpha)(\mu + \beta)\eta(\mu). \tag{6}$$

For V_E to be regular at the foci it is required that ζ^* and η^* and their first three derivatives are equal at $\lambda = \mu = -\alpha$ [19]. In practice it is therefore no loss of generality to write $\zeta^*(\lambda) = f(\lambda)$ and $\eta^*(\mu) = f(\mu)$ so that both ζ^* and η^* are the same function $f(\tau)$ — which is assumed to be at least three times differentiable — but on adjacent intervals.

The Lagrangian is $\frac{1}{2}\dot{s}^2 - V_E$ so the momenta conjugate to λ and μ are $p_\lambda = P^2 \dot{\lambda}$ and $p_\mu = Q^2 \dot{\mu}$, respectively. V_E admits two isolating integrals of motion that are quadratic in the velocities:

$$H = X + Y, \tag{7}$$

$$I = \mu X + \lambda Y, \tag{8}$$

where

$$X = \frac{p_\lambda^2}{2P^2} - \frac{\zeta(\lambda)}{P^2}, \qquad Y = \frac{p_\mu^2}{2Q^2} - \frac{\eta(\mu)}{Q^2}. \tag{9}$$

H is the Hamiltonian and is equal to the total energy E.

The equations of motion in an Eddington potential are separable. For an orbit with H=E and I=i we solve (7)-(9) for p_λ^2 and p_μ^2 to obtain

$$p_\lambda^2 = \frac{\lambda E - i + f(\lambda)}{2(\lambda+\alpha)(\lambda+\beta)} ,$$

$$p_\mu^2 = \frac{\mu E - i + f(\mu)}{2(\mu+\alpha)(\mu+\beta)} .$$

(10)

These equations can be solved by quadratures to give the coordinates λ and μ as function of time for the orbit. In practical cases the required integrations have to be performed by numerical means. Fortunately, a great deal can be learned about the orbits without doing any integrations.

Admissible motion requires that both p_λ^2 and p_μ^2 are non-negative. The orbit has turning points at the value of λ for which $p_\lambda^2=0$ and at the value of μ for which $p_\mu^2=0$. *Motion is therefore bounded by the coordinate lines.* The orbit can be thought of as the sum of two independent oscillations, one in λ and the other in μ. The frequencies of these oscillations are in general incommensurable so that the orbit is a Lissajous figure which fills the area allowed by the integrals of motion E and i.

As a result, all possible orbital shapes can be found by inspection of the coordinate lines (Figure 1). Two examples for bound orbits are shown in Figure 2. They are usually referred to as box- and tube-orbits (Ollongren [20]), or as *butterfly-* and *loop-orbits* (Binney [21]), respectively.

a)   *b)*

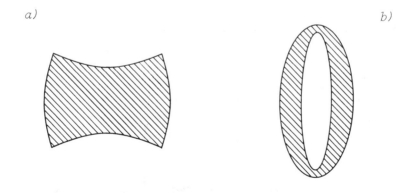

Figure 2. Two possible orbital shapes. a) Butterfly. b) Loop.

Which shapes will in fact occur in a given Eddington potential can be easily derived from equations (10) by solving $p_\tau^2=0$ for the given $f(\tau)$ ($\tau=\lambda$ or μ). For this purpose it turns out to be convenient [11] to rewrite (10) somewhat. Instead of the integral I use the integral I_1 defined by $I_1=I+\alpha H$ so that

$$i_1 = i + \alpha E. \tag{11}$$

Define $g(\tau)$ by

$$f(\tau) = (\tau+\alpha)g(\tau). \tag{12}$$

Then the equations of motion (10) may be written as

$$p_\tau^2 = \frac{T(\tau)}{2(\tau+\beta)} , \qquad\qquad \tau = \lambda \text{ or } \mu, \tag{13}$$

with

$$T(\tau) = E - \frac{i_1}{\tau+\alpha} + g(\tau). \tag{14}$$

Motion is allowed for $T(\tau)\geqslant0$ (cf. equation (2)). The boundaries of the area allowed to the orbit by E and i_1 follow from solving $T(\lambda)=0$ and $T(\mu)=0$ for λ and μ, respectively.

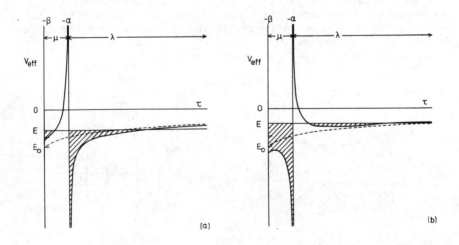

Figure 3. Orbit classification by means of the effective potential V_{eff}. The dashed line is $-g(\tau)$. The area between the lines $E=c^{st}$ and $V_{eff}(\tau)$ in the intervals where $E \geqslant V_{eff}$ is hatched.
   a) $i_1<0$. The corresponding orbit is a butterfly.
   b) $i_1>0$. The corresponding orbit is a loop.

Equation (14) can be thought of as an energy equation for one dimensional motion in an effective potential $V_{eff}(\tau)$:

$$E = T(\tau) + V_{eff}(\tau), \tag{15}$$

with

$$V_{eff} = \frac{i_1}{\tau+\alpha} - g(\tau). \tag{16}$$

In contrast with the usual cases — like the radial equation of motion in a spherical potential — here the equation should be solved in two intervals separately, one for λ and one for μ. It follows that classification of orbital shapes can be done graphically by drawing V_{eff} as function of τ for a given value of i_1 and comparing it with the horizontal lines $E=c^{st}$. This immediately gives the intervals in τ for which $T(\tau)$ is nonnegative, i.e., $E \geqslant V_{eff}(\tau)$. Two examples are shown in Figure 3. A comparison with the coordinate lines (Figure 1) then shows the area filled by the orbit.

A useful property of an orbit is its density ρ_{orb} within its allowed area. This is given by

$$\rho_{orb}(\lambda,\mu;E,i) = \iint \delta(H-E)\delta(I-i)dp_\lambda dp_\mu = \frac{p^2 q^2}{(\lambda-\mu)} \frac{1}{p_\lambda p_\mu} . \tag{17}$$

Clearly, ρ_{orb} can be simply evaluated upon substitution of (10); no (numerical) integrations are needed.
In general, the fundamental orbital frequencies are incommensurable and ρ_{orb} is the timeaveraged density of the Lissajous figure which fills the whole area allowed by E and i. If the frequencies are commensurable for a given combination of E and i then the phase difference between the λ- and μ- oscillation is a third independent isolating integral of motion. The Lissajous figure now breaks up in a continuum of periodic orbits, parametrized by the additional integral, which together fill up the whole area allowed by E and i. In this case ρ_{orb} applies to this whole collection of periodic orbits.

b. *Confocal Ellipsoidal Coordinates*

Much of the above generalizes straightforwardly to three dimensions. Let (x,y,z) be Cartesian coordinates. Confocal ellipsoidal coordinates (λ,μ,ν) are defined as the roots τ of

$$\frac{x^2}{\tau+\alpha} + \frac{y^2}{\tau+\beta} + \frac{z^2}{\tau+\gamma} = 1, \tag{18}$$

where α, β and γ are constants with $\alpha \leqslant \beta \leqslant \gamma$. The three roots satisfy the inequality

$$-\gamma \leqslant \nu \leqslant -\beta \leqslant \mu \leqslant -\alpha \leqslant \lambda. \tag{19}$$

The coordinate surfaces are confocal quadrics and are shown in Figure 4. Surfaces of constant λ are ellipsoids with the long axis in the z-direction and the short axis in the x-direction. Surfaces of constant μ are hyperboloids of one sheet lying around the x-axis. Surfaces of constant ν are hyperboloids of two sheets around the z-axis.

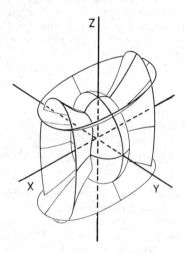

Figure 4. Confocal ellipsoidal coordinates

In each of the principal planes the coordinates are simply confocal elliptic coordinates, described in the previous section. In addition to the foci on the y-axis at $(0,\pm\sqrt{\beta-\alpha},0)$ there are two pairs of foci on the z-axis, one at $(0,0,\pm\sqrt{\gamma-\beta})$ and the other at $(0,0,\pm\sqrt{\gamma-\alpha})$.

Very close to the origin the coordinates are approximately Cartesian. At large distances from the centre the ellipsoids become progressively more nearly spherical and the hyperboloids are almost indistinguishable from their asymptotic surfaces which are planes perpendicular to the (x,y)-plane containing the z-axis and cones along the z-axis. Thus, at large radii the confocal ellipsoidal coordinates behave as ordinary spherical coordinates.

Through each point (x,y,z) there is exactly one ellipsoid $\lambda=c^{st}$, exactly one hyperboloid of one sheet $\mu=c^{st}$ and exactly one hyperboloid of two sheets $\nu=c^{st}$. These three surfaces are perpendicular to each other at this point so that (λ,μ,ν) are orthogonal curvilinear coordinates. The line element ds^2 is

$$ds^2 = P^2d\lambda^2 + Q^2d\mu^2 + R^2d\nu^2, \tag{20}$$

where the metric coefficients P, Q and R are functions of λ, μ and ν (see e.g., [19]).

The general form of the potential in which the motion is separable in these coordinates is

$$-V_E = \psi_E = \frac{\zeta(\lambda)}{P^2} + \frac{\eta(\mu)}{Q^2} + \frac{\theta(\nu)}{R^2}, \tag{21}$$

where $\zeta(\lambda)$, $\eta(\mu)$ and $\theta(\nu)$ are essentially arbitrary functions of λ, μ and ν, respectively. As in the two dimensional case V_E is often written in terms of $\zeta^*(\lambda)=4(\lambda+\alpha)(\lambda+\beta)(\lambda+\gamma)$ and $\eta^*(\mu)$, $\theta^*(\nu)$, defined in an analogous fashion. Again it is no practical loss of generality to take $\zeta^*(\lambda)=F(\lambda)$, $\eta^*(\mu)=F(\mu)$ and $\theta^*(\nu)=F(\nu)$ so that ζ^*, η^* and θ^* are the same function $F(\tau)$ on three adjacent intervals. Equation (21) then takes the form

$$-V_E = \frac{F(\lambda)}{(\lambda-\mu)(\lambda-\nu)} + \frac{F(\mu)}{(\mu-\lambda)(\mu-\nu)} + \frac{F(\nu)}{(\nu-\lambda)(\nu-\mu)}. \tag{22}$$

The Eddington potential V_E admits three independent isolating integrals of motion that are quadratic in the velocities

$$H = \quad X \quad + \quad Y \quad + \quad Z, \tag{23}$$

$$J = (\mu+\nu)X + (\nu+\lambda)Y + (\lambda+\mu)Z, \tag{24}$$

$$K = \quad \mu\nu X \quad + \quad \nu\lambda Y \quad + \quad \lambda\mu Z, \tag{25}$$

where X and Y are given in (9) and Z is defined in a similar way. H is the Hamiltonian and is equal to the total energy E.

The equations of motion can be separated and solved by quadratures. *Motion is bounded by coordinate surfaces.* Every orbit is the sum of three independent oscillations, one in each coordinate, and in general fills the volume allowed by the integrals of motion.

As in (13) we may write the equations of motion as

$$p_\tau^2 = \frac{T(\tau)}{2(\tau+\beta)}, \qquad \tau = \lambda, \mu \text{ or } \nu, \tag{26}$$

but now with (cf. (14))

$$T(\tau) = E - \frac{i_2}{\tau + \alpha} - \frac{i_3}{\tau + \gamma} + G(\tau), \tag{27}$$

where $G(\tau)$ follows from

$$F(\tau) = (\tau + \alpha)(\tau + \gamma)G(\tau), \tag{28}$$

and the integrals I_2 and I_3 are defined as

$$I_2 = \frac{\alpha^2 H + \alpha J + K}{\alpha - \gamma}, \qquad I_3 = \frac{\gamma^2 H + \gamma J + K}{\gamma - \alpha}. \tag{29}$$

We will see in §IIIc that I_2 and I_3 are related in a simple way to the classical angular momentum integrals.

Admissible motion requires (cf. equation (19))

$$T(\lambda) \geqslant 0, \qquad T(\mu) \geqslant 0, \qquad T(\nu) \leqslant 0. \tag{30}$$

The expression (27) for $T(\tau)$ may again be thought of as an energy equation (15) for one dimensional motion in an effective potential

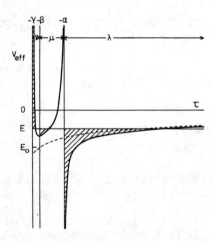

Figure 5. *Orbit classification by means of the effective potential V_{eff}. The dashed line is $-G(\tau)$. The area between the lines $E = c^{st}$ and $V_{eff}(\tau)$ in the intervals where $E \geqslant V_{eff}$ (for λ and μ) or $E \leqslant V_{eff}$ (for ν) is hatched. The corresponding orbit is a box.*

$V_{eff}(\tau)$, which is now given by

$$V_{eff} = \frac{i_2}{\tau+\alpha} + \frac{i_3}{\tau+\gamma} - G(\tau). \tag{31}$$

The equation should be solved on three intervals, one for λ, one for μ and one for ν. Orbit classification can be done graphically by drawing V_{eff} as function of τ for given values of i_2 and i_3 and comparing it with the horizontal lines $E=c^{st}$, keeping in mind (30). An example is shown in Figure 5. See also §IIIb.

Finally, the time averaged density ρ_{orb} of an orbit with H=E, J=j and K=k is given by

$$\rho_{orb}(\lambda,\mu,\nu;E,j,k) = \frac{P^2Q^2R^2}{(\lambda-\mu)(\mu-\nu)(\nu-\lambda)} \frac{1}{P_\lambda P_\mu P_\nu}. \tag{32}$$

This can be evaluated with the help of (26) and (27).

III. THE PERFECT ELLIPSOID

a. Density and Potential

Consider the density distribution given by

$$\rho = \frac{\rho_0}{(1+m^2)^2}, \tag{33}$$

where

$$m^2 = \frac{x^2}{a_1^2} + \frac{y^2}{a_2^2} + \frac{z^2}{a_3^2}, \qquad a_1 \geqslant a_2 \geqslant a_3. \tag{34}$$

The density is stratified on similar concentric ellipsoids with semiaxes ma_1, ma_2 and ma_3. The long axis of the density distribution is in the x-direction and the short axis is along the z-axis. At large distances from the centre ρ falls off as ρ^{-4}; the central density is ρ_0. The total mass M is $\pi^2 a_1 a_2 a_3 \rho_0$.

This triaxial mass model has a gravitational potential of Eddington form, i.e., separable in confocal ellipsoidal coordinates (λ,μ,ν), *irrespective of the axial ratios of the model* [11]. Separability for the oblate axisymmetric case $(a_1=a_2>a_3)$ was already known to Kuzmin [22]. One can show [23] that the density distribution (33) is the only inhomogeneous triaxial mass model in which the density is stratified on similar concentric ellipsoids, and is nowhere singular, with a separable potential. We shall call it *the perfect ellipsoid*.

Specifically, the gravitational potential of the perfect ellipsoid is given by (22) and (28) with

$$G(\tau) = \pi G \rho_0 a_1 a_2 a_3 \int_0^\infty \sqrt{\frac{(u-\beta)}{(u-\alpha)(u-\gamma)}} \cdot \frac{du}{u+\tau} \ . \tag{35}$$

The confocal coordinate system (λ, μ, ν) is defined by

$$\alpha = -a_1^2, \qquad \beta = -a_2^2, \qquad \gamma = -a_3^2. \tag{36}$$

The integral (35) can be expressed in terms of the incomplete elliptic integrals of the three kinds and is readily evaluated numerically. Figure 6 shows a graph of $G(\tau)$ for the case $a_1=1$, $a_2=5/8$ and $a_3=1/2$. These values of a_1, a_2 and a_3 lead to axial ratios that are identical to those used by Schwarzschild [6,7] for his selfconsistent numerical model of an elliptical galaxy. In order to facilitate a comparison (§IIId), all figures in the present contribution that relate to the perfect ellipsoid have been made for this choice of a_1, a_2 and a_3.

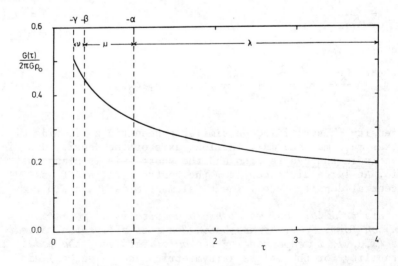

Figure 6. The function $G(\tau)$ for $a_1=1$, $a_2=5/8$ and $a_3=1/2$.

b. Orbits

The specific form of $G(\tau)$ given in (35) allows a classification of orbits in the perfect ellipsoid in terms of the integrals
of motion H, I_2 and I_3 by means of the method described in §II.
A detailed analysis is given in [11]. Here we summarize the main
results.

From symmetry it follows that a star moving in a principal
plane will remain in it, so that motion is two dimensional.
Consider first the (x,y)-plane. In this plane the coordinates
(λ,μ,ν) reduce to confocal elliptic coordinates (λ,μ). The
integral I_3 is zero. The equations of motion (26) reduce to (13)
with $g(\tau)=G(\tau)$ and $I_2=I_1$ so that motion can be analysed as in
§IIa, as expected.

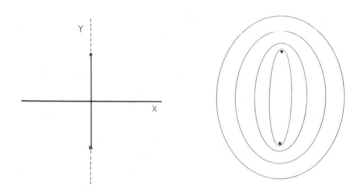

*Figure 7. Simple periodic orbits in the (x,y)-plane of the
perfect ellipsoid. The small circles denote the foci.*

The simple periodic orbits in the (x,y)-plane are drawn in
Figure 7. The two axial oscillations are solutions of (13), as
follows also from symmetry. The x-axis orbit is stable for all
energies. The y-axis orbit is stable for energies such that the
amplitude of the oscillation is smaller than $\sqrt{\beta-\alpha}$, the distance
to the foci. For larger energies the y-axis orbit is unstable.
At these energies another family of stable periodic orbits exists:
the lines $\lambda=c^{st}$ are solutions of (13). They are stable elliptic
closed orbits, and branch off the y-axis at the foci. In other
words, *the foci are the bifurcation points* on the y-axis.

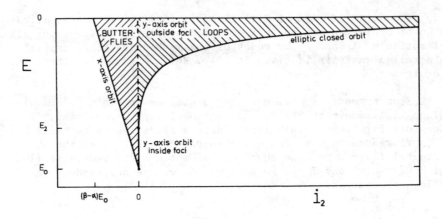

Figure 8. Classification of orbits in the (x,y)-plane of the perfect ellipsoid in terms of the integrals E and i_2.

Figure 8 illustrates the relations between the integrals E and i_2 for the three families of simple periodic orbits just mentioned. These relations form the boundary of the area in the (E,i_2)-plane that corresponds to admissible motion in the (x,y)-plane. Each point in the hatched area corresponds with a general (bound) orbit. These are divided into two families by the unstable y-axis orbit, one family having $i_2 < 0$, the other having $i_2 > 0$. From Figure 3, which was drawn for the perfect ellipsoid, it follows that the former are *butterfly orbits* and the latter are *loop orbits*. Representative shapes were shown in Figure 2. The x-axis orbit can be considered as the "parent" of the butterflies; the elliptic closed orbits are the progenitors of the loops. Note that butterflies always cross the y-axis between the foci and the loops outside the foci.

In each of the remaining principal planes motion is completely similar. Butterfly orbits are generated by the (relative) long-axis orbit, which is everywhere stable. Loop orbits have as parent the elliptic closed orbit which branches off the (relative) short-axis orbit at the appropriate foci. For amplitudes larger than the distance to the foci the (relative) short axis orbit is unstable.

In three dimensions analysis of (26) as described in §IIb (cf. Figure 5) yields the following result. There are six basic periodic orbit families. They all lie in the principal planes. Of the three axial oscillations the x-axis orbit is stable for all energies. The y-axis orbit becomes unstable in the x-direction for amplitudes larger than the distance $\sqrt{\beta-\alpha}$ to the foci. The z-axis orbit is stable for small amplitudes, becomes unstable in the y-direction for amplitudes larger than $\sqrt{\gamma-\beta}$, and unstable in the x-direction for amplitudes larger than $\sqrt{\gamma-\alpha}$ (Figure 9).

The other three simple periodic orbit families are the elliptic closed orbits, one in each principal plane. Those in the (x,y)- and (y,z)-plane are stable; the elliptic orbit around the intermediate (y-) axis is unstable to perpendicular perturbations.

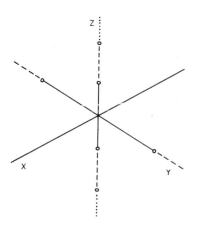

Figure 9. *Axial orbits and their stability. The small circles denote the foci.* —— *stable;* --- *unstable in one direction;* ... *unstable in two directions.*

The three stable periodic orbit families generate three families of general orbits. The x-axis orbit is the parent of the threedimensional *box orbits*, which reduce to butterflies in each of the principal planes. The two closed elliptic orbits that are stable are the progenitors of (three dimensional) *tube orbits*, one *around the long (x-) axis*, the other *around the short (z-) axis*.

It should be remarked that the above orbit classification for the perfect ellipsoid is valid for all values of $a_1 \neq a_2 \neq a_3$.

c. Degeneracies

For $a_1 = a_2 > a_3$ the mass model is an *oblate spheroid* with the
z-axis as symmetry axis. From (36) it follows that $\beta = \alpha$. This
means that the coordinates (λ, μ, ν) reduce to prolate spheroidal
coordinates (λ, ϕ, ν). In the equatorial plane the foci coincide
with the origin and the coordinates are simply polar coordinates
(R, ϕ), see Figure 10; in each plane $\phi = c^{st}$ (meridional plane) λ
and ν are confocal elliptic coordinates.

$$\beta > \alpha \qquad\qquad\qquad \beta = \alpha$$

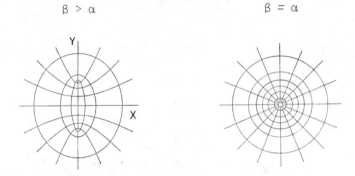

Figure 10. For $\beta = \alpha$ the confocal elliptic coordinates (λ, μ)
become polar coordinates (R, ϕ).

The function $G(\tau)$ is elementary for $\beta = \alpha$. The resulting
gravitational potential is identical to the one derived by Kuzmin
[22] along a different route. The integral I_2 reduces to a
classical integral of motion

$$I_2 = \tfrac{1}{2} L_z{}^2, \qquad\qquad\qquad \text{for } \beta = \alpha, \qquad\qquad (37)$$

where L_z is the component of the angular momentum parallel to the
symmetry axis. I_3 is the "third integral" [20,24-26] and can be
thought of as a generalization of $\vec{L}^2$, the square of the total
angular momentum.

The elliptic closed orbits in the (x,y)-plane that exist in
the triaxial case now reduce to the well known circular orbits
that exist at every radius, as can be seen from Figure 10. All
straight lines in this plane through the origin are orbits; they
are unstable. The (E, i_2)-diagram presented in Figure 8 now is a
diagram of E versus $\tfrac{1}{2} L_z{}^2$ (Figure 11). The triangular area in
Figure 8 that contains the butterflies shrinks to the vertical
line $\tfrac{1}{2} L_z{}^2 = 0$ for $\beta = \alpha$. No butterflies exist in the oblate model. In
the equatorial plane there is only one family of general orbits,
the loops around the circular orbit.

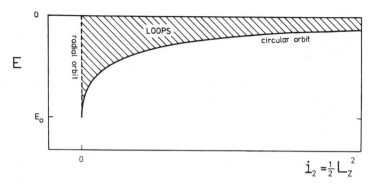

*Figure 11. Orbit classification in the equatorial plane
of the perfect ellipsoid.*

In three dimensions we have a similar result. The general
orbits are all tubes around the circular orbits in the equatorial
plane. The boxes and tubes around the long axis that exist in the
triaxial case have disappeared.

For $a_1 > a_2 = a_3$ the mass model is a *prolate spheroid* elongated
in the x-direction. The coordinates (λ, μ, ν) reduce to oblate
spheroidal coordinates and again $G(\tau)$ is an elementary function.
In this case it is I_3 that reduces to a classical integral

$$I_3 = \tfrac{1}{2} L_x^2, \qquad\qquad \text{for } \beta = \gamma, \qquad\qquad (38)$$

where L_x is the component of the angular momentum along the x-axis.
I_2 is the "third integral" and is a generalization of $\vec{L}^2$.

Just as in the oblate case motion is very simple. All general
orbits belong to one family, tubes around the circular orbit that
exists at every radius in the (y,z)-plane. Again the boxes and
one family of tubes from the triaxial model disappear.

Finally, for $a_1 = a_2 = a_3$ the mass model is a *sphere*. The
coordinates (λ, μ, ν) become ordinary spherical coordinates
(r, θ, ϕ). We find

$$I_2 + I_3 = \tfrac{1}{2}\vec{L}^2, \qquad\qquad \text{for } \alpha = \beta = \gamma. \qquad\qquad (39)$$

From the oblate and prolate cases it follows that also L_x and L_z
are integrals. As a result we have four integrals of motion:
E, L_x, L_y and L_z, as it should be. Equation (27) reduces to the
well known one for the radial coordinate in a spherical potential.
All general orbits are loops around the circular orbits that
exist at every radius in every plane through the origin.

d. Comparison with the Schwarzschild Ellipsoid

Schwarzschild [6,7] has constructed realistic models of
elliptical galaxies by means of linear programming. In the
potential of a given mass model stellar orbits and their
individual densities are calculated. The latter are added in such
a way that the mass model is reproduced, with nonnegative
occupation numbers for all orbits.
The mass model chosen by Schwarzschild is roughly ellipsoidal,
with axial ratios 1:5/8:1/2 (the Schwarzschild Ellipsoid [27]).
The radial density profile is the modified Hubble profile which
falls off as r^{-3} at large radii r, and gives a finite density in
the centre. Schwarzschild constructed first a selfconsistent
solution for the model in the absence of figure rotation, and
later with rotation.

The basic periodic orbits in the nonrotating Schwarzschild
Ellipsoid are exactly the six that occur in the perfect ellipsoid,
and with identical stability properties [10,28]. In each principal
plane there are two major families of orbits: butterflies and
loops. Two representative orbits are shown in Figure 12; they can
be compared with Figure 2.
Not surprisingly there are three major families of three dimensional orbits in the model, viz., boxes, tubes around the long
axis and tubes around the short axis. In addition to these there
are minor orbit families connected with higher order periodic
orbits, and a small fraction of stochastic orbits related to
the unstable periodic orbits [29,30].

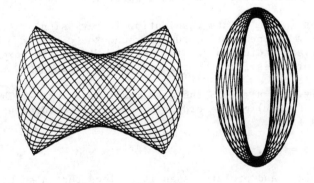

*Figure 12. Two typical orbits in the (x,y)-plane of the
            Schwarzschild Ellipsoid.*

It follows that *the perfect ellipsoid has all orbit families that are of major importance for the structure of triaxial elliptical galaxies.* This is the main point: it shows that
1. Potentials of ellipticals can be well approximated by Eddington potentials.
2. The (λ,μ,ν)'s are the natural coordinates in which to describe these systems.

IV. CONCLUDING REMARKS

The construction of selfconsistent models for density distributions that have a gravitational potential of Eddington form is straightforward by means of Schwarzschild's linear programming method. The orbital densities needed for this can be simply evaluated from (32), without even calculating the orbits themselves. Laborious numerical calculations are thus avoided. In view of the requirements for realistic models of elliptical galaxies mentioned in the introduction it is evident that a selfconsistent model based on the perfect ellipsoid will constitute the *perfect elliptical galaxy.*

Real ellipticals are nearly perfect (or, depending on one's point of view: perfect ellipticals are nearly real!). Some imperfections are:
1. In many cases the ellipticity of the isophotes varies with radius [31,32]. Also changes in the position angle of the major axis of the isophotes occur [33].
2. The radial density profile differs from that of the perfect ellipsoid.
3. Most ellipticals appear to rotate, albeit slowly, whereas the perfect ellipsoid does not. However, rotation is not very important for the dynamics [2].
4. Not all stellar orbits in an elliptical galaxy will belong to the three families present in the perfect ellipsoid. As in Schwarzschild's model, there will be minor families trapped around higher order periodic orbits, as well as a fraction of stochastic orbits. These orbits are unlikely to play an important role in the dynamics of ellipticals [10,30].
These defects detract little from the advantage of having a mass model with arbitrary axial ratios that has all important orbit families and is analytically tractable.

The mere existence of the perfect ellipsoid makes it very likely that other Eddington potentials will give rise to triaxial mass models as well. These models will not be stratified exactly on similar concentric ellipsoids (since the perfect ellipsoid is the only such one), and will have a density profile different from (33). As a result, their projected surface density will in general show twisting isophotes as well as changes in "ellipticity"

which depend both on radius and on the radial density profile.
It is therefore not unlikely that separable mass models exist
that come closer to being realistic than the perfect ellipsoid
(cf. points 1. and 2. mentioned above).

The fact that potentials that are used for models of
ellipticals are in some sense close to being separable in (λ,μ,ν)'s
leads to the question: how close? More specifically: by choosing
ζ, η and θ and the positions of the foci, how well can a given
potential be approximated by one of Eddington form? For axisymmetric potentials this problem received some attention in the
early sixties [34,35]. The triaxial problem was recently
studied by Lynden-Bell and the author [23]. They consider both
the local problem of expansions of potentials around a stable point
as well as the global problem. For the latter see also the
contribution of Wilkinson in the present volume.

The equations of motion in an Eddington potential can be
solved by quadratures. This means that one can calculate actions
J_τ and angles θ_τ ($\tau=\lambda,\mu,\nu$) for each orbit by straight integration.
The actions depend only on the integrals of motion so that the
orbits can be classified in terms of them [36]. The main advantage
of using the J_τ's is that they are adiabatic invariants. Study of
slow evolution of elliptical galaxies (e.g., [37]) is therefore
most easily done in terms of the variables (J_τ,θ_τ).

It appears that mass models with gravitational potentials
separable in confocal ellipsoidal coordinates may be useful
for the study of both the structure and the evolution of triaxial
elliptical galaxies.

It is a pleasure to thank D. Lynden-Bell for suggesting
that "it might be interesting to look at (λ,μ,ν)'s and Eddington
potentials". This work has benefitted greatly from his continued
interest and encouragement. Comments by H.C. van de Hulst and
E.F. van Dishoeck improved the manuscript. J. Audouze and J. Tran
Tranh Van are thanked for their hospitality in La Plagne, both
on as well as off the slopes. A travel grant from Het Leidsch
Kerkhoven Bosscha Fonds is gratefully acknowledged.

REFERENCES:

1. Binney, J.J. 1978a, M.N.R.A.S. 183, p. 501.
2. Binney, J.J. 1978b, Comments Ap. 8, p. 2.
3. Illingworth, G. 1981, in *The Structure and Evolution of Normal Galaxies*, ed. S.M. Fall and D. Lynden-Bell (London: Cambridge University Press), p. 27.
4. Wilkinson, A. and James, R.A. 1982, M.N.R.A.S. 199, p. 171.
5. van Albada, T.S. 1982, M.N.R.A.S. 201, p. 939.
6. Schwarzschild, M. 1979, Ap. J. 232, p. 236.
7. Schwarzschild, M. 1982, Ap. J. 263, p. 599.
8. Jeans, J.H. 1915, M.N.R.A.S. 76, p. 70.
9. Lynden-Bell, D. 1962a, M.N.R.A.S. 124, p. 1.
10. Schwarzschild, M. 1981, in *The Structure and Evolution of Normal Galaxies*, ed. S.M. Fall and D. Lynden-Bell (London: Cambridge University Press), p. 43.
11. de Zeeuw, P.T. 1983, in preparation.
12. Liouville, J. 1849, Journ. de Math. 14, p. 257.
13. Stäckel, P. 1890, Math. Ann. 35, p. 91.
14. Levi-Civita, T. 1904, Math. Ann. 59, p. 383.
15. Dall'Acqua, F.A. 1908, Math. Ann. 66, p. 394.
16. Eddington, A.S. 1915, M.N.R.A.S. 76, p. 37.
17. Weinacht, J. 1924, Math. Ann. 91, p. 279.
18. Clark, G.L. 1936, M.N.R.A.S. 97, p. 182.
19. Lynden-Bell, D. 1962b, M.N.R.A.S. 124, p. 95.
20. Ollongren, A. 1962, Bull. Astr. Inst. Neth. 16, p. 241.
21. Binney, J.J. 1982, in *Morphology and Dynamics of Galaxies*, Proc. Twelfth Adv. Course of the Swiss Society of Astr. and Astroph., ed. L. Martinet and M. Mayor, (Sauverny, Geneve Observatory), p. 1.
22. Kuzmin, C.G. 1956, Astr. Zh. 33, p. 27.
23. de Zeeuw, P.T. and Lynden-Bell, D. 1983, in preparation.
24. Oort, J.H. 1928, Bull. Astr. Inst. Neth. 4, p. 269.
25. van Albada, G.B. 1952, Proc. Kon. Ned. Akad. van Wetensch., Series B 55, No 5, p. 620.
26. Contopoulos, G. 1960, Zs. Ap. 49, p. 273.
27. Tremaine, S. 1983, in IAU Symposium No. 100, *Internal Kinematics and Dynamics of Galaxies*, ed. E.O. Athanassoula (Dordrecht: Reidel), p. 411.
28. Heiligman, G. and Schwarzschild, M. 1979, Ap. J. 233, p. 872.
29. de Zeeuw, P.T. and Merritt, D.R. 1983, Ap. J. 267, p. 571.
30. Goodman, J. and Schwarzschild, M. 1981, Ap. J. 245, p. 1087.
31. King, I.R. 1978, Ap. J. 222, p. 1.
32. Leach, R. 1981, Ap. J. 248, p. 485.
33. Williams, T.B. and Schwarzschild, M. 1979a, Ap. J. 227, p. 56. Williams, T.B. and Schwarzschild, M. 1979b, Ap. J. Suppl. 41, p. 209.
34. van de Hulst, H.C. 1962, Bull. Astr. Inst. Neth. 16, p. 235.
35. Hori, G. 1962, Publ. Astr. Soc. Japan 14, p. 353.
36. Binney, J.J. and Spergel, D.N. 1983, preprint.
37. Freeman, K.C. 1966, M.N.R.A.S. 134, p. 1.

ARE N-BODY POTENTIALS OF EDDINGTON FORM?

Althea Wilkinson

Department of Astronomy
University of Manchester
Manchester, England.

ABSTRACT

Using a formalism developed by de Zeeuw and Lynden-Bell, we are investigating whether a global fit can be made to the potential from the equilibrium phase of a triaxial n-body model. When extended from the present two-dimensional example to the fully three dimensional case, this will immediately yield the distribution function in terms of the three integrals of motion, without the necessity of integrating each orbit over many periods.

1.1 OBJECTIVES

This paper pursues one of the objectives mentioned in the previous paper (de Zeeuw, 1983), namely, that of investigating whether it is possible to make a global fit to the potential derived from a "realistic" triaxial n-body model with an Eddington potential. Phrased another way, we want to know whether such potentials are nearly separable in confocal ellipsoidal coordinates. Our reason for wanting to do this is that if the potentials can genuinely be treated in this way, it provides a simple and efficient method of understanding the n-body models in terms of the integrals of motion.

The information we should like to extract from the n-body models begins with how the overall form is constructed from the individual orbits. In order to do this, the orbits found in a particular potential must be classified, and until now this classification has been achieved in two dimensions by the tedious process of running each orbit for a long time in the fixed potential and examining the surface of section. Clearly, to investigate

315

J. Audouze and J. Tran Thanh Van (eds.),
Formation and Evolution of Galaxies and Large Structures in the Universe, 315–325.

three-dimensional potentials, and potentials which are changing,
by this method would be an unacceptably lengthy procedure. This
is also only the first step, since once the orbits are known, we
then wish to know what the integrals are, how well they are conserved, how the distribution function as a function of the integrals behaves, and the ultimate aim, how the orbits respond to
slow changes in the potential.

2.1 GLOBAL FITTING IN CONFOCAL COORDINATES

The form of the potential which is separable in confocal coordinates has been given many times (Eddington, 1916; Lynden-Bell,
1962; Lynden-Bell and de Zeeuw, 1983), and is

$$V_{Edd}(\lambda, \mu) = \frac{\zeta^*(\lambda) - \eta^*(\mu)}{(\lambda - \mu)} \qquad (2.1)$$

If $V(\lambda, \mu)$ is of Eddington form, the function

$$\chi(\lambda, \mu) = (\lambda - \mu) \; V(\lambda, \mu) \qquad (2.2)$$

should therefore have the form

$$\chi(\lambda, \mu) = \zeta^*(\lambda) - \eta^*(\mu) \qquad (2.3)$$

For an ideal potential, we should expect $\chi(\lambda_i, \mu)$ at each value of
λ_i, and $\chi(\lambda, \mu_i)$ at each value of μ_i to have the same functional
form, though possibly to be separated from each other by constant
amounts (Fig. 2.1). For a "real" potential where the foci have
not been (or cannot be globally) determined correctly, we should
expect plots of the following form (Fig. 2.2). Clearly we can use
the degree to which the functions can be made similar as a criterion for a least squares fit to the optimum foci, and this is
what we have done in practice.

2.2 N-BODY POTENTIAL FITTING

We have used a potential obtained from the later stages of
one of our n-body runs on a 32 x 32 x 32 mesh as a convenient test
example. This potential is only slightly triaxial (axial ratios
0.6:0.7:1 approximately, rather close to those used by Schwarzschild (1979) in his initial triaxial model), and is not rotating.
The radial variation is shown in Fig. 2.3.

For test purposes, we have initially considered the two-
dimensional case, taking the X-Y section (X > Y > Z) in the Z = O
plane. The geometry of confocal ellipsoids has been discussed in
the previous paper (de Zeeuw, 1983), where it has been pointed out

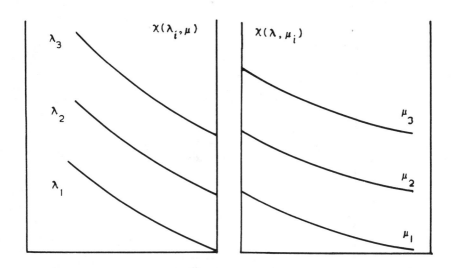

Fig. 2.1. Sections across the function $\chi(\lambda, \mu)$ at different values of λ, μ, for ideal Eddington potential.

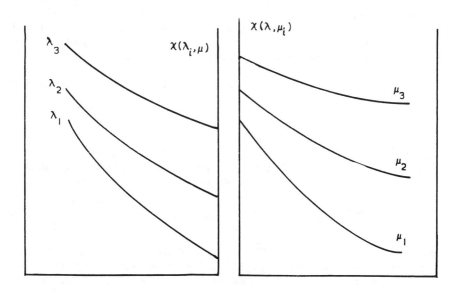

Fig. 2.2. Sections across the function $\chi(\lambda, \mu)$ at different values of λ, μ for real potential with incorrectly determined foci.

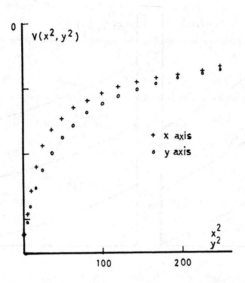

Fig. 2.3. Radial variation of the n-body potential

that the foci represent the branch points where axial orbits give
way to loop orbits. One way of determining the foci would be by
an examination of the dimensions of the smallest loop orbits and
largest y-axial orbits which are strictly in the x-y plane. However, this procedure does not define the foci accurately enough.

Lynden-Bell and de Zeeuw have devised a simple analytic method
of locating the foci. By differentiating the expression

$$(\lambda - \mu) \, V_{Edd} \, (x^2, \, y^2) = \zeta^*(\lambda) - \eta^*(\mu) \tag{2.4}$$

for the separable Eddington potential in Cartesian coordinates with
respect to λ, and equating the resulting expressions evaluated at
the origin and the focus, they obtain

$$\Delta = \frac{V_{Edd} \, (0, \, \Delta) - V_{Edd} \, (0, \, 0)}{\partial V_{Edd}(0, \, 0)/\partial x^2} \tag{2.5}$$

where $\Delta = \beta - \alpha$, which is the square of the position of the focus
along the y axis. Once $\partial V_{Edd}(0, \, 0)/\partial x^2 = g$, say, has been determined, then the intersection of the line gy with the graph of
$V_{Edd}(0, \, y^2) = V_{Edd}(0, \, 0)$ gives the required value of Δ.

The potential is given as $V(x^2, \, y^2)$ and we require it as a
function of $(\lambda, \, \mu)$. The confocal coordinates corresponding to

the Cartesian coordinates for each point at which the potential is tabulated are given by

$$\lambda = 1/2 \ (x^2 + y^2 - \alpha - \beta) + 1/2 \ ((\alpha - \beta)^2 - 2(\alpha - \beta)(x^2 - y^2)$$
$$+ \ (x^2 + y^2)^2)^{\frac{1}{2}}$$

$$\mu = 1/2 \ (x^2 + y^2 - \alpha - \beta) - 1/2 \ ((\alpha - \beta)^2 - 2(\alpha - \beta)(x^2 - y^2)$$
$$+ \ (x^2 + y^2)^2)^{\frac{1}{2}} \qquad (2.6)$$

The potential must then be interpolated onto a uniform grid in (λ, μ) space. The outer corners of the square grid, for which we do not have complete coverage in (λ, μ) have been omitted from the subsequent fit.

For the optimum foci, the sections through the function in the λ and μ directions are shown in Fig. 2.4 and 2.5.

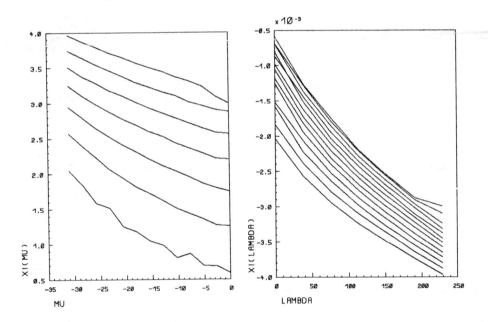

Fig. 2.4. $\chi(\lambda_i, \mu)$ at constant values of lambda

Fig. 2.5. $\chi(\lambda, \mu_i)$ at constant values of mu

We may shift the curves by an arbitrary constant without altering the form of the function, and for a perfect Eddington potential, this would result in all the curves being exactly superposed. Superposing the means of all the curves shown in Figs. 2.4 and 2.5 yields Figs. 2.6 and 2.7.

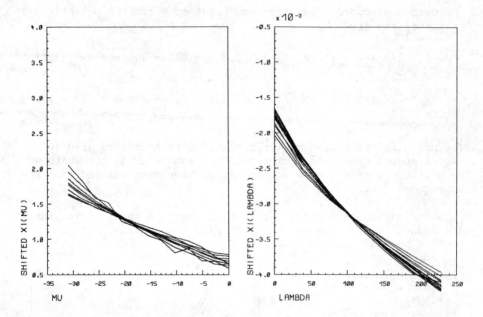

Fig. 2.6. Superposed cuts
across $\chi(\lambda_i, \mu)$ at constant λ

Fig. 2.7. Superposed cuts
across $\chi(\lambda, \mu_i)$ at constant μ

There is a slight change of form over the range of the function, but nevertheless it is clear that to first order, an
Eddington potential represents an acceptable fit.

It is reasonable to suppose that weighting the function according to the particle density might also be an allowable modification, and hence in Fig. 2.8 and 2.9 we show the effect of
weighting $\chi(\lambda, \mu)$ with the function

$$1/(constant*\tau + constant)^2 \tag{2.7}$$

where τ is either λ or μ over the appropriate range.

Fig. 2.10 and 2.11 show that we can superpose the curves
rather satisfactorily over most of the potential.

By then taking a mean over the superposed curves, we can obtain the best global estimate of the function $\chi(\lambda, \mu)$. Fig. 2.12
shows the final function, and a polynomial fit to it.

In the ideal case we should expect one continuous function
of $\chi(\tau)$ over the entire range, with $\chi(\lambda) = \chi(\tau)$ and $\chi(\mu) = \chi(\tau)$
in the appropriate adjacent intervals (cf. Fig. 6 in de Zeeuw 1983).

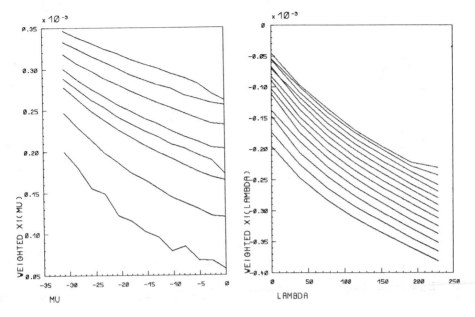

Fig. 2.8. $\chi(\lambda_i, \mu)$ at constant λ, weighted (constants 1,100)

Fig. 2.9. $\chi(\lambda, \mu_i)$ at constant μ, weighted (constants 1,100)

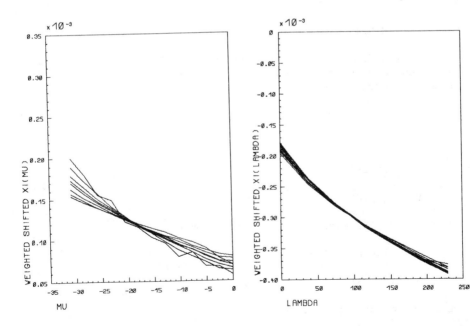

Fig. 2.10. Superposed weighted curves at constant λ

Fig. 2.11. Superposed weighted curves at constant μ

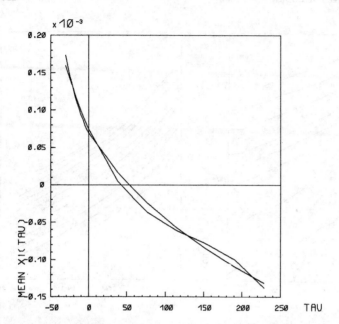

Fig. 2.12. Mean best fit to $\chi(\tau)$ and a polynomial approximation to the fit, (unweighted).

3.1 DETERMINATION OF INTEGRALS

Having fitted the potential functions $\zeta^*(\lambda)$ and $\eta^*(\mu)$, we are then in a position to be able to determine the integrals directly.   These are given by

$$H = X + Y; \quad I = X + Y \qquad\qquad (2.8)$$

where

$$X = p_\lambda{}^2/2P^2 - \zeta^*(\lambda)/(\lambda - \mu)$$
$$Y = p_\mu{}^2/2Q^2 - \eta^*(\mu)/(\lambda - \mu) \qquad\qquad (2.9)$$

P and Q are the metric coefficients given by

$$P = (\lambda - \mu)/4(\lambda + \alpha)(\lambda + \beta)$$
$$Q = (\mu - \lambda)/4(\mu + \alpha)(\mu + \beta) \qquad\qquad (2.10)$$

and the momenta in confocal coordinates are

$$p_\lambda = P^2\dot\lambda$$
$$p_\mu = Q^2\dot\mu \qquad\qquad (2.11)$$

For each orbit we have instantaneous values of x, y, $\dot{x}$, and $\dot{y}$, and hence the confocal equivalents $\dot{\lambda}$ and $\dot{\mu}$ are given by

$$2\dot{X}/X = \dot{\lambda}/(\lambda + \alpha) + \dot{\mu}/(\mu + \alpha)$$

$$2\dot{Y}/Y = \dot{\lambda}/(\lambda + \beta) + \dot{\mu}/(\mu + \beta)$$

(2.12)

and equation (2.1). Therefore by simply knowing the form of the potential at any instant, and the corresponding position and velocity for each orbit, we can immediately classify all the orbits, and represent the orbital constitution of a potential by a single H-I occupation diagram, which is just a graphical representation of the distribution function.

In practice it is almost impossible to use the n-body particle orbits because none of them are strictly two-dimensional. We have therefore illustrated the distribution functions for some families of representative test orbits in Figs. 3.1 to 3.3. These figures are to be compared with Fig. 8 of the previous paper (de Zeeuw, 1984).

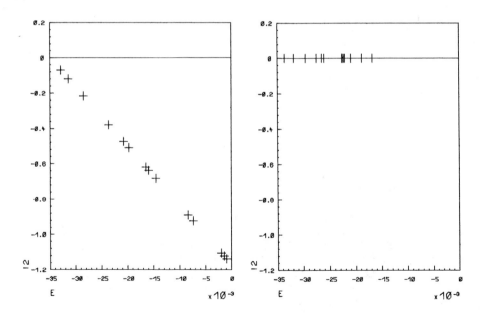

Fig. 3.1. H-I diagram for
x-axial orbits

Fig. 3.2. H-I diagram for
y-axial orbits

This marks the point at which it is necessary to move into three
dimensions - the part of the work which we are just about to under-
take. Here we shall have the advantage of a 64 x 64 x 64 grid and
three-dimensional "real" orbits for about 5000 particles, hence we
can reasonably hope for a good determination of the distribution
function.

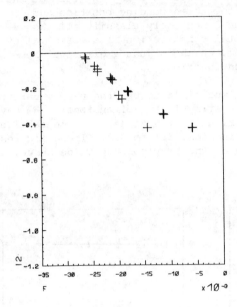

Fig. 3.3. H-I diagram for box orbits

4.1 CONCLUSIONS

     Our first attempts at fitting an n-body potential with an
Eddington potential look very promising, and have encouraged us
to press on to the three dimensional case. It seems that for at
least some astrophysically interesting models, the potentials are
such that we can make use of this neat, powerful and economical
method of extracting the information about the integrals of motion
from the experimental orbits.

     This method is complementary to the spectral analysis tech-
nique being developed by Binney and his co-workers (eg. Binney and
Spergel, 1983). For potentials where it can be applied with justification, it then involves substantially less computation than
the spectral analysis method, which does require the integration
of whole orbits. However, spectral analysis is not restricted in

the type of potential it can treat, and it is also being adapted
to cope with rotating potentials.

It therefore seems that we are very close to having the
analytical tools to achieve our goals of following slow changes in
the potential such as might happen during evolution of an isolated
system, or modification of a system during an encounter. Both
techniques will eventually yield adiabatic invariants, and study
of these during evolution of the potentials promises to yield some
very interesting results.

REFERENCES

Binney, J. and Spergel, D., 1983 (preprint).
de Zeeuw, P. T., 1984, this volume, p. 293.
Eddington, A., 1915, Mon. Not. R. astr. Soc., 76, p.37.
Lynden-Bell, D., 1962, Mon. Not. R. astr. Soc., 124, p.1.
Schwarzschild, M., 1979, Astrophys. J., 232, p.236.

SECULAR DYNAMICAL
EVOLUTION OF GALAXIES

COLIN A. NORMAN

Sterrewacht, Leiden

Institute of Astronomy, Cambridge

I. INTRODUCTION

Our knowledge and understanding of the structure of normal galaxies is growing rapidly. An important, but less studied, aspect is the slow secular dynamical evolution of galaxies over a Hubble time. As shown here, important, observable effects occur in galaxy structure and morphology. It is expected that during the Space Telescope era this can be widely studied since galaxies at a redshift of approximately unity will be observed easily.

The slow changes in potential are envisaged to occur over a large number of dynamical time scales of the system. Such changes can be caused by a variety of circumstances, some specific examples of which are: cooling and accretion of gas onto massive galaxies for example NGC 1275, where the accretion which is estimated to be $300 \ M_0 yr^{-1}$ (Fabian et al 1983); slow formation of the disks of spiral galaxies which can have a marked influence on the structure of the bulge; the non-axisymmetric driving of radial flow velocities in disks that, taken over a Hubble time, can completely alter the disk morphology; and the formation of a central massive black hole in a triaxial core producing the normal spike in the stellar density and also changing the fundamental nature of the orbits from box to tube as the sphere of influence of the hole encompasses the core.

Apart from the classical studies of the classical ellipsoids (c.f. Chandrasekhar 1969), secular evolution effects have been mainly studied in disks. As emphasized by Lynden-Bell and Kalnajs (1971), galaxies want to transfer their angular momentum outwards thereby increasing their binding energy. Schwarz (1981) has shown that an

327

J. Audouze and J. Tran Thanh Van (eds.),
Formation and Evolution of Galaxies and Large Structures in the Universe, 327–336.
© *1984 by D. Reidel Publishing Company.*

ensemble of sticky clouds driven in a disk by a bar will sweep out
the region between resonances to form annular concentrations of
gas within approximately ten dynamical times. These structures he
associated with rings. These numerical studies are modelled analytically in II.

Bar evolution is not yet clearly understood. Lynden-Bell (1979)
has argued, on the basis of an isochrone orbit analysis, that bars
form in the rising position of the rotation curve and as they
evolve and grow they become shorter and thinner. On the other
hand, Weinberg and Tremaine (1983) show that if one models bars
with planar Riemannian ellipsoids then, as they grow, they become
shorter and fatter. The final non-linear state of a bar is also
unclear from simulation work. Combes and Sanders (1981) performed
a three-dimensional simulation and showed that bars dissolved at a
sufficiently elongated axial ratio. In striking contrast, Sellwood
and coworkers (Sellwood 1980, 1981) have found no such dissolution
using long runs of their highly accurate code. The latter
calculation is, however, only two dimensional as in the work of
Efstathiou, Lake and Negroponte (1982) and it is possible that
three dimensional effects such as the firehose instability (Toomre
1966, Kulsrud, Mark and Caruso, 1972) may be important for very
radial orbits.

Combes and Sanders had a 30% mass loss but this probably does not
effect their overall conclusion. In the detailed simulation of
collapse models, van Albada (1982) found that, in the transient
initial phases, bars evolve from long and skinny to short and fat.
Sanders and van Albada (1979) have shown how a bar can drive a
significant temperature anisotropy in the disk due to preferential
heating in the plane. Stochastic effects and related heating
processes at resonances have been studied in detail by Sanders and
Teuben (1983), Contopolous (1983) and Athanassoula (1983). The
initial goal of such a study is to determine global properties of
galaxies such as shape, density profile, rotation curve and radial
velocity dispersion profile that are affected by slow changes in
the potential in quite different ways from rapid changes and the
associated violent relaxation processes.

II. FORCED RADIAL FLOWS IN GALACTIC DISKS

We study here how a population of clouds moving on ballistic
orbits with some effective drag, due to interaction with the
background disk, can undergo substantial radial motion in less
than a Hubble time. Inside corotation the clouds lose angular
momentum to a driving wave, outside corotation they gain from it.
Consequently, in the absence of an inner Lindblad resonance, gas
can be swept from inside corotation to the centre, or from outside
corotation to the outer Lindblad resonance, in the recent past.

Radial gas flows in galactic disks have often been invoked to explain metallicity gradients and, if present, are thought to have significant influence on many aspects of disk formation and evolution. Here we present a calculation of how a population of massive, presumably molecular clouds, undergoing some drag by interacting with the background disk, will evolve due to the presence of a driving non-axisymmetric disturbance such as a spiral wave or bar. Significant sweeping of the galactic disk inwards inside corotation and outwards outside corotation occurs on time-scales of ~10 rotation times. Previous numerical simulations of the bar forcing of an ensemble of colliding clouds are in agreement with our result (Schwarz, 1981). Obvious applications are to rings in S0 galaxies and gas flows in their inner regions. However, the aim of this work is to present a detailed analytic calculation of the angular momentum exchange between a spiral wave and a 'dragged' cloud.

Throughout the discussion we will focus on an average cloud from an ensemble of clouds with random positions and phases at their formation, destruction and injection. At cloud formation, we assumed the clouds are launched on ballistic, stellar, orbits with a epicyclic energy significantly larger than that of the gas (Leisawitz and Bash, 1982). The clouds feel a drag because of this difference, which we model by assuming the cloud and gas angular momentum is the same (so therefore are their angular velocity Ω and mean epicyclic frequency K), but the cloud epicyclic energy is damped such that the radial action $J_1 = \frac{1}{2} Ka^2$ is damped according to

$$\dot{J}_1 = \gamma(J_1 - J_{1gas}) = -\gamma J_{1gas}, \text{ if } J_1 \gg J_{1gas}.$$

In this limit we are, in effect, assuming that the background gas is moving on almost circular orbits and so indeed are the clouds if unperturbed by a wave. To zeroth order $J_1 \sim J_{1gas} \sim 0$. When the wave perturbation is present, we model the difference between the large epicycle of the ballistic clouds and the background by assuming in averaging over an ensemble of clouds that $\Delta J_{1gas} \sim 0$ and only the clouds responded to the perturbation.

Clouds are formed, destroyed and relaunched on timescales shorter than $1/\gamma$ so that the above inequality holds well. The drag is a simple way of describing a very complex situation involving cloud-cloud collisions, shocks and other dissipative gas dynamic effects. For example, dynamical friction estimates are given in Icke (1982), Julian (1967) and Thorne (1968) and Fokker Planck type estimates of the drag due to collisions and coagulation are discussed in Norman and Silk (1980).

However, very interesting physics can result from assuming γ is constant. We do this and leave more complex models to the future.

To first order the clouds are driven by the wave in the classical forced oscillator response where there is only a phase shift in the damped cloud response in comparison with the undamped stellar response. To ensure the accuracy of our calculation we have followed the direct second order computational technique of Lynden-Bell and Kalnajs (1971), with the exception that, because of the epicyclic damping, the first order integration of the radial action J_1 is obtain from solving

$$(\frac{d}{dt} + \gamma) \, \Delta_1 J_1 = \frac{\partial}{\partial w_1} \, \psi \, (\underline{J}, \, \underline{w}, \, t)$$

where $\Delta_1 J_1$ is the first order change in J_1 due to the perturbing potential ψ and the drag γ, and where $(J_1, \, w_1)$ are the action angle variables associated with the epicyclic motion and $(J_2, \, w_2)$ are those associated with the guiding centre. We find

$$\Delta_1 J_1 = Re \, \frac{1}{4\pi^2} \, \sum_{1,m} \psi_{\ell m} e^{i(1w_1 + mw_2 + wt)} \, \frac{1}{D(w-ij)}$$

where $D(w) = (1\Omega_1 + m\Omega_2 + w)$ and ψ, in the usual way, has been Fourier transformed with respect to the angles w_1 and w_2. The integration for $\Delta_1 J_2$ has no damping term leading to

$$\Delta_1 J_1 = Re \, \frac{1}{4\pi^2} \, \sum_{1,m} e^{i(1w_1 + mw_2 + wt)} \, \frac{m}{D(w)} \psi_{\ell m}$$

The first order corrections to the angles are

$$\Delta_1 w_1 = Re \, \frac{1}{4\pi^2} \sum_{1,m} e^{i(\ell w_1 + mw_2 + wt)} \frac{-i}{D(w)}] [\frac{1 \frac{\partial \Omega}{\partial J}}{D(w-j)} \psi_{\ell m} + \frac{m \frac{\partial \Omega}{\partial J}}{D(w)} \psi_{\ell m} - \frac{\partial \psi}{\partial J} \ell m]$$

The angular momentum behaviour of a mean cloud is obtained by averaging

$$\overset{\bullet}{h} = \Delta_2 J_2 = [\Delta_1 J_1 \frac{\partial}{\partial J_1} + \Delta_1 J_2 \frac{\partial}{\partial J_2} + \Delta_1 w_1 \frac{\partial}{\partial w_1} + \Delta_1 w_2 \frac{\partial}{\partial w_2}] \frac{\partial \psi}{\partial w_2} \ell m$$

over angles, resulting in

$$\langle \overset{\bullet}{h} \rangle = \frac{1}{(2\pi)^4} e^{-2Im(w)t} \sum_{1,m} \frac{\gamma m 1}{D(w-ij)} 2 [\frac{1}{2} \frac{\partial}{\partial J_1} \psi_{\ell m}^2 + \frac{|\psi_{\ell m}|^2}{D(w)} \frac{\partial}{\partial J_1}(1\Omega_1 + m\Omega_2)]$$

Hence further we shall consider only steady waves and $Im(w) = 0$. In the epicyclic approximation $\Omega_1 = K(h)$, the epicyclic frequency, and $\Omega_2 = \Omega(h)$, the angular rotational frequency. The partial derivative with respect to J_1 is at fixed h so

$$\langle \overset{\bullet}{h} \rangle = \frac{1}{2} \frac{1}{(2\pi)^4} \sum_{1,m} \frac{\gamma m 1}{(1K + w + m\Omega)^2 + \gamma^2} \frac{1}{Ka} \frac{\partial}{\partial a} |\psi_{\ell m}|^2$$

Using $\psi_{\ell m} = 4\pi^2 S(R_h) e^{i(1\alpha - m\Phi(R_h))} J_e(k'a)$ and $k'^2 = k^2 + \dfrac{2\Omega m}{KR^2}$,

where $J_\ell(k'a)$ is a Bessel function (Lynden-Bell and Kalnajs, 1971), we find

$$\langle \dot{h} \rangle = \tfrac{1}{2} \sum_{1,m} \frac{\gamma S^2 m1}{[\ell K + w + m\Omega]^2 + \gamma^2} \frac{1}{Ka} \frac{\partial}{\partial a} J_1^2(k'a)$$

and in the long-wavelength limit $k'a \ll 1$ only the $1 = \pm 1$ terms dominate giving our result (with $\Omega p = -w/m$)

$$\langle \dot{h} \rangle = - \frac{2m^2 \gamma k'^2 S^2(R_n)(\Omega - \Omega_p)}{[(w + m\Omega - K)^2 + \gamma^2][(w + m\Omega + K)^2 + \gamma^2]}$$

explicitly showing how the effect changes sign across corotation; losing angular momentum inside and gaining it outside.

Making order of magnitude estimates with $w \sim K \sim \Omega$ and assuming we are not near resonance then

$$\frac{\langle \dot{h} \rangle}{\Omega_m R_h^2} \sim - 2 m^2 \gamma \left(\frac{\Omega - \Omega_p}{\Omega}\right) (k'R_h)^2 \left(\frac{S(R_h)}{\Omega^2 R_h^2}\right)^2 \quad , \; k \neq 0,$$

and,

$$\sim - 2 m^2 \gamma \left(\frac{\Omega - \Omega_p}{\Omega}\right) \left(\frac{S(R_h)^2}{\Omega^2 R_h^2}\right) \quad , \; \text{if } k = 0,$$

with numerical estimates of radial inflow velocities

$$\frac{v_r}{r} \sim - 2 \gamma m^2 \left[\frac{k'R_h}{10}\right]^2 \left[\frac{S(R_n)/\Omega^2 R_n^2}{10^{-1}}\right] \left(\frac{\Omega - \Omega_p}{\Omega}\right) \quad , \; \text{if } k \neq 0$$

and

$$\sim - 2 \gamma m^2 \left[\frac{S(R_h)/\Omega^2 R_h^2}{0.3}\right] \left(\frac{\Omega - \Omega_p}{\Omega}\right) \quad , \; \text{if } k = 0$$

and with $\gamma \sim 10^9 - 10^{10} \; \text{yr}^{-1}$, $v_r \sim 1\text{-}10 \; \text{kms}^{-1}$ ar $r \sim 10$ kpc!

Lenticular SO galaxies have gas rings in the outer regions and no gas in the interior. We suggest here that they have been swept by this mechanism, then Q of the disk increased and the waves turned off. Such SO rings should always have stars and gas moving in the same direction in contrast to the infalling gas rich dwarf hypothesis (Silk and Norman 1979, Van Woerden et al 1983). Such a mechanism could be important in creating substantial radial inflow in the central parts of galaxies (Norman and Silk 1983, Lake and Norman 1983). The radial outflow outside corotation could produce observably flatter metalliaty gradient in the outer parts of disks. If bars in spirals end near corotation (Kormendy and Norman 1979) this flattening could be observed exterior to the bar. It could even applicable to self-gravitating accretion disks. Generally speaking, an unstable disk with a set of ballistic lumps with some background friction will show this interesting process of angular momentum exchange and radial flow.

III. EVOLUTION OF SPHEROIDS

The secular evolution of spheroids has been studied in immense detail by Chandrasekhar, Friedman and Lebowitz for the classical ellipsoids see Chandrasekhar 1969 for a review. Here we consider only non-classical systems with at least one additional quasi-integral. Previous studies here include the analysis of the formation and collapse phase of the Galaxy including a third integral (Yoshii and Saio 1979) and more recently numerical work on spectral stellar dynamics (Binney and Spergel 1982, 1983) who analysed tube to box transitions in slowly varying triaxial systems. We shall discuss the above problems as well as the structure of galaxies that slowly became axisymmetric from an initially triaxial state and the evolution of a galaxy's core region as a black hole is formed.

Yoshii and Saio (1979) considered the slow collapse of the Galaxy using an axisymmetric potential with a third integral given by

$$\psi\,(r,\,\xi) = \frac{GM}{b+(b^2+ r^2)^{\frac{1}{2}}} - \frac{g(\xi)}{r^2}$$

where

$$g\,(\xi) = \frac{\varepsilon^2 \tan^2(\zeta/2)}{[1 + a^2\tan^2(\zeta/2)]^{\frac{1}{2}}}$$

and ε, a and b are slowly varied, and the maximum box angle (see Fig. 1) is found from the root of the equation

$$I_3 - I_2^2(1 + \tan^2\zeta) - 2g(\zeta) = 0$$

where I_3 and I_2 are constants of the motions.

There are three adiabatic invariants observed in a slow collapse

$$J_2 = \Pi\,[\frac{2GM}{(-2E)^{\frac{1}{2}}} - I_3^{\frac{1}{2}} - (4bGM + I_3)^{\frac{1}{2}}],$$

$$J_\zeta = 4\int_0^{\zeta_{max}}[I_3 - I_2^2(1 + \tan^2\zeta) - 2g(\zeta)]^{\frac{1}{2}}d\zeta$$

and

$$J_0 = 2\Pi\,J_Z$$

These authors noted a substantial difference with respect to the rapid collapse results of Eggen, Lynden-Bell and Sandage (1962)

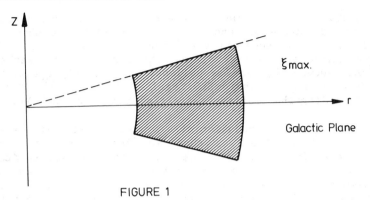

FIGURE 1

ξ_{max} is the maximum box angle defined in the text.

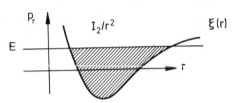

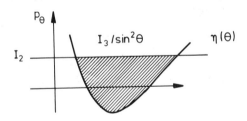

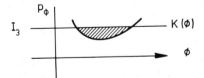

FIGURE 2

Effects of the transition of a triaxial to an axisymmetric structure (see text).

(ELS). For a rapid collapse the eccentricity changes considerably but the box angle is conserved. For a slow collapse there is little change of eccentricity (since it is related to the conserved adiabatic invariants) and the box angle decreases. Their analysis showed that a slow collapse was possibly a better fit to the data than the classic ELS results. Further work is necessary here including the important effect of a massive dark halo.

We now consider the evolution of a dynamical system from a triaxial to axisymmetric structure that may for example be caused by massive disk formation in a bulge or by the slow adiabatic spin up of a triaxial system. In a non-rotating triaxial system there are two dominant families of tube orbits: one around the largest axis (a spindle orbit called hereafter Tube 1) and the other (Tube 2) around the shortest axis. We shall see how Tube 1 dissolves as the system becomes axisymmetric and Tube 2 remains as the system slowly became axisymmetric. We model this evolution using a separable potential in spherical co-ordinates with a generic potential form (see Lake and Norman, 1983)

$$\Psi = \zeta (r) + \frac{1}{r^2} \eta (\theta) + \frac{1}{r^2 \sin^2 \theta} K (\phi)$$

where the separability condition allows direct integration to

$$p_r^2 = E - \zeta(r) - \frac{I_2}{r^2}, \quad p_\theta^2 = I2 - \frac{I_3}{\sin^2 \theta} + \eta (\theta),$$

and $p_\phi^2 = I_3 + K(\phi)$ where E, I_2 and I_3 are integrals of motion. Consider now the actions J_1

$$= \int p_r dr = \int (E - \zeta(r) - \frac{I_2}{r^2})^{\frac{1}{2}} \, dr, J_2 = \int p_\theta d\theta = \int (I_2 - \frac{I_3}{\sin^2 \theta} + \eta(\phi)^{\frac{1}{2}} d\theta, \quad \text{and}$$

$$J_3 = \int p_\theta d_\theta = \int \left(I_3 - K (\phi) \right)^{\frac{1}{2}} d\phi \text{ as displayed in Fig. 2. During slow evolution } E, I_2 \text{ and } I_3 \text{ adjust to keep the area or action instant as the potential changes.}$$

From a cursury inspection of the Figure 2 we can see the effects of a triaxial to axisymmetric transition. Tube 1 consists of a circulation in θ and a trapping in ϕ and as $K(\phi)$ goes to zero. I_3 is unchanged or increases only slightly and this tube disappears. Tube 2 is a circulation in ϕ with a trapping in θ and as $K(\phi)$ goes to zero, this tube remains. The loss of Tube 1 but not of Tube 2 leads to the conjecture that this will produce a peanut-shaped bulge (c.f. Kormendy and Illingworth 1982, Dressler and Sandage 1983) It is interesting to note here that attempts to produce this effect using disk formation in a axially symmetric bulge have not been succesful (S. White, private communication). However the three-dimensional evolving bar simulations of Combes an Sanders did produce a transient peanut-shape when viewed looking along the major axis of the bar.

The essentially graphical analysis given here is easiest to use in spherical coordinate but we note the following problems. Firstly, there is the need to fit a separable potential which can only be done strictly for non-rotating systems and these peanut-shaped bulges do rotate. Secondly, fitting separable potentials is best done in general ellipsoidal co-ordinates and the graphical method becomes more complex than in the spherical case.

Briefly turning to the growth of a central massive object such as a Schwarzschild or Kerr hole in the centre of a triaxial galaxy. The following questions arise. Is there a catastrophic instability when the sphere of influence of the hole extends past the foci of the triaxial system where the orbits change from box to tube, i.c. $GM_h/\sigma^2 \gtrsim r_{foci}$ so that the core consists of tube orbits not boxes? Do the foci smoothly change as one models the potential with a separable one as, we move outwards from the central hole. For example if α is a focus can one model the system using a WKBJ approach with $\alpha = \alpha_p \tanh (r/r_h)$ as a 'slowly' varying focus? Can one produce a stellar hydodynmaics solution to the problem using the perfect ellipsoid (de Zeeuw 1983) plus black hole?

The most obviously direct way to investigate these secular evolutionary effects is to take the self-consistent, very high dynamic-range simulations of van Albada and coworkers and slowly change the potential by growing a central massive black hole, or a bulge and disk, or by slowly torquing a triaxial configuration until it becomes axisymmetric.

Such detailed work is now a program in collaboration with A. May and T. van Albada. I have argued here that the results will be rather interesting (Norman, May and van Albada 1983).

It is a pleasure to acknowledge interesting and useful discussions with Ray Carlberg, Jerry Sellwood, Tjeerd van Albada, Andrew May, Bernard Pagel, Rosie Wyse, Simon White and Tim de Zeeuw and futher to thank Jean Audouze for his excellent hospitality at La Plagne during this stimulating workshop.

REFERENCES

Albada, T.S. van 1983, MNRAS 201, 939.
Athanassoula, E., Bienayme, O., Martinet, L. and Pfenniger, D. 1983, in IAU 100, Internal Kinematics and Dynamics of Galaxies ed. E. Athanassoula, p 209.
Binney, J.J. and Spergel, D. 1982, Ap. J. 252, 308.
Binney, J.J. and Spergel, D. 1983 preprint
Chandrasekhar, S. 1969, Ellipsoidal Figures of Equilibrium, New Haven,Yale univ. pers.
Combes, F. and Sanders, R.M. 1981, Astron. Astrophys. 96, 164.
Contopolous, G. 1983, Astron. Astrophys. 117, 89.

Dressler, A. and Sandage, A. 1983, Ap. J. 265, 664.
Efstathiou G., Lake, G. and Negroponte, J. 1982, MNRAS 199, 1069
Eggen, O.J., Lynden-Bell, D. and Sandage, A.R. 1962, Ap.J. 136, 748.
Fabian, A.C., Nulsen, P.E.J. and Canizares, C.R. 1982, MNRAS 201, 933.
Icke, V. 1982, Ap. J. 254, 751.
Julian, W.H. 1967, Ap.J. 184, 175
Kormendy, J. and Norman, C. 1979. Ap. J. 233, 539
Kormendy, J. and Illingworth, G. 1982, Ap. J. 256, 460
Kulsrud, R.M., Mark, J.W.K. and Curuso, A. 1972, in Gravitational N-Body Problem ed M. Lecar, p 180 (D. Reidel, Dordrecht-Holland).
Lake, G. and Norman, C. 1983, Ap. J. 269, in press
Leisesawitz, D. and Bash, F. 1982, Ap. J. 259, 133
Lynden-Bell, D. and Kalnajs, A. 1971, MNRAS 157, 1.
Lynden-Bell, D. 1979, MNRAS 187, 101
Norman, C. and Silk, J. 1980, Ap. J. 238, 158
Norman, C. and Silk, J. 1983, Ap.J. 266, 502.
Norman, C. May, A. and van Albada, T.S. 1983, in IAU 106, The Milky Way Galaxy ed. H. van Woerden.
Sanders, R.H. and van Albada, T.S. 1979, MNRAS 189, 791
Schwarz, M.P. 1981, Ap. J. 247, 77.
Sellwood, J.A. 1980, Astron. Astrophys. 89, 296.
Sellwood, J.A. 1981, Astron. Astrophys. 99, 362
Silk, J. and Norman, C. 1979, Ap. J. 234, 86
Teuben, P.J. and Sanders, R.H. 1983 in IAU 100, Internal Kinematics and Dynamics of Galaxies ed. E. Athanassoula, p. 211
Thorne, R. 1968, Ap. J. 151 671.
Toomre, A. 1966 'A Kelvin Helmholtz Instability', Woods Hole Summer program on Geophysical Fluid Dynamics, p 114.
Weinberg A. and Tremaine, S. 1983, preprint.
Van Woerden, H., van Driel, W. and Schwarz, U.S. 1983, In IAU 100, Internal Kinematic and Dynamics of Galaxies e.d E. Athannasousla, p. 99
Yoshi, Y. and Saio, H. 1979, P.A.S.J. 31, 339.
De Zeeuw, T. 1983, these proceedings, p. 293.

A GALAXY DISTRIBUTION FUNCTION?

Philippe Crane

European Southern Observatory

Abstract: A galaxy distribution function based on the statistical thermodynamics of gravitating objects is compared to data from the Zwicky catalogue of galaxies. The data are well represented by the proposed function. However, there remain some systematic effects to be modelled.

INTRODUCTION

Saslaw and Hamilton (ref. 1) have proposed a function to describe the distribution of galaxies. This function is based on the statistical thermodynamics of gravitationally interacting bodies. It predicts the probability of finding a given number of galaxies in a particular volume given the mean number per unit volume. The function is:

$$P(N) = \frac{\left[\overline{N}(1-b) + Nb\right]^{N-1}}{N!} \overline{N}(1-b) \exp - \left[\overline{N}(1-b) + Nb\right]$$

where N is the number of galaxies in volume V

$\overline{N}$ = nV is the average number of galaxies in V and n is the mean number of galaxies per unit volume

b = W/2K is the ratio of the correlation potential (W) to the kinetic energies (K) of the galaxies (see ref. 1 for details).

The parameter b in the equation is the interesting physical quantity. In the limit of uncorrelated motions of the test

337

J. Audouze and J. Tran Thanh Van (eds.),
Formation and Evolution of Galaxies and Large Structures in the Universe, 337–341.

objects, b = 0 and the expression for P(N) reduces to the
Poisson distribution as would be expected. In the limit b = 1
the quantity $\overline{N}(1-b)$ => cst and the expression fn P(N) describes
a system in virial equilibrium.

The derivation of the expression for P(N) given by Saslaw
and Hamilton leaves several important questions to be answered.
Among the most important are: 1) whether an equilibrium thermodynamic solution can be used to describe the distribution of
galaxies and 2) whether the particular ansatz used to derive the
result is physically correct. Perhaps the best way to test this
result is to compare it to the real world. This is done in the
next section.

APPLICATION TO THE ZWICKY CATALOGUE

The Zwicky Catalogue of galaxies (ref. 2) represents a
reasonably good sample of galaxies on which to test the proposed
galaxy distribution function. Any solid angle on the sky with
galaxy counts to a fixed limiting magnitude represents a sample
of some volume. Another equal solid angle sampled to the same
limiting magnitude is a sample of an equivalent volume with a
similar degree of incompleteness. Thus the counts of galaxies in
a given solid angle in the Zwicky catalog represent a data set
which can be compared to the proposed galaxy distribution. In
particular, galaxy counts in some arbitrary solid angle are made
for many different areas in the Zwicky catalog. The histogram of
these results is then compared to the function P(N). Figure 1
shows a typical histogram of galaxy with a plot of the fitted
function

$$F(n) = P(N) + A \exp - (N/N_0).$$

An exponential was added to the function P(N) to account
for the fact that there are far too many occurrences of areas
with no or few galaxies compared to the predictions of the formula P(N). This is most likely due to edge effects in our sample
and in the Zwicky catalog itself. In any case, this ad hoc addition of an exponential does not affect the main conclusions.

Many different samples were taken with varying solid
angles, with varying ways of choosing the shape of solid angles
of equal size, and with varying galactic latitude limits. Essentially all the derived distributions were well described by the
function P(N) with various values of the parameter "b". In this
case, "well described" means that the chi-squared values had the
expected distribution.

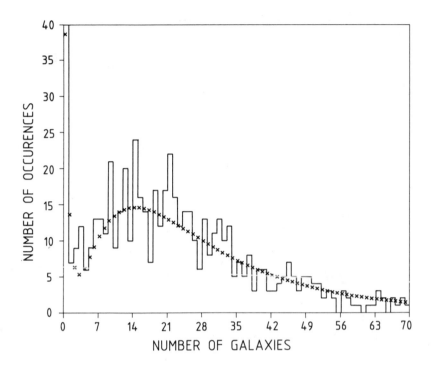

Figure 1: Histogram of the number of Zwicky catalog galaxies in
areas of 12 square degrees. All galaxies north of galactic latitude 40° are included. The crosses (+) are the fitted function
described in the text. The value of b in this case is
0.71 ± 0.05 and the chi-squared is 64.16 for 65 degrees of
freedom.

DISCUSSION:

        Two trends in the derived values of "b" were evident.
First, b depends on the galactic latitude limit chosen. Second,
b depends on the size of the solid angle. Both of these trends
have likely explanations. Nevertheless, they have not yet been
quantitatively explained, and work is in progress to model these
trends.

        The dependence of "b" on galactic latitude is shown in
Figure 2. Here, small values of b are found at low galactic latitude. Small b values correspond to more random and less correlated galaxy distributions. This result is attributed to a
random distribution of absorbing clouds at lower galactic latitudes which will make the galaxy distribution appear more
random.

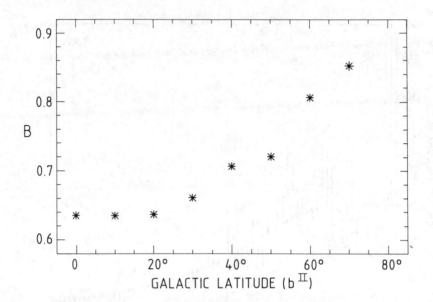

Figure 2: Plot of the dependence of the parameter "b" on
galactic latitude. These results are from data in areas of 12
square degrees.

The dependence of "b" of the size of the solid angle
chosen is shown in Figure 3. Here, small solid angles produce
small values of b. Small solid angles will only sample small
portions of nearby groups or clusters due to small angular
coverage and small portions of more distant groups and clusters
due to incompleteness. Hence small solid angles will tend not to
sample virialized or close to virialized groups and clusters in
this sample. Large solid angles will be more likely to sample
virialized groups and hence the values of b derived will tend to
be larger.

Independent of the validity of the physical reasoning that
went into deriving the formula of Saslaw and Hamilton, the
formula gives excellent agreement with the observed distributions in the Zwicky catalog. Nevertheless, further work on other
samples will be necessary to confirm the validity of the expression in general.

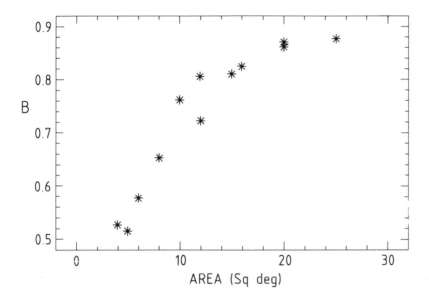

Figure 3: Plot of the dependence of the parameter "b" on the size of area chosen. These results are for data with galactic latitudes greater than 60°.

REFERENCES

1)   Saslaw, W.C. and Hamilton, A.J.S. 1983, Ap.J. (submitted).

2)   Zwicky, F., Herzog, E., Wild, P., Karpowicz, M. and
       Kowal, C.T. 1961-1968, "Catalogue of Galaxies and Clusters
       of Galaxies" (in 6 vols), Pasadena, California Institute
       of Technology.

THE THICKNESS OF GALACTIC DISKS

R.G. Carlberg

University of Toronto

I. Introduction

The local dynamical structure of disk galaxies is broadly characterized by a vertical scale height which is nearly independent of radius, the scale height being about 1/10 of the horizontal scale length (Searle and van der Kruit, 1982) and an internal population structure wherein the oldest stars have the largest random velocities (for an analysis of the data see Wielen 1977). Almost all disk galaxies contain gas and exhibit signs of ongoing star formation, indicating formation on a time scale comparable to the Hubble time. The slow and ongoing process of forming gas into stars has been shown by Sellwood and Carlberg (1983) to be vital to the continual presence of transient spiral waves. The density waves have associated potential waves which Carlberg and Sellwood (1983) and Carlberg (1983) suggest are the dominant mechanism driving secular evolution of the disk. This paper reviews the arguments for transient spiral wave heating, concentrating on the problems of wave heating in the vertical direction. The aim of this paper is to establish that the velocity dispersions in the disk are due to secular evolution, and nearly independent of cosmological conditions.

II. Why Transient Spirals?

In order that spiral waves have a significant dynamical effect they must be transient, and varying on a time scale comparable to the basic frequencies of oscillation in the disk,

343

J. Audouze and J. Tran Thanh Van (eds.),
Formation and Evolution of Galaxies and Large Structures in the Universe, 343–349.
© *1984 by D. Reidel Publishing Company.*

otherwise the adiabatic invariants insure that the disk will
return to its initial state after the waves have come and gone.
Spiral waves with a lifetime comparable to a rotation period
have been observed in many N body experiments of disk galaxies
(Lindblad 1960, Miller, Prendergast, and Quirk 1970, Hohl 1971,
Hockney and Brownrigg 1974, Sellwood and James 1978). The
associated heating is so powerful that within a few rotation
periods the disk particles acquire such high random velocities
that the disk becomes almost completely stable to spiral
instabilities. A problem then arises since most disk galaxies
are 100 rotations periods old, yet virtually all have some sort
of spiral structure.

Sellwood and Carlberg (1983) showed that dynamical cooling,
associated with dissipation in the gas and accretion of gas onto
the disk of the galaxy, continually destabilizes the galaxy
sufficiently that spirals can reform.

The spirals that are observed in the experiments have many
features in common with the swing amplification theory presented
by Toomre (1981). All spiral patterns start as weak leading
spirals, which shear around to trailing spirals growing by a
factor of 3 to 10 in amplitude. The waves then continue to wind
up on to the resonances, gradually dwindling in amplitude. The
growth and decay rates are comparable for two patterns that were
measured and the growth rates were 1/2 and 1/5 of the pattern
speeds of the associated waves. Although we do not yet have a
complete description of the evolution of spirals, the swing
amplification description is consistent with most of the
behaviour of an individual pattern.

III. Horizontal Heating by Spirals

The wave heating balances the dynamical cooling at an
equilibrium value of the velocity dispersion. The average radial
velocity dispersion, σ_u, of the disk is conveniently measured
by the dimensionless parameter $Q \equiv \sigma_u \kappa / (3.36 G \Sigma)$, where Σ is the
disk surface density and κ is the epicyclic frequency. In the N
body experiments the equilibrium Q is found to be only weakly
sensitive to the cooling rate, and to maintain a value near to
2, increasing from 1.7 to 2.3 for a factor of 10 decrease in
the cooling rate. The value in the solar neighbourhood is
estimated to be 1.5 to 2.0 (Toomre 1974), and there are
preliminary indications that other galaxies have comparable
values (Kormendy and Illingworth, private communication).

The mechanism governing the heating is well described by
considering the changes induced in a disk by a perturbation to

the axisymmetric potential. As a first approximation for the
determination of the behaviour in the plane of the galaxy the
thickness can be ignored. The change in the phase space
distribution induced by a transient spiral wave is found by
linearizing Liouville's equation for the conservation of phase
space density. The first order change phase mixes away, giving
no change in the distribution function.

The second order change in the distribution function, f,
due to a single transient spiral is an equation in the form of
a diffusion in phase space (derived in Carlberg and Sellwood
1983),

$$\Delta_2 f(\mathbf{J}, \mathbf{w}) = \frac{1}{(2\pi)^4} \sum_{\mathbf{m}} \mathbf{m} \cdot \frac{\partial}{\partial \mathbf{J}} \left({}^2 T_{\mathbf{m}} |\psi_{\mathbf{m}}|^2 \mathbf{m} \cdot \frac{\partial f}{\partial \mathbf{J}} \right). \quad (1)$$

The independent variables are chosen to be the action variables,
$J = (h, J_r)$ where h is the angular momentum, and J_r is the radial
action, which for an epicyclic of radial amplitude a is
$J_r = 1/2 \kappa a^2$. The complementary variables are the phase angles,
$\underline{w}$. The vector $\underline{m}$ is an integer vector of the Fourier indices.

The diffusion in phase space induced by the transient spirals
is governed by diffusion coefficient, which is composed of the
product of $\psi_{\underline{m}}$, which are Fourier transforms of the perturbing
potential over the unperturbed orbits, and the tuning factor
${}^2 T_m$ which indicates how well matched the perturbing potential's
frequency is to the natural frequencies of oscillation in the
disk.

A simple model of the spiral waveform with a constant pitch
angle was used to investigate the velocity dependence of the
heating rate. In the limit that the epicycles are small
compared to the distance between crests of the spiral wave the
heating is independent of the initial size of the epicycle, and
for one spiral the increase in the velocity dispersion is
approximately,

$$\Delta \sigma_u^2 \approx \frac{S^2 k'^2}{4\kappa^2}, \quad (2)$$

where k' is the effective wavenumber of the spiral. If spirals
with similar properties recur at a constant rate in time then
the increase of the velocity dispersion with age for a coeval
group of stars would be $\sigma_u^2 \sim t$, or $\sigma_u \sim t^{1/2}$. However as the
velocity dispersion increases, and the size of the epicycles

increases to a size comparable or larger than one wavelength of
the spiral wave, the heating rate diminishes to a minimum of
$\sigma_u \sim t^{1/5}$.

Heating by spiral waves therefore gives a natural explanation to the mean velocity horizontal dispersion of stars in a
disk and the shape of the age velocity dispersion relation.

IV.   Vertical Heating

Wave heating in the plane of the disk so well that one
might assume that heating in the vertical direction works
equally well. However wave heating in the vertical direction is
vexed by a number of difficulties, and at the moment no guidance
is available from N body simulations which adequately resolve
the population structure of the disk.

The basic difficulty is that heating by spirals with motions
completely in the plane is extremely slow in the vertical
direction for stars on orbits smaller than the thickness of the
perturbing layer. There are two causes of the inefficient
heating, the first being that the vertical frequency of
oscillation is much higher than the horizontal frequency of
oscillation. In the solar neighbourhood $\Omega_z \approx 3k$ and the pattern
speed of the waves is likely smaller than κ. The vertical
oscillation is so rapid that a spiral wave appears as an
adiabatic variation, which producing no permanent change in the
vertical component of the orbits. The other difficulty is
illustrated by the equation for the increase in the vertical
velocity dispersion (Carlberg and Sellwood 1983),

$$\Delta\sigma_w^2 = \frac{S^2 k_\perp^2}{64(\pi G \Sigma)^2} \sigma_w^2, \tag{3}$$

where $\kappa_\perp$ is the spiral wavenumber measured perpendicular to the
crests. The increase in the velocity dispersion increases with
initial value of the velocity dispersion. A ratio of the
heating rate (2) and (3) shows that for stars of relatively low
velocity dispersion the heating in the vertical direction falls
short of the required amount by about an order of magnitude.
The cause of this behaviour is simply that the star never feels
a vertical force if it happens to start exactly at $z = 0$, since
the perturbing layer is symmetrical about that point. Away from
the plane the perturbing force increases linearly with height
for a constant density perturbing layer.

There are two solution to the problem of heating the lowest
velocity dispersion stars, but they both come with their own
difficulties. In the pure wave heating picture the obvious thing
to do is to include vertical motions of the perturbing laryer.
There are two types of vertical motions of the spiral wave.
The perturbing layer may respond so as to keep the vertical
adiabatic invariant, proportional to $z^3\Sigma$, constant. This motion
is symmetrical about the plane and does not significantly
increase the heating.

A more promising wave motion is a bending where the plane
of the perturbing layer oscillates about the midlayer of the
stellar distribution. The heating was investigated by
integrating orbits in a disk containing the imposed spiral
potential. For a disk with structure parameters scaled to the
solar neighbourhood, i.e. a scale height of 300 pc in a disk
with a flat rotation curve, particles that are started with an
identical oscillation amplitude of 100 pc experience little
vertical heating. However the main action of the wave is to
scatter particles about their initial amplitude of oscillation,
which will widen any initial distribution of velocities, and
will lead to a net increase of the velocity dispersion. An
oscillation amplitude of 50 pc generates a ratio of the spread
in horizontal velocity amplitude to the spread in vertical
velocity amplitude of about 1/2, the observed ratio of velocity
dispersions being $\sigma_w/\sigma_u = 0.5$. If the stars start off with an
oscillation amplitude of only 50 pc the ratio of the spread
in the velocities is reduced to about 1/3. This ratio of spread
in the velocity amplitudes about their initial values is found
to be very insensitive to the details of the waves, although
the absolute magnitude of the increase of either does depend
on the strength of the perturbation and the wavenumber of the
spiral. For oscillations of the perturbing layer greater than
the amplitude of the initial orbits, no significant increase in
heating occurs.

The orbit integrations show that vertical oscillations of
the perturbing layer are capable of providing heating of the
observed magnitude. Do these vertical oscillations actually
exist? The theoretical situation is unclear, but discouraging.
The detailed analysis of Hunter and Toomre (1969, see also Shu
et al. 1983) of purely vertical bending waves found that the
waves have a very small amplitude in the inner part of the disk,
and there is no evidence for instability of modes with m > 1.
Bertin and Mark (1980, see also Mark 1983) found similarly small
amplitudes of vertical bending waves in the inner disk, but
argued that the inclusion of a spherical halo could lead to an
instability. In spite of theory (see Toomre 1983 for a
discussion, mostly of warps) observations of the galactic disk
to indicate the presence of distortions from the plane of

symmetry within the solar circle. Quiroga (1974) finds that
the HI plane has regular distortions plane with an amplitude
of 70 pc, and in the molecular cloud layer Solomon et al. (1979)
and Cohen et al. (1979) report deviations of 40 pc from the plane.

The situation for waves with a vertical component of
oscillation is therefore far from being clear. The modes of
oscillation and a source of amplification are not known. The
observations certainly indicate deviations of the regions of
maximum emission in the gas layer, but it is not known how well
this reflects deviations of the total perturbing potential.

Another possible source of the increase in the vertical
oscillations is scattering off molecular clouds. The analysis
of Lacey (1983) shows that interactions of stars and massive
gas clouds heats the stars on a long time scale, but the interactions drive the velocity ellipsoid towards an equilibrium
shape on a shorter time scale. From the data presented by Lacey
one expects that stars that are heated in the plane by any
mechanism will be scattered out of the plane on a time scale of
a few billion years. A difficulty arises if the scattering sends
the distribution all the way to equilibrium, in which case the
vertical velocity dispersion is predicted to exceed the tangential
velocity dispersion, contrary to observations.

Wave heating for stars with vertical oscillation amplitudes
exceeding the scale height of the disk is sufficient to provide
considerable heating. The increase in effectiveness is due
to the decrease in the vertical frequency of oscillation of the
stars with increasing height, which provides a better match to
the driving frequency of the perturbing spiral. However in
this case the size of the epicycles is comparable to the spiral
wavelength, and there is considerable interplay between the
heating in the two directions. At the moment the nature of
the interaction is not well enough understood to make a
prediction of the expected velocity distribution.

V.   Conclusions

Heating by transient spiral waves has a feedback mechanism
which adjusts the amplitude of the spirals to the average random
velocities of stars in a disk galaxy, and provides a natural
explanation to the mean radial velocity dispersion observed in
galactic disks, and N body experiments. The shape of the
observed age velocity dispersion relations can be predicted with
a diffusion equation, where the rate of diffusion is dependent
on the properties of the transient spirals, and the velocity
dispersion of the stars being heated.

The thickness of disk galaxies and the vertical velocity dispersion is not so easily explained. Either spiral waves with a bending motion, or scattering of orbits out of the plane by large gas clouds would be capable of providing vertical velocities of approximately the right magnitude. The cloud scattering mechanism in equilibrium predicts the wrong shape of the velocity ellipsoid, and the understanding of vertical wave oscillations and their effects on vertical motions is very incomplete.

REFERENCES

Bertin, G. and Mark, J.W.-K. 1980, Astr. Ap., 88, 289.

Carlberg, R.G. 1983, in preparation.

Carlberg, R.G. and Sellwood, J.A. 1983, in preparation.

Cohen, R.S., Tomasevich, G.R., and Thaddeus, P. 1979, in
    The Large-Scale Characteristics of The Galaxy, I.A.U.
    Symposium No. 84, ed. W.P. Burton, (Dordrecht: Reidel),
    p. 53.

Hockney, R.W. and Brownrigg, D.R.K. 1974, M.N.R.A.S., 167, 351.

Hohl, F. 1971, Ap.J., 168, 343.

Hunter, C. and Toomre, A. 1969, Ap.J., 155, 747.

James, R.A. and Sellwood, J.A. 1978, M.N.R.A.S., 182, 331.

Lacey, C.G. 1983, M.N.R.A.S., submitted.

Lindblad, P.O. 1960, Stockholm Obs. Ann., 21, 4.

Mark, J.W.-K. 1983, in Kinematics, Dynamics and Structure of
    The Milky Way, ed W.L.H. Shuter, (Dordrecht: Reidel),
    p. 289.

Miller, R.H., Prendergast, K.H., and Quirk, W.J. 1970, Ap.J.,
    161, 903.

Quiroga, R.J. 1974, Astrophys. Sp. Sci., 27, 323.

Searle, L. and van der Kruit, P.C. 1982, Astr. Ap., 110, 61.

Sellwood, J.A. and Carlberg, R.G. 1983, Ap.J., submitted.

Shu, F.H., Cuzzi, J.N., and Lissauer, J.J. 1983, Icarus, 53, 185.

Solomon, P.M., Sanders, D.B., and Scoville, N.Z. 1979, in
    The Large-Scale Characteristics of The Galaxy, I.A.U.
    Symposium No. 84, ed. W.B. Burton, (Dordrecht: Reidel),
    p. 35.

Toomre, A. 1974, in Highlights of Astronomy, Vol. 3, ed.
    G. Contopoulos, (Dordrecht: Reidel), p. 457.

Toomre, A. 1981, in Normal Galaxies, ed. S.M. Fall and D. Lynden-
    Bell, (Cambridge: Cambridge University Press), p. 111.

Toomre, A. 1983, in Internal Kinematics and Dynamics of Galaxies,
    I.A.U. Symposium No. 100, ed. E. Athanassoula, (Dordrecht:
    Reidel), p. 177.

Wielen, R. 1977, Astr. Ap., 60, 263.

HEATING OF STELLAR DISKS BY MASSIVE OBJECTS

Cedric G. Lacey

Institute of Astronomy, Cambridge, England.

ABSTRACT

I consider the possibility that the heating of stellar disks may
be due to scattering of the stars by massive objects, either
giant molecular clouds in the disk or massive black holes in the
galactic halo. I find that the molecular cloud mechanism probably
cannot be the main one because it predicts a ratio of vertical-to-
horizontal velocity dispersions in conflict with that observed.
The halo black hole mechanism can account well for the solar
neighbourhood kinematics, but the variation of disk scale-height
with radius that it predicts appears to disagree with that
observed. These results suggest that large scale perturbations
of the disk potential, such as spiral density waves, are involved
in stellar heating.

1. INTRODUCTION

A long-standing problem in galactic evolution is to explain why,
in the solar neighbourhood at least, the velocity dispersions of
disk stars increase with their age (see Wielen (1974) for a recent
observational analysis). As argued by Wielen (1977), the form of
the relation between velocity dispersion and age makes it seem
likely that it is caused by stochastic acceleration of the stars
after they are born, rather than resulting from any time-dependence
of the velocity dispersion of the gas layer from which the stars
form. This stochastic heating is of importance for at least two
aspects of the structure and evolution of galactic disks:
(i) their stability, and the development of spiral density waves
etc., depends on the stellar velocity dispersion (Toomre 1964),

351

J. Audouze and J. Tran Thanh Van (eds.),
Formation and Evolution of Galaxies and Large Structures in the Universe, 351–360.
© *1984 by D. Reidel Publishing Company.*

and (ii) the stellar scale-height, which is observed to be approximately independent of radius in other spiral galaxies (van der Kruit & Searle 1981a,b), depends on the vertical component of the velocity dispersion.

One possible explanation for the stochastic heating that is worth investigating is that it results from the scattering of stars by much more massive objects. As is well known, the timescale for two-body encounters between stars to be effective is roughly 10^{14} years (Chandrasekhar 1960), but this timescale can be much shorter for encounters between stars and more massive objects, varying as $1/nM^2$ for objects with mass M and number density n. There are two populations of objects which may have sufficiently large masses and number densities to produce a significant effect over lifetime of the galaxy, having quite different dynamical properties:
(i) Giant molecular clouds in the disk. These are observed to have masses in the range $10^5 - 10^6$ $M_\odot$ (Solomon & Sanders 1980) and are dynamically "cold" in the sense that they are on nearly circular orbits in the disk with smaller velocity dispersions than the stars. Their possible role in disk heating was first discussed by Spitzer & Schwarzschild (1951).
(ii) Massive black holes that might constitute the "dark matter" in galactic halos. This population would be dynamically "hot", being supported against gravity mainly by its velocity dispersion, which would greatly exceed that of the disk stars.

2. CALCULATION OF THE STELLAR HEATING: COMMON FEATURES

The calculations of stellar heating by molecular clouds and by halo black holes are similar. This section describes the common features.

2.1 Assumptions.

The principal assumptions I make are the following:
(i) The orbits of the stars in the background galactic potential are described by epicyclic theory.
(ii) The perturbers (molecular clouds or black holes) are much more massive than the stars.
(iii) The effective duration of a typical encounter is short compared to an epicyclic period.
(iv) The perturbation of stellar velocities is dominated by the effects of the many distant, weak encounters rather than by a few close, strong encounters so that the process can be treated as a diffusion in velocity space.

2.2 Epicyclic Approximation.

In cylindrical polar coordinates (R,θ,z), the velocity components (u,v,w) relative to the local circular velocity $(0,V_c(R),0)$ are

$$u = \dot{R}, \tag{1}$$

$$v = R\dot{\theta} - V_c(R), \tag{2}$$

$$w = \dot{z}. \tag{3}$$

In the epicyclic approximation (e.g. Chandrasekhar 1960) the energies E_e and E_z of horizontal and vertical epicyclic oscillations are separately conserved:

$$E_e = 1/2 (u^2 + \beta^2 v^2), \tag{4}$$

$$E_z = 1/2 (w^2 + \nu^2 z^2), \tag{5}$$

where

$$\beta = 2\Omega/\kappa, \tag{6}$$

and Ω, κ and ν are the frequencies of circular motion and of horizontal and vertical epicyclic oscillations respectively. For real disks, the value of β ranges between 1 (solid-body rotation) and 2 (Keplerian). The other constant of the motion is the angular momentum J.

2.3 Treatment of Encounters.

In an encounter between a star and a massive perturber, the epicyclic energies are changed by amounts

$$\Delta E_e = u\Delta u + 1/2(\Delta u)^2 + \beta^2 (v\Delta v + 1/2(\Delta v)^2), \tag{7}$$

$$\Delta E_z = w\Delta w + 1/2(\Delta w)^2. \tag{8}$$

According to assumptions (iii) and (iv), the statistical properties of the velocity perturbations are given by the velocity diffusion coefficients derived by Chandrasekhar (1960). The non-vanishing moments of first and second order of the velocity change in a short time interval Δt due to scattering of the star by a population of perturbers having mass M, local number density n and an isotropic Maxwellian velocity distribution with one-dimensional dispersion σ, relative to which the star is moving a velocity $\underline{V}_{rel}$, are:

$$<\Delta V_{\#}> = - \frac{4\pi G^2 n\ M^2\ \ln\Lambda}{V_{rel}^2}\ G_1(\gamma)\ \Delta t\ , \qquad (9)$$

$$<(\Delta V_{\#})^2> = \frac{8\pi G^2 n\ M^2\ \ln\Lambda}{V_{rel}}\ G_2(\gamma)\ \Delta t\ , \qquad (10)$$

$$<(\Delta V_{\perp})^2> = \frac{8\pi G^2 n\ M^2\ \ln\Lambda}{V_{rel}}\ G_3(\gamma)\ \Delta t\ , \qquad (11)$$

where $\#$ and $\perp$ denote components parallel and perpendicular to $\underline{V}_{rel}$ respectively, and assumption (ii) has also been used. In the above,

$$\gamma = V_{rel}/\sqrt{2}\ \sigma\ , \qquad (12)$$

and

$$G_1(\gamma) = \Phi(\gamma) - \gamma\Phi'(\gamma)\ , \qquad (13)$$

$$G_2(\gamma) = [\Phi(\gamma) - \gamma\Phi'(\gamma)]/2\gamma^2\ , \qquad (14)$$

$$G_3(\gamma) = [(2\gamma^2-1)\ \Phi(\gamma) + \gamma\Phi'(\gamma)]/2\gamma^2\ , \qquad (15)$$

where

$$\Phi(\gamma) = (2/\pi^{\frac{1}{2}}) \int_0^\gamma \exp(-x^2)\ dx\ . \qquad (16)$$

Equations (9) – (11) can be used to derive $<\Delta u>$, $<(\Delta u)^2>$ etc. and thus $<dE_e/dt>$ and $<dE_z/dt>$, the expected instantaneous rates of change of E_e and E_z, using equations (7) and (8).

2.4 Evolution of a Population of Stars due to Encounters.

To derive the rates of change of the mean energies for a population of stars, the expressions for $<dE_e/dt>$ and $<dE_z/dt>$ must be averaged over the stellar distribution. I assume that the distribution function is always in a quasi-steady state, so that the phase space density obeys $f = f(E_e, E_z, J)$ and further that f is approximately exponential in the epicyclic energies and so (by equations (4) and (5)) Gaussian in the velocities:

$$f \propto \exp\ [-(\frac{E_e}{<E_e>} + \frac{E_z}{<E_z>})]$$

$$= \exp\ [-(\frac{u^2}{2\sigma_u^2} + \frac{v^2}{2\sigma_v^2} + \frac{w^2}{2\sigma_w^2} + \frac{z^2}{2h_s^2})]\ . \qquad (17)$$

The J-dependence of f has been suppressed because it makes no
difference to the calculation if the epicyclic amplitude is small.
Then the velocity dispersions and scale-height are related to the
mean energies by

$$\sigma_u^2 \equiv <u^2> = <E_e> , \tag{18}$$

$$\sigma_v^2 \equiv <v^2> = <E_e>/\beta^2 , \tag{19}$$

$$\sigma_w^2 \equiv <w^2> = <E_z> , \tag{20}$$

$$h_s^2 \equiv <z^2> = <E_z>/\nu^2 . \tag{21}$$

3. HEATING BY MOLECULAR CLOUDS

3.1 Theory.

I assume the velocity dispersion of the clouds to be negligible
compared to that of the stars. This calculation thus generalizes
that of Spitzer & Schwarzschild (1953), which only considered
motions in the plane of the disk.

The number density of clouds (of mass M_c) is assumed to vary as

$$n = (N_c(R)/(2\pi)^{\frac{1}{2}}h_c) \exp(-z^2/2h_c^2) . \tag{22}$$

The expressions for $<dE_e/dt>$ and $<dE_z/dt>$ derived as explained in
Section 2.3 are averaged over the distribution function (17) and
over the epicyclic phases. The final result is

$$\frac{d(\sigma_u^2)}{dt} = \frac{2G^2 N_c M_c^2 \ln\Lambda}{\sigma_u(h_s^2+h_c^2)^{\frac{1}{2}}} K(\alpha,\beta) , \tag{23}$$

$$\frac{d(\sigma_w^2)}{dt} = \frac{2G^2 N_c M_c \ln\Lambda}{\sigma_u(h_s^2+h_c^2)^{\frac{1}{2}}} L(\alpha,\beta) , \tag{24}$$

where

$$\alpha = \sigma_w/\sigma_u , \tag{25}$$

and $K(\alpha,\beta)$ and $L(\alpha,\beta)$ are integrals over the epicyclic phase.
They are plotted as functions of α for various values of β in Fig.1.

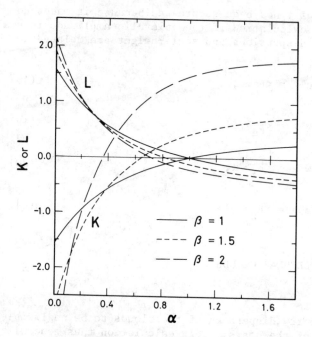

Figure 1. Dependence of rates of change of epicyclic energies on epicyclic energy ratio ($\alpha = \sigma_w/\sigma_u$) and rotation curve shape ($\beta = 2\Omega/\kappa$), as given by $K(\alpha,\beta)$ and $L(\alpha,\beta)$.

The evolution of the velocity dispersions given by equations (23) and (24) can be divided into two distinct phases:

(i) <u>Transient Relaxation</u>. The ratio of vertical-to-horizontal velocity dispersions, σ_w/σ_u or σ_w/σ_v, relaxes to a steady final value. The shape of the velocity ellipsoid is then determined entirely by the value of $\beta = 2\Omega/\kappa$:

$$\sigma_v/\sigma_u = 1/\beta , \tag{26}$$

$$\sigma_w/\sigma_u = \alpha_s(\beta) . \tag{27}$$

The equilibrium values of σ_v/σ_u and σ_w/σ_u are plotted as functions of β in Fig.2.

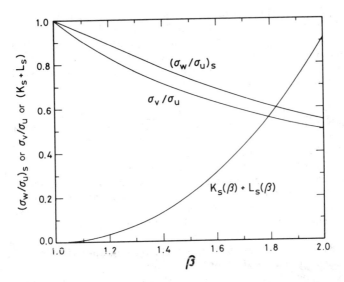

Figure 2.   Dependence of velocity dispersion ratios σ_w/σ_u and σ_v/σ_u and heating rate (as given by $K_s(\beta) + L_s(\beta)$) on rotation curve shape in steady heating phase.

(ii) Steady Heating. The velocity dispersions continue to increase at fixed ratios (except for the case $\beta = 1$ - solid-body rotation - when no heating occurs). For most disk stars, $h_s \gtrsim h_c$, so the total velocity dispersion σ obeys

$$\frac{d\sigma^2}{dt} \approx \frac{D}{\sigma^2} \quad , \tag{28}$$

where D depends on N_c, M_c β etc. but not on σ. (The β-dependence of the rate of change of total epicyclic energy is given by $K_s(\beta) + L_s(\beta)$, plotted in Fig.2). If D is constant, then for stars all born at t=0,

$$\sigma(t) \approx (\sigma^4(o) + 2Dt)^{1/4} . \tag{29}$$

3.2 Comparison with Observations.

Three independent comparisons of the theoretical predictions with observations are possible:

(i) Shape of the Velocity Ellipsoid. If scattering by clouds is important, the velocity dispersion ratios should obey equations (26) and (27). These predict $\sigma_w > \sigma_v$, whereas virtually all observational determinations given $\sigma_w < \sigma_v$ (e.g. Delhaye 1965). In more detail, estimates of β from measurements of the rotation curve and of σ_v/σ_u locally give $1.3 \lesssim \beta \lesssim 1.8$, for which the theory predicts $0.6 \lesssim \sigma_w/\sigma_u \lesssim 0.8$, while for typical disk stars the observed value is $\sigma_w/\sigma_u = 0.52 \pm 0.05$ (Wielen 1974). This comparison appears to rule out the cloud mechanism as being important unless there is a large distortion of the local stellar kinematics by a density wave.

(ii) <u>Age-dependence of the Velocity Dispersions</u>. If the cloud masses and number densities are independent of time, the equation (29) predicts $\sigma \sim t^{1/4}$, whereas the observations imply $\sigma \sim t^{1/2}$ or $t^{1/3}$ (Wielen 1974). In fact, the cloud number density at least is likely to be a function of time. The simplest assumption is that D in equation (28) is proportional to the star formation rate, in which case the discrepancy between the predicted and observed time-dependence remains (see Lacey & Fall 1983). One can also try to predict the magnitude of the heating using the observed cloud properties. Taking molecular cloud data from Liszt, Xiang and Burton (1981), I predict for the velocity dispersion of the oldest disk stars in the solar neighbourhood $13 \lesssim \sigma(T_D) \lesssim 54$ kms^{-1}, for an initial velocity dispersion of 10 kms^{-1}, where the uncertainties allow for a factor of 2 error either way in the cloud masses and for the time-averaged value of D being up to a factor of 4 greater than the present one. The observed value is $\sigma(T_D) \approx 60\text{-}80$ kms^{-1} (Wielen 1974). Given the uncertainty in cloud masses, this comparison does not rule out the mechanism.

(iii) <u>Radial Dependence of the Disk Scale-height</u>. The scale-height of the disk varies as $h_D \propto \sigma_D^2/\mu_D$, where σ_D is an effective average velocity dispersion and μ_D is the surface density. The radial variation of σ_D for heating by clouds can only be predicted if one makes some assumption about how the cloud properties vary with radius and time. If one assumes that D is proportional to the star formation rate, then $\sigma_D \propto \mu_D^{1/4}$ (see Lacey & Fall 1983), so

$$h_D \propto \mu_D^{-1/2} , \tag{30}$$

and for an exponential disk, $\mu_D \propto \exp(-\alpha R)$,

$$h_D \propto \exp(\alpha R/2) . \tag{31}$$

This is not compatible with the observations of van der Kruit & Searle (1981a,b), who find the disk scale-heights in other spiral galaxies to be approximately independent of radius.

I conclude that the cloud mechanism in the form assumed here probably cannot be the dominant disk heating mechanism. The best evidence for this comes from the ratio σ_w/σ_u, since the other tests depend on additional, uncertain, assumptions about cloud masses and about the variation of cloud properties with time and with radius. It is possible that the mechanism might work if some of the assumptions were modified e.g. to allow for clustering of the clouds or induced stellar wakes (c.f. Julian & Toomre 1966).

4. HEATING BY HALO BLACK HOLES

4.1 Theory.

The approximation made in this case is that the stellar velocity dispersion is negligible compared to both the velocity dispersion of the halo σ_H and the relative velocity of the disk and halo $(V_c - \overline{V}_H)$, where $\overline{V}_H$ is the mean rotational velocity of the halo. I derive

$$\frac{d\sigma_u^2}{dt} = \frac{2\pi G^2 \, n_H \, M_H^2 \, \ln\Lambda}{(V_c - \overline{V}_H)} \, (G_3(\gamma) + 2\beta^2 \, G_2(\gamma)), \tag{32}$$

$$\frac{d\sigma_w^2}{dt} = \frac{2\pi G^2 \, n_H \, M_H^2 \, \ln\Lambda}{(V_c - \overline{V}_H)} \, G_3(\gamma) \quad , \tag{33}$$

with

$$\gamma = (V_c - \overline{V}_H)/\sqrt{2} \, \sigma_H \, . \tag{34}$$

Therefore the time-dependence of the velocity dispersions is of the form

$$\sigma(t) = (\sigma^2(o) + D't)^{\frac{1}{2}} \, , \tag{35}$$

while the ratio σ_w/σ_u tends to a value given by the ratio of the heating rates in equations (32) and (33):

$$\sigma_w/\sigma_u = (1 + 2\beta^2 \, G_2(\gamma)/G_3(\gamma))^{-\frac{1}{2}} \tag{36}$$

4.2 Comparison with Observations.

(i) Shape of the Velocity Ellipsoid. The predicted σ_w/σ_u is consistent with that observed for $\beta \gtrsim 1.7$, for a slowly-rotating halo $(\overline{V}_H/V_c \lesssim 1/3)$, which is within the observational range of $(1.3 \lesssim \beta \lesssim 1.8)$ mentioned previously.

(ii) Age-dependence of the Velocity Dispersions. Equation (35) gives $\sigma \sim t^{\frac{1}{2}}$, which agrees well with the observations. The magnitude of the heating cannot be predicted since M_H is unknown. Rather, we can use the observations to determine M_H. To obtain a total velocity dispersion $\sigma \cong 80$ kms^{-1} for the oldest disk stars, one requires $n_H M_H^2 \approx 2\times10^4$ $M_\odot^2$ pc^{-3}, or $M_H \approx 2\times10^6$ $M_\odot$ for a halo density $n_H M_H \cong 10^{-2}$ $M_\odot$ pc^{-3}.

    (iii) <u>Radial Dependence of the Disk Scale-height.</u> $\sigma_D \propto \rho_H^{\frac{1}{2}}$ where ρ_H is the halo density (assuming M_H is everywhere the same), so that

$$h_D \propto \rho_H / \mu_D \ . \tag{37}$$

Then if the halo density varies as $\rho_H \propto 1/(R^2 + R_c^2)$ and the disk is exponential,

$$h_D \propto \exp(\alpha R)/(R^2 + R_c^2) \ . \tag{38}$$

This does not appear to be consistent with the observations of van der Kruit & Searle, even allowing for projection effects and letting R_c be a free parameter. Therefore it seems likely that halo black holes are not responsible for heating the disk, and the value of M_H derived is best viewed as an upper limit.

A fuller account of the calculation of heating by molecular clouds is contained in a paper submitted to Monthly Notices of the Royal Astronomical Society. The calculation of heating by black holes was part of a collaborative project with J.P. Ostriker.

REFERENCES

Chandrasekhar, S. 1960, "Principles of Stellar Dynamics" (Dover).
Delhaye, J. 1965, in "Galactic Structure", eds. Blaauw, A. & Schmidt, M. (Chicago University Press), p.61.
Julian, W.H. & Toomre, A. 1966, Astrophys. J. 146, p.810.
Lacey, C.G. & Fall, S.M. 1983, Mon. Not. Roy. astr. Soc. 204.
Liszt, H.S., Xiang, D. & Burton, W.B. 1981, Astrophys. J. 249, p.532.
Solomon, P.M. & Sanders, D.B. 1980, in "Giant Molecular Clouds in the Galaxy", eds. Solomon, P.M. & Edmunds, M.G. (Pergamon), p.41.
Spitzer, L. & Schwarzschild, M. 1951, Astrophys. J. 114, p.385.
Spitzer, L. & Schwarzschild, M. 1953, Astrophys. J. 118, p.306.
Toomre, A. 1964, Astrophys. J. 139, p.1217.
van der Kruit, P.C. & Searle, L. 1981a, Astron. Astrophys. 95, p.105.
van der Kruit, P.C. & Searle, L. 1981b, Astron. Astrophys. 95, p.116.
Wielen, R. 1974, Highlights of Astron. 3, p.395.
Wielen, R. 1977, Astron. Astrophys. 60, p.263.

THE ORIGIN OF ROTATION IN GALAXIES AND CLUSTERS

G. Efstathiou[1] and J. Barnes[2]

[1]Institute of Astronomy, Madingley Road, Cambridge, U.K.
[2]Astronomy Department, University of California, Berkeley,
California, U.S.A.

In this article we review recent observations of the rotation-al properties of galaxies and clusters. We investigate whether these observations can be explained if galaxies and clusters were set in rotation by the tidal torques of their neighbours. We present new results on the angular momentum–mass relation and on the dependence of angular momentum on initial conditions. These results are discussed within the context of the White-Rees model for galaxy formation. A new test of tidal torques is proposed.

1. INTRODUCTION

The dimensionless spin-parameter

$$\lambda = J|E|^{\frac{1}{2}} G^{-1} M^{-5/2} \tag{1}$$

is a convenient quantity with which to parameterise the angular momentum content of galaxies and clusters. In equ. (1) J, E and M are respectively the total angular momentum, energy and mass of the system and G is the gravitational constant.

If a proto-system collapses and reaches virial equilibrium without dissipating binding energy so that J, E and M are conserved during collapse, the dimensionless spin-parameter will also be conserved. Clusters and groups of galaxies are unlikely to have dissipated much binding energy because their radiative cooling times exceed a Hubble time (Rees and Ostriker 1977). Measure-ments of λ in clusters and groups may, therefore, provide a direct estimate of the pre-collapse value of λ for comparison with theoretical predictions.

361

J. Audouze and J. Tran Thanh Van (eds.),
Formation and Evolution of Galaxies and Large Structures in the Universe, 361–377.
© *1984 by D. Reidel Publishing Company.*

If a proto-system collapses and dissipates binding energy, so that only J and M are conserved, the observed value of λ may be compared with theoretical predictions to compute the ratio of the pre-collapse to final half-mass radius. Dissipation has clearly played an important role in the formation of the disc components of spiral and lenticular galaxies but it is less clear whether this is the case for ellipticals and the bulges of disc galaxies. Indeed, several authors (eg. Sandage, Freeman and Stokes 1970, Gott and Thuan 1976) have argued that the amount of binding energy dissipated during the collapse may determine whether a proto-galactic cloud becomes an elliptical or a spiral galaxy. The rotational properties of galaxies provide an important test of this idea.

2. OBSERVATIONS

2.1 Ellipticals.

If ellipticals were rotationally flattened oblate spheroids with apparent ellipticity ε, the ratio of the peak rotational velocity (v_m) to the peak velocity dispersion (σ_p) should obey the relation (Binney 1978)

$$\frac{v_m}{\sigma_p} \simeq \frac{\pi}{4} \left[\frac{(3-2e^2) \; \sin^{-1}e \; - \; 3e(1-e^2)^{\frac{1}{2}}}{e(1-e^2)^{\frac{1}{2}} \; - \; (1-e^2) \; \sin^{-1}e} \right]^{\frac{1}{2}} \tag{2}$$

where $e^2 = 1 - (1-\varepsilon)^2$. In the last few years rotation curves and velocity dispersion profiles have been measured for a large number of bright $(M_B < -19)$[1] elliptical galaxies and it is now well established that most giant ellipticals rotate more slowly than is predicted by equ. (2) (see e.g. Illingworth 1981 and references therein). This result may be understood if the flattening of giant ellipticals is due to anisotropic velocity dispersions rather than rotation (Binney, 1978).

Recently, Davies et al. (1983) have measured rotation curves and velocity dispersion profiles for 11 faint ellipticals $(M_B > -19)$. They find that the rotational properties of ellipticals correlate with luminosity, with faint ellipticals rotating nearly as rapidly as predicted by equ. (2) (Figure 1). A rough approximation to their results is the power-law relation

$$v_m/\sigma_p \simeq (0.27\pm0.06) \; (L/L^*)^{-0.38\pm0.04}, 0.06 \lesssim L/L^* \lesssim 10 \tag{3}$$

where we have taken $L^* = 1.1 \times 10^{10} \; L_\odot$. This relation is shown in Figure 2.

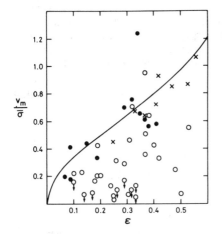

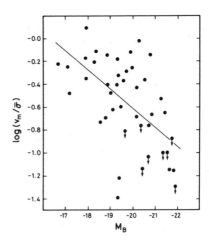

Fig. 1. v_m/σ plotted against ellipticity. Open circles show ellipticals with $M_B<-19$ and filled circles show ellipticals with $M_B>-19$. Bulges are plotted as crosses. The solid line is given by equation (2).

Fig. 2. $\text{Log}(v_m/\sigma)$ for ellipticals plotted against absolute magnitude. The solid line shows the power law relation given by equation (3)

For a spherical galaxy with a de-Vaucouleurs $r^{1/4}$ surface density profile and a flat rotation curve, we expect $\lambda \cong 0.43\ v_m/\sigma_p$. Thus, equation (3) implies

$$\lambda \sim 0.1\ (L/L^*)^{-0.4}\ ,\quad 0.06 \lesssim L/L^* \lesssim 10\ . \tag{4}$$

2.2. Bulges of Disc Galaxies.

The bulges of several spiral and lenticular galaxies have been studied in some detail by Kormendy and Illingworth (1982) and Illingworth and Schechter (1982). Most bulges have absolute magnitudes fainter than −19 and they are found to rotate rapidly, consistent with the predictions of the oblate isotropic model (equ. 2) and with the rotational properties of faint ellipticals. The results for bulges are shown as the crosses in Figure 1.

2.3. Disc Galaxies.

The surface brightness profiles of the disc components of spiral galaxies may be roughly parameterised by the form,

$$I(r) = I_o\ \exp(-\alpha r) \tag{5}$$

with $I_0 \cong 145 L_\odot pc^{-2}$ (Freeman 1970, Boroson 1981). Rotation
curves have been measured for a large number of disc galaxies
using both 21 cm and optical emission lines (see e.g. Rubin 1983,
Bosma 1983 and references therein). In most cases the rotation
curves are very nearly flat out to the limits of detectability,
which usually corresponds to $\lesssim$ the Holmberg radius $r_H \sim 4.5 \, \alpha^{-1}$.
The rotation curves, especially of late type discs, may be roughly
parameterised by the relation

$$v(r) = v_m \left\{ \frac{r^2}{r^2 + r_m^2} \right\}^{\frac{1}{2}} \tag{6}$$

with αr_m typically within the range 0.1 to 0.8 (Fall and Efstathiou
1980).

Flat rotation curves have usually been interpreted as evidence
in support of dark haloes with density profile $\rho(r) \cong v_m^2/4\pi G r^2$
(see eg. Faber and Gallagher 1979 for a review). However,
because of the lack of coordinated photometric and kinematic
studies, this result should be viewed with some caution. Kalnajs
(1983) has used detailed surface photometry of four disc galaxies
to predict their rotation curves under the assumption of a constant
mass-to-light (M/L) ratio. The results agree remarkably well with
the observed rotation curves. Kalnajs' results illustrate the
dangers of using mean relations such as equations (5) and (6) to
infer the mass distribution in the outer parts of discs. The
existence of dark haloes is of crucial importance to the present
discussion. As we will show in Section (4) it seems that massive
haloes are necessary if disc angular momentum was acquired by
tidal torques.

An additional uncertainty involves the amount of random
motion in disc galaxies. We will assume that discs are cold
($\sigma \ll v_m$) in analogy with the kinematics of stars in the local solar
neighbourhood ($\sigma \cong 40$ km sec^{-1}, $v_m \cong 250$ km sec^{-1}).

2.4. The Coma Cluster.

There are few rich clusters with enough measured redshifts
to make a search for rotation worthwhile, so we will confine the
discussion to Coma, the best studied of all the rich clusters.
Several authors have investigated the central regions ($\theta \lesssim 180$
arcmin) but no rotation has yet been detected (Peebles 1971, Rood
et al. 1972, Gregory 1975). We have re-examined this problem
using the extensive list of redshifts given by Kent and Gunn
(1982). We compute the quantities (Peebles 1971),

$$\ell_x = \frac{1}{N} \sum_i x_i (v_i - \bar{v}) \quad , \quad \ell_y = \frac{1}{N} \sum_i y_i (v_i - \bar{v}) \quad , \tag{7}$$

where the sums extend over N galaxies located within a radius θ_{max} of the centre (taken to be NGC 4874). Here $\bar{v}$ is the mean radial velocity of the Coma cluster. We eliminate foreground and background galaxies using the following iterative technique. We reject galaxies with radial velocities outside the range $\bar{v}\pm2.5\sigma_v$ where σ_v is the measured cluster velocity dispersion. We then recompute $\bar{v}$, σ_v ℓ_x and ℓ_y and repeat the background-foreground check. This procedure converges very rapidly for $\theta_{max} < 180'$ (cf. Figure 1 of Kent and Gunn 1982). The results are summarised in Table 1.

TABLE 1. Test for rotation in the Coma cluster

θ_{max} arcmin	N	ℓ_x	σ_{xv} arcmin km sec^{-1}	ℓ_y	σ_{yv}	$(\ell_x{}^2+\ell_y{}^2)^{\frac{1}{2}}$
60	194	990	1390	480	1140	1100
120	265	-1770	2130	1660	1840	2430
180	292	-2760	2720	980	2140	2930

In Table 1,

$$\sigma_{xv} = \sigma_v \, (\sum_i x_i{}^2)^{\frac{1}{2}}/N \quad , \quad \sigma_{yv} = \sigma_v \, (\sum_i y_i{}^2)^{\frac{1}{2}}/N$$

are the expected variances of ℓ_x and ℓ_y respectively under the assumption that (x_i,v_i) and (y_i,v_i) are uncorrelated. To check this assumption and to interpret these results in terms of the λ-parameter we have computed a series of Monte-Carlo simulations of the Coma cluster. The surface density profile (Kent and Gunn 1982 Figure 6) is well fitted by a de-Vaucouleurs' $r^{1/4}$ law with effective radius $r_e = 65.3$ arcmin. We have used this model to generate spherical clusters of 400 points with isotropic velocity dispersions calculated from the equations of stellar hydrodynamics. The peak projected velocity dispersion was fixed at 1100 km sec^{-1} and each cluster was set in cylindrical rotation with a constant rotational velocity and a randomly oriented spin axis. The clusters were projected along the line of sight and analyzed as described above. The results are summarized in Table 2.

TABLE 2. Monte-Carlo simulations of rotating clusters

θ_{max} arcmin	$\lambda = 0.04$ $(\ell_x{}^2+\ell_y{}^2)^{\frac{1}{2}}$	$\lambda = 0.08$ $(\ell_x{}^2+\ell_y{}^2)^{\frac{1}{2}}$ arcmin km sec^{-1}	$\lambda = 0.12$ $(\ell_x{}^2+\ell_y{}^2)^{\frac{1}{2}}$	$0.47(\sigma_{xv}^2+\sigma_{yv}^2)^{\frac{1}{2}}$
60	2050±1020	2860±1190	4030±1340	1020
120	2660±1240	4430±1500	6490±1710	1440
180	3140±1470	5470±1870	8280±2290	1750

The errors on $(\ell_x{}^2 + \ell_y{}^2)^{\frac{1}{2}}$ represent one standard deviation and are close to the predicted variances $0.47\ (\sigma^2{}_{xv} + \sigma^2{}_{yv})^{\frac{1}{2}}$. Only 1 in about 20 Monte-Carlo models with $\lambda \cong 0.08$ gives values for $(\ell_x{}^2 + \ell_y{}^2)^{\frac{1}{2}}$ as low as those observed for the Coma cluster. Thus Coma probably has $\lambda \lesssim 0.08$, unless its angular momentum vector points very nearly along the line of sight.

We have also checked for rotation around the apparent minor axis. Coma is highly flattened, with $\varepsilon = 0.47 \pm 0.07$ and a major axis position angle of $74° \pm 4°$ for $\theta \sim 1$ arcmin (Carter and Metcalfe 1980, Schipper and King 1978). Fitting a flat rotation curve we find $v_{rot} = 60 \pm 90$ km sec^{-1} (positive values correspond to NE receeding) so there is no evidence for rotation around the apparent minor axis, in agreement with previous work.

Note that if Coma were oblate with isotropic velocity dispersions, equ. (2) predicts $v_m/\sigma_p \cong 0.95$ which is clearly excluded by the observations. If Coma were more nearly prolate with isotropic velocity dispersions we could get the observed results only if the angular momentum vector were very nearly parallel to the line of sight. Assuming that clusters have the same ellipticity distribution as elliptical galaxies (cf. Carter and Metcalfe 1980) we can use Figure 4 of Binney (1978) to assign a probability of $\lesssim 2\%$ for getting the observed result. It therefore seems likely that the flattening of Coma is due to anisotropic velocity dispersions rather than rotation.

2.5. The Local Supercluster

Several authors have searched for rotation in the Local Supercluster (Rubin 1951, de Vaucouleurs 1958, 1972, Stewart and Sciama 1967). In the most recent study, Aaronson et al. (1982) use the infra-red Tully-Fisher relation to test for infall of nearby galaxies towards the Virgo cluster centre. They find an infall velocity of 250 ± 64 km sec^{-1} at the position of the Local Group. In addition, they have tested for differential rotation assuming spheroidal rotation of the form

$$v(r) = v_{rot} \left(\frac{r}{R_{LG}} \right) \exp \left(- \left(\frac{r}{R_{LG}} \right)^2 + 1 \right) \tag{8}$$

(de Vaucouleurs 1958) where R_{LG} is the distance of the Local Group to the centre of Virgo. Interestingly, they find $v_{rot} = 180 \pm 58$ km sec^{-1}.

It is not possible to convert this result into a unique value of λ for the Local Supercluster since we can add an arbitrary amount of solid body rotation to equ. (8) without altering the predicted line of sight velocity of a galaxy. Nevertheless, to compare with the theoretical predictions of the next Section and

to see what sort of rotational motions we might expect in a
structure like the Local Supercluster we have calculated λ for
a system with a flat rotation curve and an r^{-2} density profile,

$$\lambda \cong 0.05 \left(\frac{v_{rot}}{200 \text{ km sec}^{-1}} \right) (\bar{\Omega \delta})^{-\frac{1}{2}} (R/15 \text{ Mpc})^{-1} , \qquad (9)$$

where Ω is the ratio of the mean matter density to the critical
density in an Einstein–de Sitter model, $\bar{\delta}$ is the mean overdensity
of material within a radius R from the centre and we have neglected
the effect of material at radii greater than R.

3. TIDAL TORQUES

      One of the consequences of the gravitational instability
picture of galaxy formation is that individual proto-structures
would have been set in rotation by the tidal torques from their
neighbours. Hoyle (1949) was the first to suggest that this
mechanism might account for the rotation of galaxies and the idea
was refined further by Peebles (1969). Peebles used linear
perturbation theory to estimate the r.m.s. angular momentum in
randomly placed spheres and matched the results to a simple nonlinear model in which the torque on a proto-system was calculated
as the product of its quadrupole moment with the tidal field of
neighbouring point masses. Assuming that density perturbations
at recombination can be approximated as a random Gaussian process
with a power-spectrum

$$|\delta_k|^2 \propto k^n , \qquad (10)$$

Peebles derived $\lambda \cong 0.08$ for n = -1/2 with a theoretical uncertainty of about a factor of three.

      A more accurate estimate may be obtained from N-body simul-
ations (Peebles 1971, Efstathiou and Jones 1979, hereafter EJ).
In particular, the latter authors studied the tidal torque process
using 1000 particle models with $\Omega=1$ and Poisson initial conditions.
For bound clumps they find the median value of λ to be $\lambda_{me} = 0.06$.
The distribution of λ shows a large scatter; the 90 and 10 per-
centile points of the distribution lie at $\lambda = 0.11$ and $\lambda = 0.03$
respectively.

      Recently we have run a large numerical experiment using the
numerical scheme described by Efstathiou and Eastwood (1981).
The model contains 20000 particles which were initially distributed
at random positions inside a cube of unit volume. The cosmological
density parameter was set to $\Omega=1$. Figure 3 shows a projection of
the particle distribution after the system had expanded by a

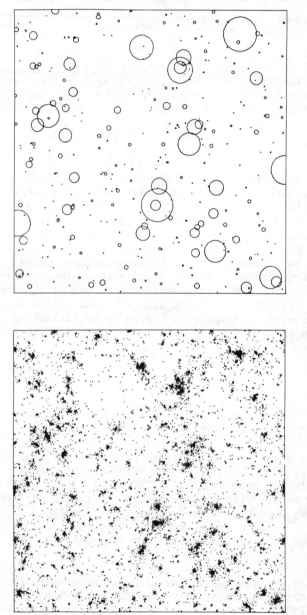

Fig. 3. Projection of the particle distribution in an N-body simulation with Poisson initial conditions and $\Omega = 1$ after the system had expanded by a factor of 28.11. Groups identified using a density contrast criterion $\delta\rho/\rho = 6$ and containing more than 8 particles are shown schematically. To avoid overcrowding in the figure, we have scaled the radius of each group so that $r = 10^{-3}$ M in units in which the box length is unity and M is the number of group members.

factor of 28.11 together with a schematic representation of groups
identified using a density contrast criterion $\delta\rho/\rho$ = 6.

   The local median value of λ is plotted against mass in
Figure 4. In matching the results at different epochs we have
used the similarity scaling,

$$\lambda_{me}(m,t) = \lambda_{me}(m/m^*) \quad m^* \propto t^{4/(3+n)} \tag{11}$$

which should apply in an $\Omega=1$ model (see e.g. Efstathiou, Fall and
Hogan 1979). The median value of λ derived from this simulation
(λ_{me} = 0.065) and the distribution of λ are in excellent agreement
with the results of EJ. Further, the new simulation shows that
Poisson initial conditions lead to a relation between λ_{me} and mass
in which λ_{me} is nearly independent of mass at small masses and
declines slowly with increasing mass at large masses. The dashed
line in Figure 4 shows the empirical relation

$$\lambda = 0.07 \; (1 + (m/m_c)^2)^{-1/4} \; . \tag{12}$$

Note that whilst only 8% of the identified groups are more massive

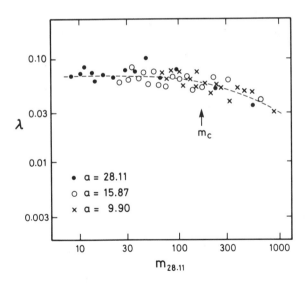

Figure 4. Local median value of λ plotted against mass. $m_{28.11}$
refers to the masses of groups identified when the system had
expanded by a factor of 28.11. Masses for groups identified at
earlier times were scaled according to equ. (11). The dashed
line shows equ. (12).

than m_c, they contain 41% of the total number of particles.

Another problem that we have begun to study concerns the dependence of the tidal torque results on the power-spectrum index n (equ. 10). The linear calculation of Peebles (1969) does not help to answer this question because the r.m.s. angular momentum $<J^2>^{\frac{1}{2}}_x$ in a randomly placed sphere of radius x involves a summation which couples long-wavelength modes with short-wavelengths where the clustering is highly non-linear. If the power-spectrum (10) is extrapolated to arbitrarily high k one finds that $<J^2>^{\frac{1}{2}}_x$ diverges if n lies outside the range -1<n<0. The quadrupole model of Thuan and Gott (1977) suggests that tidal-torques should be relatively insensitive to the primordial mass-distribution, though it is doubtful that this model is sufficiently realistic to provide more than a rough guide.

The results of a series of small (N $\sim$ 1000) simulations run with Ω=1 are summarized in Table 3. The first column gives the initial particle distribution. The cells distribution is set up by placing one particle at a random location in each of N contiguous cubical volumes and corresponds to n = +2 on scales larger than the interparticle separation. The rods distribution is set up by placing a number of rods, with particles at fixed spacing along each rod, at random orientations within the comput- ational volume. This distribution corresponds to a power-spectrum with n = -1 on scales larger than the interparticle separation (Aarseth, Gott and Turner 1979). The third column gives the mean value of λ for the groups with density contrast $\delta\rho/\rho$ = 10 con- taining more than 10 particles which were identified after each model had expanded by a factor of 8.4. The fourth column gives the mean value of the dimensionless parameter κ defined as

$$\kappa = J M^{-2} G^{-1} P^{-1} |E|$$ (13)

TABLE 3. Results from N-body simulations

Distr.	n	$<\lambda>$	$<\kappa>$	λ_{me}(<0.5)	λ_{me}(>0.5)	ε(J/M$\propto$M$^{\varepsilon}$)
Cells	+2	0.040±0.005	0.22±0.04	0.04	0.03	-0.41±0.72
Poisson	0	0.065±0.006	0.28±0.04	0.06	0.05	0.65±0.19
Rods	-1	0.064±0.005	0.26±0.05	0.07	0.05	0.40±0.20

where P is the total linear momentum of each group. This parameter measures the ratio of rotational velocity to the clusters centre of mass velocity ($\kappa \cong v_{rot}/4 v_{cm}$). $<P^2>^{\frac{1}{2}}$ should diverge if the spectrum (10) extends to k=0 with n<-1. There is no divergence in the rods experiments because the spectrum is truncated due to the finite size of the numerical simulations. The fifth and sixth columns give the median value of λ for groups with masses less than and greater than the median group mass respectively. The final column gives the exponent of the specific angular momentum-mass relation.

The results from the cells and rods experiments agree qualitatively with those from the Poisson models. The median value of λ for high mass groups is slightly smaller than that for low mass groups in agreement with the trend seen in Figure 4. The median value of λ is not particularly sensitive to initial conditions; note that the median value of λ for the cells experiments may be too low because our criterion that groups should contain more than 10 members means that we sample only the richest groups (containing $\cong$ 10% of the total mass) whereas in the Poisson and rods experiments we sample a wider range of group masses.

We have also tested the sensitivity of these results to the overdensity criterion used to identify groups. The median value of λ is extremely insensitive to the density contrast (cf. also EJ). The test is reliable only for $\delta\rho/\rho \lesssim 30$ in the 20000 body experiment, since the softening of the gravitational potential significantly influences estimates of the potential energy if $\delta\rho/\rho \gtrsim 30$. In the experiments run with Aarseth's code we used a harder potential which allows a reliable test to $\delta\rho/\rho \cong 50$.

4. DISCUSSION

4.1 Clusters of Galaxies

The results of the previous section suggest that clusters of galaxies should rotate slowly, with $\lambda \cong 0.07$. This is certainly consistent with the observational results described in Sections 2.4 and 2.5 for Coma and the Local Supercluster. Further, the κ parameter (equ. 13) shows that $v_{cm} \sim v_{rot}$ if $2 > n > -1$. Thus the agreement between the infall velocity derived by Aaronson et al. (1982) and our motion with respect to the microwave background radiation does not necessarily exclude the possibility that the rotation of the Local Supercluster was caused by tidal torques.

If the tidal torque theory is correct, the flattening of relaxed galaxy clusters must be due to anisotropic velocity dispersions. It would be interesting to test in detail whether tidal interactions between neighbouring protostructures can account for the observed distribution of cluster flattenings. Theoretical calculations by Binney and Silk (1979) indicate that tidal interactions lead to prolate structures with typical ellipticities $\varepsilon \sim 0.2 - 0.3$.

4.2 Disc Galaxies

        Recent applications of the cosmic virial theorem indicate
$\Omega = 0.2 \times 1.5^{\pm 1}$ under the assumption that galaxies are good
tracers of the mass distribution (Davis and Peebles 1983, Bean
et al. 1983). This implies a mean mass-to-luminosity ratio of M/L
$\cong 300$ ($\Omega/0.2$) which is compatible with the M/L values deduced from
studies of rich clusters and groups but far in excess of the M/L
values inferred within the optical radii of elliptical and spiral
galaxies (M/L $\sim$ 3-15). Any theory of galaxy formation should
take into account the large amounts of dark material implied by
these observations and should explain why the M/L values for the
inner parts of galaxies are so much smaller than the mean.

        White and Rees (1978) propose the following model. Most
of the material in the universe is assumed to be in a dark
component which can cluster gravitationally but does not dissipate
energy. Any of the usual candidates for the dark component such
as low mass stars, the remnants of supermassive stars or massive
(>1 keV) weakly interacting particles would be acceptable.
Primordial gas accounts for only a small fraction of the mean mass
density ($\sim$1-10%) but can radiatively cool and dissipate binding
energy to form the dense luminous components of galaxies. Each
galaxy would have formed within the potential well of a dark halo
and the high overdensities and low M/L ratios of galaxies can be
explained if the gas collapsed by a factor of $\sim$10 in radius before
fragmenting into stars. The requirement that the gas should cool
on a time-scale shorter than the Hubble time leads to an upper
mass limit of $\sim$10^{12} $M_\odot$ which agrees with the masses of large
galaxies.

        In this theory, it is reasonable to expect that the
specific angular momentum of the gas should be equal to that of
the dark halo material,

$$J_D/M_D = J_H/M_H .$$
                                                                    (14)

This is because all the material in the protosystem experiences
the same external torques before separating into two distinct
components. (However, note that if the primordial gas is ionized,
it would stay locked to the comoving frame by Compton drag for
$z \gtrsim 140$ $(\Omega)^{1/5}$ in which case J_D/M_D could be less than J_H/M_H, Hogan
1979). The numerical experiments described in the previous section
predict that a typical protosystem will have a pre-collapse spin
parameter $\lambda \sim 0.07$. We now wish to compute the factor by which the
gas must collapse to produce a centrifugally supported disc with
an exponential surface density profile (equ.5) and a flat rotation
curve with amplitude v_m out to some limiting radius R_H. If the
disc's self-gravity is neglected, the halo profile is

$$\rho_H(r) = v_m^2/4\pi G \ r^2 \ , \qquad\qquad r < R_H \qquad\qquad (15a)$$

and we will assume that

$$\rho_H(r) = 0 \ . \qquad\qquad r > R_H \qquad\qquad (15b)$$

The disc's angular momentum is

$$J_D = 2 \ G^{\frac{1}{2}} \ Y \ M_D^{3/2} \ \alpha^{-\frac{1}{2}} \ , \quad Y = v_m/(\alpha M_D \ G)^{\frac{1}{2}} \ . \qquad (16)$$

Equs. (1), (15) and (16) give

$$J_H = G \ \lambda_H \ E_H^{-\frac{1}{2}} \ M_H^{5/2} \ , \quad M_H/M_D = Y^2 \ \alpha R_H \qquad (17)$$

and, since the halo component does not dissipate binding energy we can compute E_H from (15). Equs. (14), (16) and (17) then give

$$\alpha R_H = \sqrt{2}/\lambda_H \qquad\qquad (18)$$

which agrees well with the more detailed computations of Fall and Efstathiou (1980). If $\lambda_H = 0.07$, we find $\alpha R_H \cong 20$ which is compatible with observational arguments which suggest that the disc collapsed by a factor of ~ 10 (Eggen, Lynden-Bell and Sandage 1962). In this model, the ratio M_H/M_D depends on the parameter Y (equ. 17) which cannot be directly determined from observations of disc galaxies. Theoretical arguments concerning the stability of disc galaxies suggest that $Y \approx 1.1$ (Efstathiou, Lake and Negroponte 1982) in which case $M_H/M_D \sim 20(0.07/\lambda_H)$ which agrees roughly with the parameters required by the White-Rees theory.

If there are no heavy haloes, the protocloud must collapse by an excessively large factor if it is to form a centrifugally supported disc. If we approximate the protocloud as a uniform sphere of radius R_i and total energy $E_i = -3/5 \ GM^2/R_i$, we get $\alpha R_i = 150 \ (0.07/\lambda_i)^2$. For a disc of scale length $\alpha^{-1} = 3$ kpc and mass $10^{11} \ M_\odot$, the free-fall time is $t_{ff} \cong \pi(R_i^3/8GM)^{\frac{1}{2}} \sim 2 \times 10^{10}$ yrs which is unreasonably long. Thus, if dark haloes do not exist, it is difficult to see how the tidal torque theory could account for the origin of disc angular momentum.

Several authors (Faber, 1982, Burstein and Sarazin 1983, Fall 1983, and ourselves) have noticed that, under certain special assumptions, the arguments summarized in equs. (14)-(18) can lead to relations similar to those inferred from the infra-red Tully-Fisher relation ($v_m^4 \propto L_D$) and Freeman's law ($I_0 = $ const, equ. 5):

$$v_m^4 \propto L_D \left(\frac{M_D}{L_D}\right)\left(\frac{M_H}{M_D}\right) M_H R_H^{-2} \, , \tag{19a}$$

$$I_0 \propto \lambda_H^{-2}\left(\frac{L_D}{M_D}\right)\left(\frac{M_D}{M_H}\right) M_H R_H^{-2} \, . \tag{19b}$$

To explain the observations requires that $M_H/M_D \cong$ const, $M_H \propto R_H^2$, $(M_D/L_D) \cong$ const, $\lambda_H \cong$ const or some other special relations between these quantities. Since some of these relations are highly un-certain we do not propose to give a detailed discussion here. Instead, we wish to consider the following point. We have mentioned in Section (3) that the tidal-torque theory predicts a broad spread in λ, so if the model for disc formation sketched above is correct, we should see this scatter reflected in the properties of present day spirals. Now we know that the scatter in the infra-red Tully-Fisher relation is small (eg. Aaronson et al. 1982) and this also seems to be true in the B-band if Hubble type is used as an indicator of M_D/L_D (Rubin 1983). These results suggest that the combination $(M_H/M_D) M_H R_H^{-2}$ in equ. (19a) has a fairly small scatter. If M_H/M_D is fairly constant, equ. (19b) says that we should see the scatter in λ_H reflected as a large spread in I_0. As a test of this idea we have compared the distribution of $I_0^{-\frac{1}{2}}$ for the disc galaxies studied by Boroson (1981) with the distribution of λ measured in the N-body experiments (Figure 5). The distributions agree quite well once we have matched the area under each curve and the position of the median values. Of course,

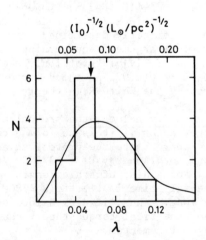

Figure 5. Comparison of the distribution in central surface bright-ness of disc components with the distribution of λ found in N-body experiments with n = 0. The arrow shows Freeman's value $I_0^{-\frac{1}{2}} = (145 \ L_\odot/pc^2)^{-\frac{1}{2}}$.

our main purpose in presenting this graph is to illustrate what
might turn out to be an interesting test of the tidal torque
theory given more photometry and an independent indicator of
M_D/L_D (perhaps Hubble type or colour). It may even be possible
to test for the mass dependence of λ predicted in Section 3.

4.3 Elliptical Galaxies

As discussed above, the low M/L values in the inner parts
of elliptical galaxies and their high overdensities argue against
dissipationless collapse models. Consistent with this, the λ
values inferred for ellipticals with $L < L^*$ exceed the median
values found in the numerical simulations. One possibility is
that ellipticals formed by the dissipative collapse of gas in
dark haloes (Kashlinsky 1982, Faber 1982). If the protosystem is
approximated as a uniform sphere of radius R_i with spin parameter
λ_i, the requirement that $J_E/M_E = J_H/M_H$ (here the subscript E refers
to "elliptical") leads to the following relation for the ratio of
rotational velocity to central velocity dispersion of the luminous
component,

$$v_m/\sigma_p \cong 1.4\ \lambda_i\ (\frac{M_H+M_E}{M_E(1+f)})^{\frac{1}{2}}(R_i/r_e)^{\frac{1}{2}} \qquad (20)$$

Here f is the ratio of dark mass to luminous mass interior to the
effective radius r_e and we have assumed that the luminous component
rotates on cylinders with a flat rotation curve and a flat
dispersion profile. If the relaxed halo component approximates
an isothermal we must have $r_e/R_i \cong M_E\ f/2\ M_H$. Equ. (20) then gives

$$v_m/\sigma_p \cong 1.7\ (\lambda_i/0.06)(M_H/20\ M_E)\ [2/(1+f)f]^{\frac{1}{2}} \qquad (21)$$

where we have inserted values for the parameters in rough agreement with those expected in the White-Rees theory. If equ. (21)
is compared with the observations (equ.3) we see that ellipticals
must have formed in protosystems with unusually low values of λ
($\cong 0.01$) or low values of M_H/M_E. The main problem with this model
lies in accounting for the observation that ellipticals occur
predominantly in rich clusters of galaxies for it is difficult to
see how λ or M_H/M_E could be correlated on scales of galaxy clusters
at the time of galaxy formation.

Other possibilities, such as galaxy mergers and tidal
stripping are discussed by Davies et al. (1983) and Fall (1983).
But it is tempting to speculate on whether the properties of
elliptical galaxies could be explained if primordial fluctuations
were non-Gaussian (Peebles 1983).

Acknowledgements.   We thank the SERC for providing computing time.
G.E. thanks King's College, Cambridge, for financial support.

1. In this article we assume a Hubble constant H_o = 100 km sec^{-1}
Mpc^{-1}. All magnitudes and luminosities refer to the B_T system
defined in the Second Reference Catalogue of Bright Galaxies (de
Vaucouleurs, de Vaucouleurs and Corwin 1976).

REFERENCES

Aaronson, M., Huchra, J., Mould, J., Schechter, P.L. and Tully,
    R.B., 1982, Astrophys. J., 258, 64.
Aarseth, S.J., Gott, J.R. and Turner, E.L., 1979, Astrophys. J.,
    228, 664.
Bean, J., Efstathiou, G., Ellis, R.S., Shanks, T. and Peterson,
    B.A., 1983, Mon. Not. R. astr. Soc., in press.
Binney, J.J., 1978, Mon. Not. R. astr. Soc., 183, 501.
Binney, J.J. and Silk, J., 1979, Mon. Not. R. astr. Soc., 188, 273.
Boroson, T., 1981, Astrophys. J. Suppl. Ser., 46, 177.
Burstein, D. and Sarazin, C.L., 1983, Astrophys. J., 264, 427.
Bosma, A., 1983 in "Internal Kinematics and Dynamics of Galaxies"
    IAU Symp. No.100, ed. E. Athanassoula, Reidel, Dordrecht, p.11.
Carter, D. and Metcalfe, N., 1980, Mon. Not. R. astr. Soc. 191, 325.
Davies, R.L., Efstathiou, G., Fall, S.M., Illingworth, G. and
    Schechter, P.L., 1983, Astrophys. J., 266, 41.
Davis, M. and Peebles, P.J.E., 1983, Astrophys. J., 267, 465.
de Vaucouleurs, G., 1958, Astronom. J., 63, 253.
de Vaucouleurs, G., 1972, in "External Galaxies and Quasistellar
    Objects" IAU Symp. No.44, ed. D.S. Evans, Reidel, Dordrecht,
    p.353.
de Vaucouleurs, G., de Vaucouleurs, A. and Corwin, H.R., 1976,
    "Second Reference Catalogue of Bright Galaxies", University
    of Texas Press, Austin.
Eggen, O.J., Lynden-Bell, D. and Sandage, A., 1962, Astrophys. J.,
    136, 748.
Efstathiou, G. and Jones, B.J.T., 1979, Mon. Not. R. astr. Soc.,
    186, 133.
Efstathiou, G. and Eastwood, J.W., 1981, Mon. Not. R. astr. Soc.,
    194, 503.
Efstathiou, G., Fall, S.M. and Hogan, C., 1979, Mon. Not. R. astr.
    Soc., 189, 203.
Efstathiou, G., Lake, G. and Negroponte, J., 1982, Mon. Not. R.
    astr. Soc., 199, 1069.
Faber, S.M., 1982, in "Astrophysical Cosmology", eds. H.A. Brück,
    G.V. Coyne and M.S. Longair, Pontificia Academia Scientarium,
    Vatican, p.219.
Faber, S.M. and Gallagher, J.S., 1979, Ann. Rev. Astr. Astrophys.,
    17, 135.

Fall, S.M., 1983, in "Internal Kinematics and Dynamics of Galaxies"
    IAU Symp. No.100, ed. E. Athanassoula, Reidel, Dordrecht,
    p.391.
Fall, S.M. and Efstathiou, G., 1980, Mon. Not. R. astr. Soc.,
    193, 189.
Freeman, K.C., 1970, Astrophys. J., 160, 811.
Gregory, S.A., 1975, Astrophys. J., 199, 1.
Gott, J.R. and Thuan, T.X., 1976, Astrophys. J., 204, 649.
Hogan, C.J., 1979, Mon. Not. R.astr. Soc., 188, 781.
Hoyle, F., 1949, in "Problems of Cosmological Aerodynamics", eds.
    J.M. Burgers and H.C. van de Hulst. Central Air Documents.
    Ohio, p.195.
Illingworth, G., 1981, in "The Structure and Evolution of Normal
    Galaxies", eds. S.M. Fall and D. Lynden-Bell, Cambridge
    University Press, Cambridge, p.27.
Illingworth, G. and Schechter, P.L., 1982, Astrophys. J., 256, 481.
Kalnajs, A., 1983, in "Internal Kinematics and Dynamics of Galaxies"
    IAU Symp. No.100, ed. E. Athanassoula, Reidel, Dordrecht,p.87.
Kashlinsky, A., 1982, Mon. Not. R. astr. Soc., 200, 585.
Kent, S.M. and Gunn, J.E., 1982, Astronom. J., 87, 945.
Kormendy, J. and Illingworth, G., 1982, Astrophys. J., 256, 460.
Peebles, P.J.E., 1969, Astrophys. J., 155, 393.
Peebles, P.J.E., 1971, Astron. Astrophys., 11, 377.
Peebles, P.J.E., 1983, Astrophys. J., in press. (The Sequence of
    Cosmogony and the Nature of Primeval Departures from Homo-
    geneity).
Rees, M.J. and Ostriker, J.P., 1977, Mon. Not. R. astr. Soc.,
    179, 541.
Rood, H.J., Page, T.L., Kitner, E.C. and King, I.R., 1972,
    Astrophys. J., 175, 627.
Rubin, V.C., 1951, Astronom. J., 56, 47.
Rubin, V.C., 1983, in "Internal Kinematics and Dynamics of Galaxies"
    IAU Symp. No. 100, ed. E. Athanassoula, Reidel, Dordrecht, p.3.
Sandage, A., Freeman, K.C. and Stokes, N.R., 1970, Astrophys. J.,
    160, 831.
Schipper, L. and King, I.R., 1978, Astrophys. J., 220, 798.
Stewart, J.M. and Sciama, D.W., 1967, Nature, 216, 748.
Thuan, T.X. and Gott, J.R., 1977, Astrophys. J., 216, 194.
White, S.D.M. and Rees, M.J., 1978, Mon. Not. R. astr. Soc.,
    183, 341.

ROTATION, DISSIPATION AND ELLIPTICAL GALAXIES

Rosemary F.G. Wyse * & Bernard J.T. Jones **

* Princeton University Observatory &
  Astronomy Department, U.C. Berkeley.

** Observatoire de Meudon.

ABSTRACT We discuss a new correlation between the mean surface brightness within the effective radius of elliptical galaxies, and their level of rotational support using published data. The latter parameter is described by the ratio of rotation velocity to velocity dispersion, normalised to the value required to stabalise an isotropic, oblate galaxy of the observed flattening. A stronger correlation results when the velocity dispersion is corrected for second parameter effects. The sense of the correlations is that rotation is more important for higher surface brightness ellipticals. We understand the correlations as resulting from an increased importance of dissipation during the formation and early evolution of the higher surface brightness galaxies. We envisage that ellipticals form by dissipative collapse in dark halo potentials, the amount of dissipation depending on the initial gas fraction.

1 INTRODUCTION

Models for elliptical galaxy formation have assigned a varied importance to dissipation and to the role of dark matter. The observational evidence is not clear cut, but there have been some major advances in recent years. Early work revealed correlations of line strength, colour (1) and velocity dispersion (2) with absolute magnitude. These indicated that knowledge of one parameter – luminosity – was enough to specify an elliptical completely. Terlevich et al (3), however, discovered a second parameter, manifest in correlated deviations from the mean

379

J. Audouze and J. Tran Thanh Van (eds.),
Formation and Evolution of Galaxies and Large Structures in the Universe, 379–387.

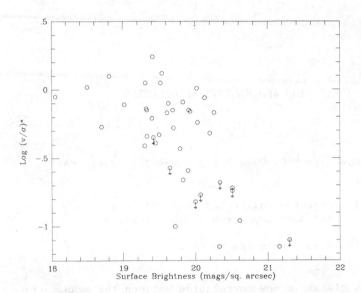

Figure 1 Mean surface brightness in magnitudes within the half
light radius against rotation parameter (v/σ)*, for the sample of
43 ellipticals with necessary data (5).

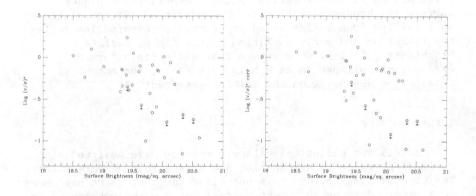

Figure 2 (a) As Figure 1, but for the 34 galaxies with colour
data. (b) As (a), but with the second parameter corrected
velocity dispersions.

relations with luminosity. The second parameter is probably
related to the chemistry or star formation history of the
galaxy.

The observations that high luminosity ellipticals are
generally (though not always) pressure supported (4, and refs
therein) while lower luminosity ellipticals are rotationally
supported (5) were unexpected. The dependence of rotation on
luminosity has been investigated (5) and used to test theories of
elliptical galaxy formation. The slope of the correlation seems
equally consistent, within the unfortunately large uncertainties,
with the predictions from clustering hierarchies (with and
without dissipation), merging of stellar discs and dissipative
pancake or blast wave scenarios. The amplitude of the relation
is more difficult to explain (7).

2 ROTATION AND SURFACE BRIGHTNESS

2.1 The Correlation

Elliptical galaxy samples with a large luminosity baseline
have surface brightness increasing with luminosity (5,6). Since
surface brightness is distance (and distance error) independent,
unlike luminosity, a plot of surface brightness against rotation
parameter should contain interesting information, to constrain
theories of elliptical galaxy formation.

Figure 1 shows mean surface brightness within the effective
radius, versus $(v/\sigma)^*$, the observed rotation velocity: velocity
dispersion ratio, divided by the value corresponding to an oblate
isotropic model (8). 43 galaxies have the necessary data (5).
There is clearly a trend in that galaxies of higher surface
brightness have higher rotation parameter. This is in the sense
expected from the luminosity correlation (5). The strength of
the correlation could not have been predicted, however. Treating
the upper limits as detections, the Spearman rank correlation
coefficient is -0.59, significant to 0.05%.

There is considerable scatter, galaxies with the same
rotation parameter differing in surface brightness by up to a
factor ten. A quick investigation shows that such galaxies
differ in colour too, an indication that second parameter effects
may be operative. The overall distribution suggests the presence
of a lower envelope.

2.2 The Second Parameter Corrected Correlation

We undertook a second parameter analysis of the sample by
investigating the deviations from the (b-V), M correlation mean

line (colours from ref (9)) and from the velocity dispersion, M
mean line. The log σ and (b-V) residuals are indeed correlated,
allowing a second parameter corrected velocity dispersion to be
defined (see ref (3) for discussion of the second parameter
correction procedure). 34 galaxies have colours and hence
corrected dispersions. Figure 2 shows the $(v/\sigma)*$, surface
brightness plots for this sample - 2(a) shows raw $(v/\sigma)*$; 2(b)
shows the relation using corrected dispersions. There is an
improvement in the correlation, the regression coefficient
increasing from -0.53 to -0.58. The major feature of the plot
remains a lower envelope. Here and in Figure 1 the envelope
corresponds to

$$(v/\sigma)* \; \alpha \; \text{Surface brightness} = L/r^2$$

There is no a priori reason why points should not fall in
the region below this envelope, unless constrained by real
physical effects due to the galaxy formation process. Hence this
envelope tells us about how ellipticals formed. (Note that since
the envelope falls on the high surface brightness limit, it
should not be the result of a simple bias in the observations;
spurious constant surface brightness correlations have been
discussed in (10)).

One might think that the trends shown in Figures 1 and 2
could be due to inclination effects - as an oblate galaxy is
viewed in an increasingly pole-on orientation, its surface
brightness and apparent rotation velocity both decrease - the
overall trend of the envelope. Here however, the plots refer not
simply to v/σ but to the shape-normalised quantity $(v/\sigma)*$, which
varies in a complicated way with inclination. Binney (11) has
shown that for oblate galaxies, $(v/\sigma)*$ at an edge-on orientation
is the minimum possible. The largest increase in $(v/\sigma)*$ as the
galaxy is tilted towards face-on orientations is for anisotropic,
intrinsically rather flat galaxies, though even isotropic
flattened galaxies (around E5) can be observed to have
$(v/\sigma)* \geqslant 1.4$. (Note that anisotropic galaxies increase $(v/\sigma)*$
towards unity, while isotropic galaxies increase $(v/\sigma)*$ above
unity.) Thus, since the surface brightness is a maximum in the
edge-on orientation, projection effects will cause a scatter
roughly perpendicular to the lower envelope in Figures 1 and 2,
and cannot produce the correlation; the envelope slope represents
the inherent (edge-on) correlation. This lower envelope is not
obvious in the luminosity : $(v/\sigma)*$ plane, and it is less easy to
isolate the effects of inclination uncertainties on this latter
distribution.

Evidence that the spread is at least in part due to
inclination comes from noticing that the distribution of
(observed) ellipticity on the $(v/\sigma)*$ plane is non-uniform.

There is a tendency to find the rounder galaxies in the top right
hand quadrant of the plane, and only galaxies observed flatter
than E2.5 along the lower envelope. The crucial point is the
lack of round ellipticals along and near to the lower envelope
where we would predict the edge-on galaxies should lie. The
triangular shape of the distribution is partly due to the fact
that projection effects favour $(v/\sigma)*$ unity, and environmental
effects discussed below.

3 THEORETICAL MODELS

3.1 Existing Theories

The slope of the lower envelope can be used to test galaxy
formation theories, in a manner analogous to ref (5).

It has been proposed (12, 13) that normal ellipticals formed
from the merger of stellar disc galaxies. Most simulations of the
merging process have simplified the situation to identical
elliptical building blocks (14 and refs within). Simple energy
conservation arguments indicate that the merger remnant will have
mean surface brightness within the half mass (light) radius
decreased by a factor two from the progenitors. Neglecting the
orbital contribution, assuming random orientations of progenitor
spin axes suggest that (v/σ) for the remnant will be a factor
$N^{-3/2}$ smaller than that for each of the N progenitors. Thus the
simplest merger theory using small elliptical building blocks
would predict a slope 0.70 for the rotation, surface brightness
correlation. This is reasonably close to the observed $(v/\sigma)*$
relation, but there is the added complication of the essentially
unknown ellipticity distribution expected in the merger remnants,
which depends rather crucially on the assumed initial orbital
parameters. In any case this simple merger model faces
difficulties with the observed luminosity-velocity dispersion
relation. A picture with spiral building blocks faces the
immediate problem that the surface density of the progenitors is
already lower than observed ellipticals.

The dissipationless formation of ellipticals in a clustering
hierarchy does not produce nearly a strong enough dependence of
rotation on mass, or a high enough amplitude (5). The occurence
of dissipation leads to more scaling relations than free
parameters and allows a consistency check of the theoretical
predictions against the observations (c.f. ref 5). Using the
slope of the envelope of the surface brightness - rotation
correlation, we find that the slopes cannot easily be made
consistent (see ref (15)).

The rotational properties of galaxies formed in dissipative

pancake and blast wave scenarios have not yet been investigated
in sufficient detail to be constrained strongly, but a simple
suggestion that the initial turbulence is independent of scale
(18) seems inconsistent with the envelope slope.

3.2 A Simple Model

The underlying key to the problem is evidently the role
played by dissipation of random energies in the rotation plane,
in the gas that formed the luminous material, since this
increases both the surface brightness and the importance of
rotational velocities, and also modifies the chemistry and star
formation rates within the proto galaxy. From Figures 1 and 2,
high surface brightness, rotating ellipticals – which are also
low luminosity ellipticals – dissipated more than their high
luminosity counterparts.

The idea that disc galaxies formed from gas which collapsed
in a dark halo potential is fairly well established (refs (17),
(18) & (19)). Simplicity arguments suggest that ellipticals
formed in the same way. Unfortunately, the most straightforward
extension of dissipative collapse in a hierarchy of haloes has
been shown to face severe problems in explaining the existence of
slow rotation in ellipticals which have large collapse factors
(small effective radii) compared to disc galaxies of the same
luminosity. However, the presence of a dark halo introduces
another parameter into the problem – the fraction of gas
initially present in the perturbation. The square root of this
appears as a multiplicitive factor in the expression for the
final rotation velocity : velocity dispersion ratio in terms of
the collapse factor (ref 20). The fraction of gas present is not
likely to remain constant once galaxy formation commences, even
though simplicity arguments may favour a universal value at the
earliest epochs. A variable fraction of gas can explain the
increasing importance of a pressure supported component (19).

We favour a scheme in which lower mass systems remain
predominantly gaseous longer. In our model, the luminous
components of ellipticals form from gas collapsing in the
potential wells of aspherical dark haloes (c.f. (17) and (19)).
The asphericity of the potential results in a preferential
collapse axis (the shortest) allowing shock heating, compression
and cooling of gas to occur before much collapse has yet occured
in the other directions. Stars formed in this initial collapse
phase come from gas which has not dissipated its energy in the
rotation plane. Stars formed after violent relaxation of the
potential will be more rotationally supported. Thus the rotation
parameter observed for a galaxy depends on the relative amounts
of stars formed prior to and after relaxation of the potential.
There exists a critical shock temperature and hence infall height

from which gas can cool rapidly. The ratio of cooling to
collapse times is governed principally by the gas fraction, with
contributions from the asphericity and density profile of the
dark halo. The trends of Figures 1 and 2 then reflect a sequence
of initial gas fraction - lower surface brightness (higher
lumnosity) ellipticals form from perturbations with a higher
initial gas fraction, and have a lower mass to light ratio.
Perturbations destined to form higher surface brightness, lower
luminosity ellipticals contain a smaller fraction of gas, at
lower densities.

The assumption of a simple scaling relationship between the
gas fraction and the initial perturbation mass, with index κ,
allows an estimate of κ to be made from consistency requirements.
Define n as the power spectrum slope for the hierarchy of halo
density perturbations. For n in the range compatible with galaxy
correlation function observations (ref 7) we find that $\kappa \cong 0.25$
(n=0); $\kappa \cong 0.05$ (n=-1.5). It is important that κ is constrained
to positive values, since we envisage κ positive in our models
(higher luminosity ellipticals forming from higher mass
perturbations with a higher initial gas fraction).

The shape of the distribution in Figures 1 and 2 is then
explained as follows. The lower envelope results from a variable
gas fraction. The spread away from the envelope is caused in
part by observational errors (the effective radius estimates are
the most likely source of error) but mostly by projection
effects. The large spread in rotation parameter for the lower
surface brightness galaxies can be explained as a real physical
effect - the low surface brightness galaxies which have a high
rotation parameter tend to be the dominant member of groups of
galaxies, and so are perhaps stationary with respect to the group
potential well. This aids accretion of intergalactic gas, and
inhibits dynamical removal of gas in the early phases of galaxy
formation. The fact that cD galaxies rotate slowly perhaps
argues against the plausibility of this explanation, but cDs are
atypical ellipticals.

The distinguishing features between spirals and ellipticals
then arise from both nature and nurture. The increased
predominance of a centrifugally supported component results from
initial conditions of lower gas fraction and decreasing
asphericity of the halo potential. A cold, gas disc is a
secondary structure, formed by ejecta from the pressure supported
component, and gas only now cooling. High density environments,
as favoured by present day ellipticals, (refs (21) and (22)) can
lead to dynamical removal of gas in the outer regions of the
initial perturbation, resulting in no rotationally supported
stellar system. This is because this outlying gas is shock
heated to highest temperatures during the collapse and hence is

destined to cool and form stars only after the relaxation is
completed. Continuing gas removal inhibits secondary disc
formation. It is also likely that in proto-ellipticals the
angular momentum axis induced by tidal torques by nearest
neighbours is not generally perfectly aligned with the principal
plane of the potential. This leads to an additional thickening
of the relaxed stellar system (c.f. ref (23)) and can explain the
observation that there exist no ellipticals as flattened as disc
galaxies.

RFGW is grateful to the English Speaking Union for support
through a Lindemann Trust Fellowship.

REFERENCES

(1) Faber, S.M., 1973. Ap.J., 179, pp731-754.
(2) Faber,S.M. & Jackson, R.E., 1976. Ap.J., 204, pp668-683.
(3) Terlevich, R.J., Davies, R.L., Faber, S.M. & Burstein, D.,
    1981. M.N.R.A.S., 196, pp381-395.
(4) Illingworth, G., 1981. In "Normal Galaxies" eds S.M. Fall &
    D. Lynden-Bell; pp27-41. (CUP).
(5) Davies, R.L., Efstathiou, G., Fall, S.M. Illingworth, G. &
    Schechter, P., 1983. Ap.J., 266, pp41-57.
(6) Tonry, J.L. & Davis, M., 1981. Ap.J., 246, p680.
(7) Fall, S.M., 1983. In IAU Symp. #100, ed E. Athanassoula;
    pp391-399. (Reidel)
(8) Binney, J.J., 1978. M.N.R.A.S., 183, pp501-514.
(9) Sandage, A. & Visvanathan, N., 1978. Ap.J., 225,
    pp742-750.
(10) Disney, M.J., 1976. Nature, 263, pp573-575.
(11) Binney, J.J., 1983. In "Morphology & Dynamics of Galaxies"
     eds L. Martinet & M. Mayor; pp1-112. (Geneva Observatory)
(12) Toomre, A., 1977. In "Evolution of Galaxies & Stellar
     Populations" eds B.M. Tinsley & R.B. Larson; pp401-416.
(13) Fall, S.M., 1979. Nature, 281, pp200-202.
(14) White, S.D.M., 1983,. In "Morphology & Dynamics of Galaxies"
     eds L. Martinet & M. Mayor; pp289-420 (Geneva Observatory).
(15) Jones, B.J.T. & Wyse, R.F.G., in preparation.
(16) Doroskevich, A.G., Shandarin, S.F. & Saar, E., 1978.
     M.N.R.A.S., 184, pp643-660.
(17) White, S.D.M. & Rees, M.J., 1978. M.N.R.A.S., 183,
     pp341-358.
(18) Fall, S.M. & Efstathiou, G., 1980. M.N.R.A.S., 193,
     pp189-206.
(19) Jones, B.J.T. & Wyse, R.F.G., 1983. A. & A., 120,
     pp165-180.

(20) Efstathiou, G. & Jones, B.J.T., Comments on Astrophysics, 8, pp169-176.
(21) Davis, M. & Geller, M.J., 1976. Ap. J., 208, pp13-19.
(22) Dressler, A., 1980. Ap. J., 236, pp351-365.
(23) Aarseth, S. & Binney, J.J., 1978. M.N.R.A.S., 185, pp227-243.

VI

CHEMICAL EVOLUTION OF GALAXIES

NUCLEOSYNTHESIS CONSTRAINTS ON EARLY GALACTIC EVOLUTION

James W. Truran

University of Illinois

This paper identifies abundance patterns observed in metal-deficient stars which serve to distinguish them from stars of solar composition and reviews the defining characteristics of the nucleosynthesis mechanisms which are believed to be responsible for their formation. We then seek to identify constraints that these combined observational and theoretical considerations impose upon models of the early history and evolution of our galaxy and cosmology.

1. INTRODUCTION

Studies of chemical evolution quite generally seek to account for the distribution of abundances of the elements observed in the stars and interstellar gas in galaxies. The details of such studies have been reviewed quite extensively by Audouze and Tinsley (1976) and Tinsley (1980) and will not be elaborated here. There are a variety of factors which must necessarily enter into any considerations of the early chemical evolution of the galaxy. In particular, it is necessary to provide some measure of the rate of star formation at early times (relative to dynamic timescales), of the fraction of gas processed through early stellar generations, of the initial mass function of the first stellar generations, of the evolutionary characteristics of metal-deficient stars, and of the consequences of nucleosynthesis in metal-poor stars. Such factors may strongly influence the relative abundances of heavy elements produced in early generations of stars. In turn, scrutiny of the abundance patterns observed in metal-deficient stars can allow inferences to be drawn concerning, and constraints to be imposed upon, the char-

J. Audouze and J. Tran Thanh Van (eds.),
Formation and Evolution of Galaxies and Large Structures in the Universe, 391–399.
© *1984 by D. Reidel Publishing Company.*

acteristics of early stellar generations.

My aim in this paper will be to identify such constraints as
studies of this nature can impose upon the early evolutionary
history of our galaxy and upon cosmology. In the following
section, we identify groups of elements which are believed to
have common origins in astrophysical environments, review the
distinguishing characteristics of the nuclear processes which
are assumed to be responsible for their formation, and summarize
existing observations of the abundances of these elements in the
most metal-deficient (and thus, presumably, the oldest) stars in
our galaxy. Significant abundance patterns and trends are then
discussed and possible implications for the early history of
stellar activity in our galaxy are noted.

2. NUCLEOSYNTHESIS PROCESSES AND ELEMENT ABUNDANCES

Our aim in this paper is to identify constraints on the early
evolutionary history of our galaxy which are imposed by observations of abundance patterns in metal-deficient stars. In order
to provide a framework within which such observations may be interpreted, we first present, in this section, a review of the
basic mechanisms of nucleosynthesis (see also Arnett, 1973;
Truran, 1973; Trimble, 1975) which calls attention to critical
dependences upon the characteristics of the underlying stellar
or supernova environments. A brief survey of significant abundance trends in metal-deficient stars, involving the products of
specific nucleosynthsis processes, is also included. In these
discussions, we will distinguish specific classes of heavy elements with common nucleosynthesis histories.

2.1. The Light Elements H, He, Li, Be, and B

The nuclei H, D, ^{3}He, ^{4}He, and ^{7}Li constitute the expected products of nucleosynthesis accompanying the cosmological big bang,
as is discussed in other contributions to these proceedings.
The relative concentrations of these various constituents are of
course dependent upon the conditions prevailing in the early
universe, and observations can be utilized to impose constraints
on cosmology. The abundances of H and ^{4}He in the halo stars are
compatible with this model. The situations for ^{3}He and ^{7}Li are
complicated by the fact that these nuclei can be both formed and
destroyed in different stellar environments.

The observed abundances of ^{6}Li, ^{9}Be, ^{10}B, and ^{11}B are consistent
with the view that they have been formed by the interactions of
cosmic rays with the constituents of the interstellar medium
over the history of the galaxy (see the review by Reeves et al,
1973). The production of sufficient concentrations of the domi-

nant isotope of lithium, ^{7}Li, does not occur in this environment. Alternative possible sites for ^{7}Li synthesis include the
cosmological big bang, supermassive stars, red giant stars, and
novae. Unanswered questions associated with the detailed opera-
tion of these mechanisms and with the extent of ^{7}Li destruction
in stellar envelopes render predictions of ^{7}Li synthesis rather
uncertain.

The abundance of lithium in unevolved halo stars has recently
been determined by Spite and Spite (1982a,b). They deduce a
primordial value for the abundance of lithium of $N_{Li}/N_H = 11.2$ x
10^{-11}. This result clearly has important implications for cosmology and thus for the composition of the gas out of which the
galaxy formed. The substantial uncertainties associate with
theoretical predictions of lithium formation (and destruction)
in stars unfortunately limit the effectiveness of the use of
lithium as a probe of early galactic evolution.

2.2. The Elements Carbon, Nitrogen and Oxygen

^{12}C and ^{16}O represent the primary products of helium burning in
stars. There is some indication (Arnett, 1978; Weaver, Zimmerman and Woosley, 1978) that the ^{16}O/^{12}C ratio characterizing
matter processed through massive stars is high relative to that
of solar system matter. Intermediate mass stars may alternatively provide the source of ^{12}C necessary to meet galactic requirements (Iben and Truran, 1978). ^{14}N is formed as a bypro-
duct of CNO-cycle hydrogen burning. This is thought to occur in
the hydrogen burning shells of red giant stars where the ^{14}N
thus formed can subsequently be transported to the surface by
convection and enrich the interstellar medium as a consequence
of mass loss or planetary nebula ejection. In general, all primordial CNO nuclei in the shell will be converted to ^{14}N due to
these burning sequences: the ^{14}N thus formed therefore represents a secondary nucleosynthesis product. The formation of
substantial ^{14}N in first generation (zero metal) stars would demand some mixing of the products of shell helium burning into
the overlying hydrogen shell during the course of red giant evolution (Truran and Cameron, 1971; Iben and Truran, 1978; Renzini
and Voli, 1981), thus in essence forming nitrogen as a primary
nucleosynthesis product.

Trends in CNO abundances have been reviewed most recently by
Tinsley (1979). She argued that the high [O/Fe] ratios characterizing halo stars were generally compatible with the site of
origin being massive stars (Arnett, 1978; Weaver, Zimmerman and
Woosley, 1978). High O/Fe ratios are also observed in globular
cluster stars (Pilachowski, Sneden, and Wallerstein, 1983). The
situation for carbon reveals [C/Fe] $\approx$ 0 even for very iron-
deficient stars. This is consistent with the view that much of

the carbon comes from less massive stars than those which produce oxygen: presumably intermediate mass stars (Iben and
Truran, 1978). Unfortunately, the available data on nitrogen in
extreme metal-deficient stars is not sufficient to allow any
existing abundance trends to be clearly identified. There is
considerable variation in values determined for halo dwarfs and
for stars at the same point in the color-magnitude diagram for
globular clusters, and trends in C and N abundances can be
complicated (see, for example, Kraft, 1979; Freeman and Norris,
1981).

2.3. The Elements Neon to Nickel

It is generally agreed that most isotopes in the mass range $20 \leq$
$A \leq 60$ represent products of explosive nucleosynthesis (Arnett
1973; Truran 1973). Successive exoergic stages of burning of
hydrogen, helium, carbon, oxygen, and silicon fuels define the
presupernova evolution of massive stars. When the ashes of
these burning epochs are subsequently subjected to high temperatures and densities accompanying their ejection in supernova
events, further thermonuclear processing yields elemental isotopic abundance patterns resembling those of solar system matter. The products of these explosive burning processes are both
temperature and composition dependent. Explosive carbon and
oxygen burning at temperatures $2 \times 10^9 \leq T \leq 3.6 \times 10^9$ K form
nuclei in the mass range $20 \leq A \leq 44$, while the burning of carbon, oxygen or silicon fuels at temperatures $4 \times 10^9 \leq T \leq 6 \times$
10^9 K produces nuclei in the iron peak $48 \leq A \leq 62$. The relative contributions will clearly be somewhat dependent upon the
temperature-density structure of the presupernova star so that
the ejecta of supernova of differing masses (or stellar populations) may be expected to differ in their relative concentrations of, for example, Ne, Mg, Si, S, Ar, Ca and Fe. Furthermore, calculations indicate that, for metal-poor stars, the resulting concentrations both of odd-Z nuclei and of the neutron-
rich isotopes of even-Z nuclei may be significantly reduced
(Truran and Arnett, 1971); some odd-even effects in Z may thus
be expected for extremely metal-deficient stars.

Perhaps the most complete and systematic survey of element abundances in halo population stars is that provided by the ongoing
study of extremely metal-deficient red giants by Luck and Bond
(1981; 1983). Their published results for a sample of 21 metal
poor stars with derived [Fe/H] ratios ranging from -1.4 to -2.7
reveal several interesting trends. The elements Mg, Si, Ca, and
Ti are generally found to be enriched by roughly 0.5 dex relative to Fe in red giants with [Fe/H] < -2. Moreover, the ratio
[Ni/Fe] is found to rise approximately linearly, from 0 to +1,
over the range in [Fe/H] from -2 to -3. For Fe concentrations
[Fe/H] $\gtrsim$ -2, the relative abundances of Mg, Si, Ca, Ti, Fe and

Ni are all compatible with those of solar system matter. The
data may also possibly suggest the existence of a mild odd-even
effect, in the sense that elements containing odd numbers of
protons show somewhat greater deficiencies at given [Fe/H], but
the uncertainties are such that a definite trend is not established. Interestingly, Pilachowski, Sneden and Wallerstein
(1983) have studied a sample of globular clusters with derived
[Fe/H] ratios ranging from -0.9 to -2.2 and found trends similar
to those characteristic of halo stars: the elements Mg, Si, Ca
and Ti are enriched relative to Fe by approximately 0.5 dex.

Note that all of these abundance variations involve elements
which are the expected products of a single nucleosynthesis
mechanism: explosive charged-particle nucleosynthesis in supernova environments. Perhaps the most straightforward possible
interpretation is that the trends in [Mg, Si, Ca, and Ti/Fe]
reflect differences in the density structures of the progenitor
stars of supernova of Population II. The situation, unfortunately, is more complicated. One can argue alteratively (Luck
and Bond, 1981) that massive stars preferentially synthesized
Mg, Si, Ca, and Ti relative to Fe early in the history of the
galaxy and that an independent source of iron was later provided
when intermediate-mass progenitors of (Type I) supernovae appeared. This alternative interpretation, involving two stellar
or supernova sources of Fe, necessarily introduces dependences
on the structure and time dependence of the initial mass function and upon the rate and efficiency of star formation early in
the galaxy's history. The problem is further complicated by the
fact that, since the formation of iron peak nuclei in massive
stars occurs in the innermost regions ejected from the vicinity
of the neutronized core, predictions of the relative abundances
of iron in supernova ejecta are sensitive to details of supernova hydrodynamic models and are therefore quite uncertain.

2.4. The Heavy Elements

Elements beyond the iron abundance peak (A > 60) are formed in
nature primarily by means of neutron capture processes. The observed abundance patterns in the heavy element region indicate
that at least two distinct neutron fluxes have contributed. Depending upon whether the characteristic lifetimes against neutron capture are longer than (s-process) or shorter than (r-
process) those for beta decay, the resulting neutron capture
path and therefore the nuclei formed by these reaction sequences
can differ dramatically. Typically, the r-process capture path
lies off the valley of beta stability and gives rise to the production of more neutron-rich isotopes of heavy elements while
the s-process forms isotopes which lie on or near the valley of
beta stability. The problem of distinguishing s-process and r-
process patterns in stars is complicated by the facts that: (1)

most heavy elements receive contributions from both s-process
and r-process nucleosynthesis; (2) there exist only a few elements, like europium, which are formed primarily by the r-
process; and (3) the generation of neutron fluxes compatible
with s-process nucleosynthesis in different stellar environments
can give rise to distinctly different abundance patterns (Truran,
1980; Ulrich, 1982).

Data regarding heavy element abundances in metal deficient stars
has recently been provided by several authors (Peterson, 1976;
Spite and Spite, 1978; Luck and Bond, 1981; Sneden and Partha-
sarathy, 1983). The data confirms the existence of depletions
in the abundances of the designated s-process elements Sr, Y,
Zr, Ba, La, and Ce relative to iron in stars of low Fe/H; in
particular, these trends are evident for [Fe/H] $\lesssim$ -1.5. The existence of systematic differential or aging effects in abundance
trends, first emphasized by Pagel (1968) and more recently reviewed by Tinsley (1979) and others, is substantially confirmed
for Sr and Ba. Luck and Bond (1981) note a strong aging trend
in Ba, with [Ba/Fe] rising from -1.6 in the most iron-deficient
giants to approximately zero at [Fe/H] $\approx$ -2.0 to -1.5; Y and Zr
are found to behave similarly, but with somewhat lower relative
depletions.

At first glance, the interpretation of these trends as reflecting the secondary character of the s-process of nucleosynthesis
seems straightforward: the abundance levels of the heavy elements lag behind that of Fe during the early history since primordial Fe concentrations are demanded in the stars in which the
s-process nuclei are formed. This would suggest that the stars
in which we see heavy elements are very likely third generation
objects - first generation (pure hydrogen and helium) stars
formed iron while second generation stars processed iron to heavy elements - and perhaps strengthens arguments for the existence of a Population III. This straightforward interpretation
nevertheless has problems. Tinsley (1979) called attention to
the fact that the observed trends in [Y/Fe] and [Ba/Fe] with
[Fe/H] are not consistent with the theoretical behavior predicted for secondary elements. While it is possible to generate
ad hoc assumptions which allow these observational trends in s-
process abundance formally to be understood, such attempts to
date have not been entirely satisfactory.

Truran (1981) proposed an alternative interpretation of these
abundance trends as being due to the fact that the heavy element
abundances in extreme metal-deficient stars are, rather, products of r-process nucleosynthesis. This view is suggested by
the observations of Spite and Spite (1978) that the abundance of
europium (an element whose abundance in nature is due primarily
to r-process nucleosynthesis) relative to iron is essentially

solar even for stars of [Fe/H] ~ -2.6. The determination by
Luck and Bond (1983) that the pronounced "aging effects" observed for Ba and Sr are not evident for Nd, La, Ce and Pr is
also consistent with the r-process model.

While questions remain concerning the character and origin of
the heavy element abundances in metal-deficient stars, the util-
ity of these abundance trends as probes of the rates of star
formation and nucleosynthesis during the early history of the
galaxy is greatly restricted.

3. DISCUSSION AND CONCLUSIONS

We wish now to consider whether the distinctive abundance trends
observed in metal-deficient stars impose interesting constraints
on the character of the earliest stages of galactic evolution.

The significant abundance trends that have been identified may
be summarized as follows: (1) the ratio [O/Fe] is high (~ 0.5)
for extreme halo stars while [C/Fe] ≈ 0; (2) values of [Mg/Fe],
[Si/Fe], [Ca/Fe], [Ti/Fe], and [Ni/Fe] of ~ 0.5 characterize
stars for which [Fe/H] ≲ -2; (3) systematic depletions of the s-
process elements Sr, Y, and particularly Ba, relative to iron,
are observed for stars of [Fe/He] ≲ -1.5; (4) the r-process element europium is present in solar proportions relative to iron
in extremely metal-deficient stars. Of these, the systematic
overabundances of O, Mg, Si, Ca, and Ti relative to Fe are consistent with the predictions of models of supernovae involving
massive stars (Arnett, 1978; Weaver, Zimmerman and Woosley,
1978). If we adopt Tinsley's (1979) assumption of a second
source of iron, we can understand the subsequent approach of
these ratios to their solar values as the lower mass stars which
form iron evolve on longer timescales. The fact that [C/Fe] remains compatible with solar even for the oldest stars suggests
that the source of carbon, like that of iron in Tinsley's model,
may again be low or intermediate mass stars.

The trends in s-process elements are of particular interest
here. As we have noted previously, the presence of s-process
nuclei in stars of metal concentrations as low as [Fe/H] = -2.6
alone holds implications for galactic evolution. The simplest
interpretation of these as secondary elements implies that the
matter we are seeing in these stars has been processed through
at least two prior stellar environments. This might reasonably
be viewed as supporting evidence for the existence of a "Population III." The alternative interpretation of these abundances
as produces of r-process nucleosynthesis (Truran, 1981) imposes
less severe demands on early galactic evolution. In principle,
the formation of r-process nuclei can occur in a first genera-

tion, zero metal star: the existence of primordial concentrations of seed Fe nuclei is not absolutely demanded.

It is important to emphasize that it is the timescale of element
formation which is the critical factor. If it were possible to
identify a nucleosynthesis product of the evolution of very low
mass stars whose lifetimes far exceeded the dynamic timescale of
the galactic halo, then a clear need for a distinct prior stellar generation might be established. Alternatively, the identification of significant concentrations of secondary elements,
whose presence in metal deficient stars demands that the matter
has been processed sequentially through two previous stars,
would impose severe restrictions as well. In this regard, the
fact that the Sr and Ba concentrations in metal deficient stars
can be interpreted alternatively as primary (r-process) elements, reduces their usefulness as probes of early galactic evolution. Increased and improved data regarding nitrogen in metal
deficient stars could also prove extremely important, although
here again both primary and secondary production of nitrogen is
possible (Truran and Cameron, 1971; Iben and Truran, 1978).

In order to place this problem in perspective, it may be instructive to call attention to the fact that the observed abundance patterns in the most extreme metal-deficient stars can all
be interpreted as involving primary elements which can be formed
in massive stars of short lifetimes in the very earliest stages
of galactic history. In particular, the published results of
Arnett (1978) for stars in the range 35-50 M_θ (core masses 16-24
M_θ) give approximately the following ratios: [C/Fe] $\approx$ 0,
[O/Fe] $\sim$ 0.5, and [(Ne + Mg + Si/Fe] $\sim$ 0.5. The site of r-
process nucleosynthesis has not yet been firmly established. If
we further assume the r-process to occur in these massive stars,
associated with the expansion and ejection of highly neutronized
matter from the outer extremities of the core, the observed ratio [Eu/Fe] $\sim$ 0, [Sr/Fe] $\sim$ -0.5 and [Ba/Fe] $\sim$ -1 can also be
understood (Truran, 1981). Thus, virtually every feature of the
seemingly anomalous abundance patterns characterizing extreme
metal-deficient stars can result from nuclear process occurring
in a single, "first generation," massive star. Note also that a
single such massive star ejecting 10 M_θ or more of heavy elements can contaminate 10^6 M_θ of gas to a level [Fe/H] $\sim$ -3.
This may be relevant both to questions concerning the source of
the abundances of heavy elements in globular cluster stars and
to the fact that globular cluster stars exhibit very similar
abundance patterns to those of extreme halo population stars.

4. REFERENCES

Arnett, W.D. 1973, Ann. Rev. Astron. Astrophys. 11, pp. 73-94.

Arnett, W.D. 1978, Astrophys. J. 219, pp. 1008-1016.

Audouze, J. and Tinsley, B.M. 1976, Ann. Rev. Astron. Astrophys. 14, pp. 43-79.

Freeman, K.C. and Norris, J. 1981, Ann. Rev. Astron. Astrophys. 19, pp. 319-356.

Kraft, R.P. 1979, Ann. Rev. Astron. Astrophys. 17, pp. 309-343.

Iben, I. Jr. and Truran, J.W. 1978, Astrophys. J. 220, pp. 980-995.

Luck, R.E. and Bond, H.E. 1981, Astrophys. J. 244, pp. 919-937.

Luck, R.E. and Bond, H.E. 1983, Astrophys. J., in press.

Pagel, P.E.J. 1968, in L.H. Ahrens (Ed.), "Origin and Distribution of the Elements," Pergamon Press, Oxford, pp. 195-204.

Peterson, R.C. 1976, Astrophys. J. 206, pp. 800-808.

Pilachowski, C.A., Sneden, C. and Wallerstein, G. 1983, Astrophys. J., in press.

Reeves, J., Audouze, J., Fowler, W.A. and Schramm, D.N. 1973, Astrophys. J. 179, pp. 909-919.

Renzini, A. and Voli, M. 1981, Astron. Astrophys. 94, pp. 175-194.

Sneden, C. and Parathasarathy, M. 1983, Astrophys. J. 267, pp. 757-778.

Spite, M. and Spite, F. 1978, Astron. Astrophys. 67, pp. 23-31.

Spite, F. and Spite, M. 1982a, Astron. Astrophys. 115, pp. 357-366.

Spite, M. and Spite, F. 1982b, Nature 297, pp. 483-485.

Tinsley, B.M. 1979, Astrophys. J. 229, pp. 1046-1056.

Tinsley, B.M. 1980, Fund. Cosmic Phys. 5, pp. 287-388.

Trimble, V. 1975, Rev. Mod. Phys. 47, pp. 877-976.

Truran, J.W. 1973, Space Sci. Rev. 15, pp. 23-49.

Truran, J.W. 1980, Nukleonika 25, pp. 1463-1475.

Truran, J.W. 1981, Astron. Astrophys. 97, pp. 391-393.

Truran, J.W. and Arnett, W.D. 1971, Astron. Space Sci. 11, pp. 430-442.

Truran, J.W. and Cameron, A.G.W. 1971, Astrophys. Space Sci. 14, pp. 179-222.

Ulrich, R.K. 1982, in C.A. Barnes, D.D. Clayton, and D.N. Schramm (Eds.), "Essays in Nuclear Astrophysics," Cambridge University Press, Cambridge, pp. 301-324.

Weaver, T.A., Zimmerman, G.B., and Woosley, S.E. 1978, Astrophys. J. 225, pp. 1021-1029.

STOCHASTIC STAR FORMATION IN DWARF IRREGULAR GALAXIES

C. Chiosi[1,2]   and   F. Matteucci[3]

1) Institute of Astronomy, University of Padova, Italy
2) International School for Advanced Studies, Trieste, Italy
3) Istituto Astrofisica Spaziale, C.N.R., Frascati, Italy

Summary

The observational properties (fractionary mass of gas, metallicity and colours) of a sample of dwarf irregular and magellanic type galaxies are discussed in the light of the stochastic self propagating star formation model of Gerola et al. (1980). To this aim, one-zone model galaxies are presented which take into account both the bursting mode of star formation and infall of unenriched gas. The comparison with the observational data shows that the high dispersion in the properties of these galaxies cannot be explained by the stochastic mechanism alone. Three possible ways out are indicated, namely variations in the metal yield, variations in the ratio of the star formation to mass accretion rates, and finally variations in the ratio of star formation to galactic wind rates.

Introduction

Dwarf irregular galaxies are a distinct class of objects, ranging from red low surface brightness to high brightness blue compact systems. These latter are of particular interest, because they are characterized by the occurrence of very intense stellar activities at the present time, blue colours, high hydrogen contents and low metal abundances. However, a few objects like IZW 18 (Lequeux and

J. Audouze and J. Tran Thanh Van (eds.),
Formation and Evolution of Galaxies and Large Structures in the Universe, 401–415.
© *1984 by D. Reidel Publishing Company.*

Viallefond, 1980) and IZW 36 (Viallefond and Thuan, 1982) seem not
to obey the above schematization, as they posses both low metallicities and low hydrogen to total mass ratios at the same time.
As already suggested by Searle et al. (1973), these galaxies are
likely to suffer from sporadic episodes of strong stellar activity,
as continuous star formation all over the galaxy life at the present rate, would completely exhaust their gas and overproduce metals. Although the nature of the physical process causing the burst
of star formation is not yet clear, current theoretical understanding of this phenomenon indicates that both dynamical stimulation
in interacting systems (Larson and Tinsley, 1978) and/or the stochastic self propagating mechanism of Gerola et al. (1980) in general are good candidates. In this paper, we discuss the main properties of a selected sample of irregular galaxies (blue compact and
magellanic irregulars) in the light of Gerola's et al. (1980) theory, because it naturally provides the bursting mode of stellar
activity in systems of relatively small size. In particular, we
derive the gas content, the chemical abundances and colours of
model galaxies with a different number of bursts of star formation
and compare them with the observational data.

1. The Observational Material

Basic properties for some 45 galaxies have been collected from various sources in the literature. These are reported in Table 1,
which displays in columns the galaxy identification with references
(1) and (2) respectively, the distance in Mpc, (3), the (U-B) and
(B-V) colours, (4) and (5), the total mass M_T in solar units, (6),
the gas mass M_g in solar units, (7), the gas to total mass ratio,
(8), the mean metal abundance Z, (9), the Holmberg diameter, (10),
which is taken here as size indicator. This sample is quite inhomogeneous in terms of galaxy morphology and quality of the data.
In particular, masses and distances are reported as given in the
original sources but all scaled to the same Hubble constant of
100 Km/ sec/ Mpc. Furthermore, the gas mass determined from the

Table 1

Name	Ref	D	(U–B)	(B–V)	M_T^*	Mg^*	Mg/M_T^*	Z	R^{**}
Mkn 600	1	10.3	–0.26	0.44	8.55	8.17	–0.38	0.0018	2.5
IIZW40	1,7,10,4	6.9	–0.11	0.69	8.87	8.39	–0.48	0.0035	2.1
IZW18	1,4,7,9	8.0	–0.61	0.09	8.81	7.79	–1.02	0.0004	1.4
DDO64	1	4.6	–0.44	0.33	8.75	8.07	–0.68	0.0018	3.5
A1116+51	1,8	14.2	–0.66	0.10	8.27	8.01	–0.26	0.0009	1.1
IC3258	1	11.6	–0.39	0.34	9.18	8.27	–0.91	0.0072	5.6
A1228+15	1	11.6	–0.48	0.11	8.50	7.42	–1.08	0.0011	1.3
IC3453	1	11.6	–0.53	0.47	9.03	8.00	–1.03	0.0029	4.4
Mkn450	1	9.1	–0.44	0.52	8.66	7.99	–0.67	0.0043	3.9
A2228–00	1	17.7	–0.34	0.29	8.15	7.86	–0.29	0.0011	1.5
NGC55	2	2.5			10.63	9.73	–0.89	0.0058	16.7
NGC1613	2	0.6		0.55	8.38	7.80	–0.58	0.0019	2.3
NGC1156	2	4.8		0.44	9.40	8.64	–0.76	0.0052	4.4
NGC1569	2	4.9		0.46	9.55	8.94	–0.61	0.0038	3.9
NGC2366	2	2.7	–0.40	0.42	9.31	8.76	–0.55	0.0024	5.1
NGC2574	2	3.5		0.33	9.93	9.36	–0.57	0.0027	11.1
NGC4214	2	3.1	–0.36	0.39	10.12	9.06	–1.04	0.0057	7.0
NGC4236	2	2.3		0.25	10.32	9.00	–1.32	0.0055	10.5
NGC4449	2	3.7	–0.32	0.33	10.37	9.60	–0.77	0.0055	5.3
NGC4653	2	3.5	–0.29	0.25	10.11	9.00	–1.11	0.0022	10.3
NGC6822	2	0.5			9.19	8.20	–0.99	0.0049	2.9
IC10	3	3.0		0.70	9.73	9.11	–0.61	0.0040	4.4
IIZW70	3,4,5	13.3	–0.71	0.25	8.96	8.23	–0.73	0.0031	5.8
IIZW71	3,5	13.3	–0.14	0.49	9.89	8.69	–1.20		7.3
Mkn19	4,10		–0.69	0.45				0.0080	
Mkn35	4,10	10.2	–0.37	0.41	9.58	8.62	–0.96	0.0050	
Mkn36	4,10	5.9	–0.68	0.34	8.52	7.14	–1.38	0.0015	
Mkn59	4,10		–0.67	0.37	10.30	9.68	–0.62	0.0032	
Mkn108	4,10		–0.65	0.43				0.0030	
Mkn156	4,10		–0.34	0.50				0.0025	
Mkn171	4,10		–0.39	0.44				0.0003	
VIIZW403	6		–0.26	0.39	8.30	7.72	–0.58	0.0014	
Mkn67	4	76.5	–0.40	0.45	10.50	9.21	–1.32	0.0029	
LMC	3	0.05		0.43	9.78	8.85	–0.92	0.0083	12.0
SMC	3	0.07		0.36	9.17	8.81	–0.37	0.0026	8.0
IZW36	11	3.46	–0.51	0.51	8.30	7.72	–0.57	0.0022	
NGC1800	12	6.0	–0.23	0.48	9.44	8.07	–1.37	0.0140	5.8
NGC2146	12	11.1	0.17	0.76	11.20	9.82	–1.42	0.0230	28.2
Haro22	12	13.9	–0.32	0.32		8.33		0.0080	6.0
NGC3274	12	4.8	–0.11	0.39	9.91	8.73	–1.18	0.0071	4.9
NGC3310	12	10.5		0.32	11.30	9.60	–1.68	0.0090	17.2
NGC3510	12	6.6	–0.23	0.38	10.30	8.78	–1.49	0.0080	11.6
NGC3738	12	2.7	–0.17	0.43	9.53	8.00	–1.53	0.0092	3.5
NGC5253	12	3.4	–0.23	0.40	9.38	8.27	–1.11	0.0058	7.6
DDO168	12	1.7	–0.24	0.44	8.56	7.88	–0.67	0.0214	3.8

*) Logarithm ;   **) in Kpc

21 cm line data is uncertain by the contribution of molecular hydrogen, a quantity which very difficult to determine, whereas the
total mass, with the exception of the few cases for which both
21 cm line maps and rotation curves exist, is derived from the
indicative mass of Fisher and Tully (1975). The gas mass M_g is 1.3
M_{HI}, M_{HI} being the atomic hydrogen mass from 21 cm line observations,
to take He into account (Lequeux et al. 1979). The metallicity Z
is obtained from the ratio N(O)/N(H) with the aid of the relation
Log Z = 1.42 + Log N(O)/N(H) due to Lequeux et al. (1979). The
oxygen to hydrogen ratios used to derive Z are as given in the original sources and no attempt is made to make them uniform as far as
the derivation method is concerned. In the following, we will briefly discuss some of the observational properties that are relevant
to subsequent model construction:

i) The Z vs Log M_g / M_T relation

With the aid of the data of Table 1, we derive the relation between
Z and M_g/M_T which is customarily used to test models of chemical
evolution. This is shown in Fig. 1, together with the similar re-

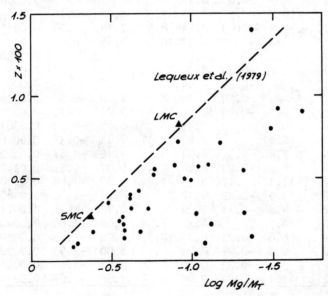

Fig. 1   Observational Z vs Log M_g/M_T relation for the galaxies of
          Table 1. The dashed line indicates the Z vs Log M_g/M_T
          relation of Lequeux et al. (1979)

lation found by Lequeux et al. (1979) for a less numerous sample of galaxies. A distinct feature of this diagram is that our galaxies seem to evenly populate the region below the line of Lequeux et al. (1979), which corresponds to $Z = 0.004 \ln M_T/M_g$ of the simple closed model. It is soon evident that unless the scatter is entirely attributable to observational uncertainties, the simple closed model cannot account for the distribution of our galaxies in the Z vs Log M_g/M_T plane. It is however worth recalling that M_T is uncertain by a factor of about four, whereas no straightforward estimate of the uncertainty in M_g and Z can be given.

ii) The rate of star formation

Observational determinations of the rate of star formation in these galaxies are difficult to obtain. The luminosity to total mass ratio, L/M_T, which is a measure of the intensity of the stellar activity, turns out to be much greater in compact galaxies than in irregular galaxies (> 1 compared to 0.2 to 0.3 L_Θ/M_Θ, .Searle and Sargent, 1972; Lequeux et al., 1979). Furthermore, the rate of star formation per unit mass of gas (normalized to the solar vicinity value and expressed by the ratio of the number of massive stars to the mass of hydrogen) is seen to vary from 1 to 0.03 (Lequeux, 1979a), and to be anticorrelated to the average volume density of gas (hydrogen) contrary to the expectation from the Schmidt's law of star formation. In addition to this, the recent studies of Hunter (1982) and Hunter et al. (1982) indicate that the rate of star formation in these galaxies can be much higher than the local galactic value (up to 10^3 times), and that no correlation exists with global gas and/or abundance parameters. They also point out that the fractional involvement of the galaxy in the present star forming activity decreases with increasing rate of star formation and decreasing time scale of gas consumption. Finally, Hunter (1982) finds that distinct cells of stellar activity can be isolated whose typical size is of the order of 100 - 400 pc. Noteworthy, the occurrence of star formation in cells is compatible with the stochastic star formation process of Gerola et al. (1980). As a

last remark, Hunter's (1982) data suggest that the process of star
formation in irregular galaxies is not entirely random, but a global mechanism is likely to exist which regulates star formation
with time.

iii) Evidence of Infall

Extensive HI observations have indicated that blue compact and irregular galaxies are surrounded by extended HI haloes, whose dimensions are much greater than those of the optical structures (Gordon
and Gottesman, 1981, and references therein). It has been suggested
that the existence of such HI haloes could be taken as an evidence
of primordial gas still collapsing on an already formed central
structure. This view is compatible with the theory of galaxy formation of Larson (1977), according to whom the slower is the collapse
the later is the morphological type of the formed galaxy.

2. Theoretical Models

We present models of chemical evolution of irregular and compact
galaxies, which incorporate the stochastic self propagating star
formation mechanism of Gerola et al. (1980), and as much as possible of the observational hints we have summarized in the foregoing
section. We assume a single zone description and complete instantaneous mixing of gas. We also release the instantaneous recycling
approximation, thus following in detail the time evolution of several important elements (H, He, C+O, Si+Fe) due to stellar nucleosynthesis, stellar mass ejection and inflow of unenriched gas. The
usual set of differential equations describes the time variation
of the fractionary mass of gas G_i in form of a given element i:

$$\dot{G}_i(t) = -X_i(t)\,B(t) + \int_{M_L}^{M_u} B_M(t-\tau_M)\,Q_i(t-\tau_M)\,dM + \dot{G}_{ia} \,, \quad (1)$$

where $B(t)$ is the total rate of star formation, $B_M(t) = B(M)\,\Phi(M)$,
where $\Phi(M)$ is the stellar mass function, $Q_i(t,M)$ is the fraction
of mass ejected by a star of mass M in form of an element i.

$G_{ia}(t)$ is the infall rate. τ_M, M_u and M_L are the stellar lifetime, the upper and lower cutoff masses of $\Phi(M)$. By definition, the abundance by mass X_i of an element i is $X_i = G_i/G$, where G is the total fractionary mass. Following Chiosi (1980), the infall rate is

$$\dot{G}_{ia} = (X_i)_a \exp(-t/\tau) \tau^{-1} (1 - \exp(-t_g/\tau))^{-1} , \qquad (2)$$

where $(X_i)_a$ stands for the primordial abundance of the species i, τ is the time scale of mass accretion, t_g is the galaxy life. In the course of this paper, we adopt $t_g = 12 \times 10^9$ yr. Relation (2) is particularly suited for studies of chemical evolution as it can be related to the dynamical nature of the problem of galaxy formation. The fractions $Q_i(t,M)$ incorporate the nucleosynthesis results of Arnett (1978) and Iben and Truran (1978) for massive and intermediate mass stars respectively. The stellar mass function is expressed as a power law of the mass $\Phi(M) \propto M^{-x}$, where x may vary with the range of mass. The following normalisation condition is used

$$\int_1^{M_u} \Phi(M) \, dM = \zeta \qquad (3)$$

ζ being the fraction of the stellar mass function in stars more massive than 1 $M_\odot$. The rate of star formation is given by

$$B(t) = \nu \, \eta(t) \, G(t) \qquad (4)$$

where $\eta(t)$ represents the fraction of the system undergoing star formation at any given time, and it contains the bursting mode of the process as in Gerola's et al. (1980) numerical experiments. $G(t)$ is the current fractionary mass of gas and it may reflect effects of infall or other global mechanisms. Finally ν is an efficiency parameter to be fixed by the comparison with the observational data. To derive an estimate of the parameter ν we make use of the properties of the Magellanic Clouds, for which a fluctuating but never ceasing star formation rate is predicted by the theory. We get $\nu = 4$. If the rate of star formation per unit mass of gas participating to the star forming process, ν, is an universal property of galaxies, we would expect that their major observable features (Z and M_g/M_T) can be obtained by varying only the number of bursts according to the galaxian size in the manner predicted by Gerola's et al. (1980) theory, while keeping ν constant. The number of burst ranges from 1 to 20 when the size increases from 0.4 to 1 Kpc.

3. Model Results

An extensive grid of models has been calculated for different values of ν, τ and two choices for $\Phi(M)$, namely the Salpeter (1955) mass function, $x = 1.35$, and a two slope case with $x = 1.35$ for $M < 2\ M_\odot$ and $x = 2$ above. This latter choice somehow takes into account the recent determination of Lequeux (1979b). The two mass functions are thereinafter indicated by S and L respectively. Table 2 summarizes the characterizing parameters for each set of models calculated with different number of bursts. Fig. 2 shows the fundamental relationship between Z and Log G. It is soon evident that this relation is mostly governed by the mass function through the parameters x and ζ, whereas a secondary role is played by ν and τ. The latter in particular has little effect on model results, whereas the higher is ν the lower is G and the higher is Z at any given time. As SMC and LMC are likely to be the best studied galaxies in our sample, we make use of their properties (Z and G) to choose among the various models, and in consequence to fix ν. It turns out that set C ($\nu = 4$) matches the properties of SMC, whereas set D ($\nu = 10$) those of LMC. The rate of star formation per unit mass of gas predicted for SMC and LMC

Table 2

Set	ν	τ^*	Φ	ζ
A	4	1	S	0.50
B	10	1	S	0.50
C	4	1	L	0.25
D	10	1	L	0.25
A'	4	6	S	0.50
B'	10	6	S	0.50
C'	4	6	L	0.25
D'	10	6	L	0.25
A"	4	1	S	0.25
B"	10	1	S	0.25

* in units of 10^9 yr

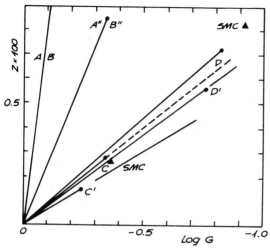

Fig. 2 Theoretical Z
       vs G relationships
for the sets of models of
Table 2. All the models
are given at the present
age. The full dots show
the location of models
with fluctuating but ne-
ver ceasing star formation.
They are aimed t· ·epre-
sent LMC and ⁀⁀

is 0.3 and 2.0 times the solar vicinity value. To the latter we
assign the value of 0.75 10^{-9} yr^{-1} on the basis of the estimate of
Miller and Scalo (1979) of the rate of star formation, and the
estimate of Pagel and Patchett (1975) of the surface mass density
of gas. Let us examine now whether ν has to be greater in LMC than
SMC, or other explanations can be put forward. In fact, it is easy
to understand that this finding would somewhat weaken the universa-
lity of the stochastic mechanism, in that other physical processes
should be invoked to justify the increase of ν from SMC to LMC.
An obvious way out of this difficulty is that the two galaxies have
not the same chemical age, in the sense that star formation in SMC
started much later than in LMC. If this is the case, an unique
value of ν is possible, as model D, which at the age of 12×10^{9} yr
matches LMC, may even reproduce SMC when about 5×10^{9} yr old. Of
course this does not mean that star formation could not occur in
SMC at earlier epochs, but only that significant star formation
started at relatively recent times. The study of Barbaro (1982),
indicating that extremely few clusters older than 6×10^{9} yr seem to
exist in SMC, somehow supports this view. In such a case, the parameter ν assumes the value of 10. Nevertheless, models D that
are seen to reproduce the main features of LMC and SMC are apparently unable to account for the wide spread properties of the whole

sample of galaxies. In fact, the models independently of the number
and intensity of the star forming events, are expected to be located along a line in the Z vs Log G plane, fixed by the stellar mass
function alone. Therefore, they cannot reproduce galaxies with the
same Z but different M_g/M_T at the same time. In the reasonable
hypothesis that at least part of the scatter is real and not due
to the sole uncertainty in the observational quantities, it is worth
looking for plausible ways out of the difficulty encountered by
our models. This is studied by means of the simple model of galactic
evolution which is known to approximate fairly well this type of
galaxies.

a) Effect of the chemical yields

Although compelling evidences are still missing, nevertheless the
possibility that the stellar mass function and at less extent the
stellar nucleosynthetic processing, may vary from galaxy to galaxy
cannot be excluded. This can be tested in terms of the global yield
p_Z. Fig. 3 shows the Z vs Log G plane for both closed and infall

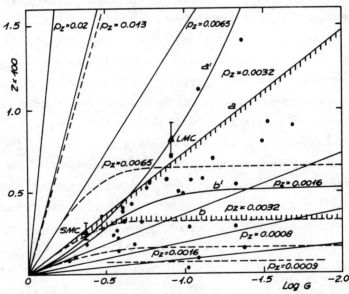

Fig. 3 Theoretical Z vs Log G plane for models with different
yields. The vertically hatched area visualizes the region
where the infall rate is lover than the star formation rate

models (whereby the infall rate equals the star formation rate) for
different values of p_Z. In the case of the closed model, p_Z should
vary from 0.0004 to 0.004 to encompass the whole range of metallicities. Similarly for the infall models.

b) Effect of Infall

To analyse the effect of infall at different rates, we calculate
models in which the rate of star formation is assumed to be proportional to the infall rate. Denoting with Ψ and f the rate of star
formation and gas infall in the standard notation (Tinsley, 1980),
we assume f = (1- R)Ψ Λ , where R is the return fraction per stellar
generation and Λ is a free parameter. With this assumption, the
usual system of equations describing the time variation of Z
(Tinsley, 1980) can be solved analytically. Solutions for several
values of Λ and an unique value of p_Z are shown in Fig. 4

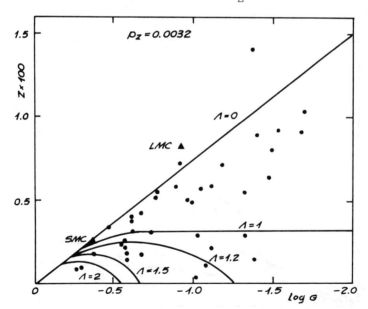

Fig. 4   Theoretical Z vs Log G relationship for simple models with
          infall. The infall rate is assumed to be proportional to
the star formation rate as described in the text. Only the particular case of p_Z = 0.0032 is shown. Full dots and triangles are the
observational data

The suggestion arises that the large scatter shown by our test
galaxies may be due to the different importance of the star formation rate relative to the infall rate. In such a case the topology
of Fig. 4 may give some indication of the parameter Λ for each
individual (or group of) galaxy.

c) Effect of Galactic Winds

Galactic winds energized by supernova explosions are likely to be
a distinct feature of these galaxies as they are relatively loosely
bound and they may occasionally suffer from very intense bursts
of star formation. To explore the point, we assume the rate of
gas loss W to be proportional to the rate of star formation.
According to the standard notation we write $W = (1 - R) \Psi \lambda$, where λ is a free parameter. With this assumption, the usual equations
(Tinsley, 1980) can be solved analytically. Fig. 5 shows these
solutions for different values of λ and an unique value of p_Z.

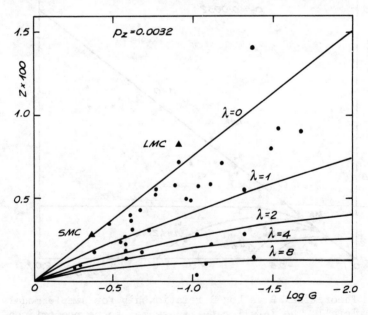

Fig. 5   Theoretical Z vs Log G relationship for simple models
            evolved in presence of a galactic wind. The wind rate
is taken to be proportional to the star formation rate. Only the
particular case of p_Z = 0.0032 is displayed. Full dots and triangles
are the observational data

Although highly speculative, the results of Fig. 5 indicate that
galactic winds may concur to determine the wide scatter in the
observational data.

4. Photometric Evolution

In order to test whether our models galaxies reproduce the observed
distribution in the (U-B) vs (B-V) plane, integrated colours are
calculated with the same method as in Searle et al. (1973), however
adapted to incorporate more recent stellar models of massive stars.
With the aid of burst ages and star formation rates of our numerical
models, we derive the two colour plots shown in Fig. 6.

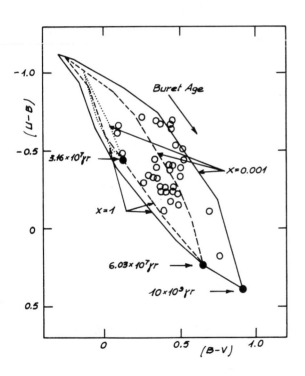

The areas comprised
between the three con-
tour lines indicate the
overall effect of bursts
of different age and
intensity, when applied
to red, blue and very
blue galaxies. These re-
sults agree with those of
Searle et al. (2973),
Huchra (1977) and Larson
and Tinsley (1978).
They however differ in
the stellar mass function
and definition of burst
intensity. Here this
quantity is defined as
    x = η (t) / < η (t) >.
Although there is in
general no unique set of
burst parameters for a
given point in the two-
colour plot, some limits

Fig. 6 Two-colour plot of burst
        models. The open circles
are the observational data. The
burst age, intensity and age of
the underlying galaxy are indicated

can be derived.

5. Conclusions

In this paper we present models based on the stochastic self propagating star formation theory of Gerola et al. (1980), which are
aimed to provide a guide line for a quantitative comparison of the
theory with observations. The major conclusion of this analysis is
that the wide spread scatter of the data in the Z vs Log M_g/M_T
plane cannot be interpreted by the stochastic mechanism alone.
This in the reasonable assumption that besides the observational
uncertainties, part of the scatter is intrinsic. Three ways out are
indicated, namely i) variation in the yield of metals; ii) different
importance of the infall rate relative to the star formation rate;
iii) different importance of the galactic wind rate relative to the
star formation rate. This last option is very attractive as galactic
winds are likely to occur in galaxies of this type, which are
loosely bound and occasionally may suffer from very intense bursts
of star formation. Future implementation of data on metallicities,
gas and total masses are highly desirable to cast light on this
problem.

This work has been financially supported by the Italian National
Council of Research (C.N.R.).

References

Alloin, D., Bergeron, J., Pelat, D., 1978, (10), Astron.
          Astrophys. 70, 141
Arnett, D.W., 1978, Astrophys. J. 219, 1008
Arp, H., O'Connell, R.W., 1975, (8), Astrophys. J. 197, 191
Balkowski, C., Chamaraux, P., Weliachew, L., 1978, (5), Astron.
          Astrophys. 69, 263
Barbaro, G., 1982, Astrophys. Space Sci. 83, 143

Bergeron, J., 1977, (7), Astrophys. J. 211, 62

Chiosi, C., 1980, Astron. Astrophys. 86, 206

Fisher, J.R., Tully, R.B., 1975, Astron. Astrophys. 44, 151

French, H.B., 1980, (4), Astrophys. J. 240, 41

Gerola, H., Seiden, P.E., Schulman, L.S., 1980, Astrophys. J. 242,517

Gordon, D., Gottesman, S.T., 1981, Astron. J. 86, 2

Huchra, J.P., 1977, Astrophys. J. 217, 298

Hunter, D.A., 1982, preprint

Hunter, D.A., Gallagher, J.S., Rautenkranz, D., 1982, (12),
        Astrophys. J. Suppl. 49, 53

Kinman, T.D., Davidson, K., 1981, (1), Astrophys. J. 243, 127

Iben, I. Jr., Truran, J.W., 1978, Astrophys. J. 220, 980

Larson, R.B., 1977, Am. Scientist 65, 189

Larson, R.B., Tinsley, B.M., 1978, Astrophys. J. 219, 46

Lequeux, J., 1979a, Rev. Mex. Astron. Astrophys. 4, 325

Lequeux, J., 1979b, Astron. Astrophys. 80, 35

Lequeux, J., Peimbert, M., Rayo, J.F., Serrano, A., Torres Peimbert,
        S., 1979, (3), Astron. Astrophys. 80, 155

Lequeux, J., Viallefond, F., 1980, (9), Astron. Astrophys. 91, 269

Miller, G.E., Scalo, J.M., 1979, Astrophys. J. 41, 513

Pagel, B.E.J., Patchett, B.E., 1975, Monthly Notices Roy. Astron.
        Soc. 172, 13

Salpeter, E.E., 1955, Astrophys. J. 160, 405

Searle, L., Sargent, W.L.W., 1972, Astrophys. J. 173, 75

Searle, L., Sargent, W.L.W., Bagnuolo, W.G., 1973, Astrophys. J.
        179, 427

Talent, D.L., 1981, (2), Ph. D. Thesis, Univ. of Texas, Houston

Tinsley, B.M., 1980, in Fundamentals of Cosmic Physics, Vol. 5, 287

Tully, R.B., Boesgaard, A.M., Dyck, H.M., Schempp, W.v., 1981,
        (6), Astrophys. J. 246, 38

Viallefond, F., Thuan, T.X., 1982, (11), preprint

THE OXYGEN ANOMALY IN METAL POOR STARS

C. Chiosi[1,2]   and   F. Matteucci[3]

1) Istituto di Astronomia, Università di Padova, Padova, Italy
2) International School for Advanced Studies, Trieste, Italy
3) Istituto Astrofisica Spaziale, C.N.R., Frascati, Italy

Summary

In this paper we present updated chemical yields of C, O and Fe
for both Pop I and Pop III stars. These yields are used to study
the history of production of C, O and Fe in the galactic disk.
We find that the yields and yield ratios must have varied during
the disk life in order to account for the observed abundances of
C, O and Fe (Clegg et al 1981). The temporal variation of the
yield of C, O and Fe consistent with the observed abundances is
derived in the context of a simple model of disk evolution. The
resulting yields of O and Fe cannot be easily explained by standard
prescriptions for the evolution of massive stars. The suggestion
arises that O and Fe are produced by stars in different ranges of
mass, whereas C is ejected by both low-intermediate mass and massive stars. Finally, constraints to the chemical make-up of the
very early stellar generations are derived.

Introduction

In recent years, evidence has been accumulating that the nucleosynthetic production of several important elements such as C, N,
O and Fe has varied during the galaxy life. Sneden et al. (1979)

417

J. Audouze and J. Tran Thanh Van (eds.),
Formation and Evolution of Galaxies and Large Structures in the Universe, 417–435.
© 1984 by D. Reidel Publishing Company.

have clearly shown that the $|O/Fe|$ ratio in the sun is different
from that in metal poor halo stars, indicating that the first stars
in the disk formed out of gas already O-rich relative to Fe.
Clegg et al. (1981) analysing a sample of unevolved F and G disk
stars, pointed out that while the abundance of C and N closely
follow that of Fe, the O abundance varies much more slowly than
Fe. More precisely, the following relationships were established:
$|O/Fe| = 0.6$ for $-2.5 \leq |Fe/H| \leq -1$, $|O/Fe| = -0.48 |Fe/H|$ for
$-1 \leq |Fe/H| \leq 0.4$, $|C/Fe| = -0.16 |Fe/H|$ and $|N/Fe| = 0$, these
latest two all over the range of Fe. An important new dimension
to this problem has been added by Bessel and Norris (1982) who
found two extremely metal poor stars (HD74000 and HD160617) with
$|Fe/H| = -2$, in which N is significantly overabundant, $1.7 \leq |N/Fe|$
≤ 2. The most natural implications of these observations are that
the production of O and N during the first stellar generations has
proceeded decoupled from Fe, and that in these stars N was produced
as a primary element. On the other hand, N production in the stars
of Clegg et al. (1981) is consistent with the notion of a secondary
process, even though there is at present a growing evidence that
part of N ought to be of primary origin (Pagel and Edmunds, 1981
and references therein). The understanding of the relative elemental abundances is a subject of great complexity and uncertainty
owing to its relationships to many aspects of galactic evolution,
such as stellar nucleosynthesis, star formation and stellar mass
function, which are still points of contention. The most recent
study of this problem is by Twarog and Wheeler (1982) who discussed
the data of Clegg et al.(1981) in the context of standard nucleosynthesis prescriptions and a simple model of galactic evolution.
Their chemical yields are based on the mass function of Miller and
Scalo (1979) and the stellar elemental production of Arnett (1978),
whereas the chemical model of galactic evolution is as in Twarog
(1980). Major conclusions of their study were: i) Clegg's et al.
(1981) data are consistent with the assumption of the yield ratios
being constant with time; ii) Standard yields (Arnett' data plus

Miller's and Scalo's stellar mass function) lead to overproduction
of O and C relative to Fe;   iii) The high |O/Fe| ratio for the most
metal poor stars strongly indicates that O was truly overproduced
over the very early stages of galactic evolution (presumably the
halo phase). This likely requires stellar yields different from
Arnett (1978) and/or a mass spectrum for these stars skewed in
favour of very massive objects. Nevertheless, the recent developments in stellar models all over the range of masses, in the stellar
mass function for massive stars and finally in the chemical yields
fully justify a new examination of the problem at the hand. In
this paper , first we endeavour to show that Clegg's et al. (1981)
data imply a variation of the yields during the disk lifetime.
Secondly, we put forward a plausible way out of the difficulties
encoutered by Twarog and Wheeler (1982) by assuming that the yields
have changed in the course of galactic evolution. Finally, we
derive some constraints to the chemical make-up and stellar mass
function holding for the first stellar generations.

1. Properties of stellar models relevant to the yields
In the following, we briefly summarize those advancements in stellar models that may be relevant to yield calculations.
i) Low and intermediate mass stars ($1\ M_{\odot} \leq M \leq 9\ M_{\odot}$)
Renzini and Voli (1981) have followed the evolution of the surface
abundances of He, C^{12}, C^{13}, and O^{16} from the main sequence phase,
till the stage of envelope ejection. The following physical processes have been taken into account: convective dredge up, by which
the external convection brings to the surface nuclearly processed
material; nuclear burning in the deepest layers of the convective
envelope itself; mass loss by stellar wind during the red giant
and asymphotic giant branch and envelope ejection leading to the
formation of planetary nebulae. These calculations show that intermediate mass stars are important sources of C^{12}, He and primary N^{14},
this latter via the hot bottom burning mechanism. Furthermore, they

also constitute an important source of C^{13}.

ii)Massive stars (9 $M_\odot \leq M \leq$ 120 $M_\odot$)

Since the bare core calculations of Arnett (1978), new important
results have been achieved during the past few years. It is however worth recalling that Arnett's (1978) models, in which initially pure He stars of mass M_α have been evolved up and through the
very latest stages of Si-burning, electron capture and thermal
disintegration, have been commonly used to determine the stellar
yields of heavy elements. These yields were given as a function
of M_α and in turn, once a $M(M_\alpha)$ relationship is assigned, as a
function of the initial mass of the star. One of the major uncertainties was the $M(M_\alpha)$ itself, which in Arnett (1978) was derived
from constant mass models. The effect of mass loss by stellar
wind was pointed out by Chiosi and Caimmi (1979) and Dearborn and
Blake (1979). The elemental yields were expected to decrease with
incresing mass loss rate, due to the different $M(M_\alpha)$ relation for
mass losing models. However, some assumptions justified in the
above studies turned out to be inadequate in the light of subsequent results on stellar nucleosynthesis and mass loss by stellar
wind. In particular, a new dimension to elemental nucleosynthesis
is given by the work of Woosley and Weaver (1982 and references
therein), whereas the role played by mass loss has been recently
focussed by Maeder (1981 and references therein). As for the latter
point, the effect of deep external convection in the red supergiant
phase and of mass loss during the so called WR stages has been
clarified. Nevertheless, due to the many uncertainties still
affecting the evolution of massive stars in presence of mass loss,
the net amounts of various elements ejected at the end of a massive
star life is very uncertain. An amalgamation of Arnett's (1978)
stellar yields of heavy elements and main properties of models
evolved in presence of mass loss is given by Maeder (1981).

iii) Very massive stars (120 $M_\odot \leq M \leq$ 300 $M_\odot$)

While the mechanism whereby stars lighter than 100 $M_\odot$ become super-

novae has always been controversial, the pair instability is commonly thought of to be responsible of explosion in more massive
stars. The main difficulty with these very massive stars is that
very few, if any at all, are believed to be forming nowadays.
However there are reasons for believing that very massive stars
existed in the early evolution of galaxies (Carr et al. 1982),
providing the very first burst of chemical enrichment. Another
important question, is the mass range for which pair instability
leads to supernova explosion. The numerical models of Woosley
and Weaver (1982), El Eid et al. (1982), Ober et al. (1982) and
the semianalytical analysis of Bond et al. (1982) set this range
of mass from about 120 $M_\odot$ to about 300 $M_\odot$, even though the upper
limit is found to depend on the rate of mass loss during core H-
and He-burning phases. Stars more massive than about 300 $M_\odot$ are
still pair unstable, but they likely collapse into black holes.
The numerical models of very massive stars of Pop I have been carried out at constant mass (Woosley and Weaver, 1982), while on the
contrary mass loss is likely to be important. Conversely, models
of very massive stars pertinent to the first stellar generation
(Pop III) have been evolved in presence of mass loss, thereby
supposed to be driven by the nuclear energized vibrational instability (Ober et al. 1982; El Eid et al. 1982). The most distinct
feature of these stars is that O is synthesized and ejected in
great amounts, whereas no significant amounts of elements heavier
than Ca^{40} are produced.

2. Chemical yields

With the aid of the nucleosynthesis results recalled in the foregoing section and the formalism of Talbot and Arnett (1973), we
have computed the yields per stellar generation of several important
elements (He, C^{12}, O^{16}, Si-Fe and Si-Ca-S) for a variety of stellar
mass functions. In particular, the yields are presented for stars
of Pop I and Pop III (zero initial metallicity) chemical composition.
Talbot's and Arnett's (1973) formalism, particularly suited for

studies of chemical evolution, describes the ejected nucleosynthesis products from stars in terms of a production matrix which involves all elemental species. According to their notation the yield
of an element i is

$$Y_i = \frac{\sum_{i \neq j} X_j \int_{M_L}^{M_u} Q_{ij}(M) \, \Phi(M) \, dM}{1 - \sum_j X_j \int_{M_L}^{M_u} Q_{ij}(M) \, \Phi(M) \, dM} \qquad (1)$$

where $Q_{ij}(M)$ specifies for a star of mass M the fraction of the
star initially present in form of the species j with abundance X_j
ejected as species i. $\Phi(M)$ is the stellar mass function, whereas
M_u and M_L stand for low and upper mass limit of born stars. The
stellar mass function obeys the following normalization condition

$$\int_1^{M_u} \Phi(M) \, dM = \zeta \qquad (2)$$

where ζ denotes the fraction of mass of $\Phi(M)$ in stars more massive than 1 $M_\odot$. Two different stellar mass functions have been used
assuming $\Phi(M) \propto M^{-x}$. The first one is the Salpeter (1955) mass
function of slope x = 1.35 which incidentally is very similar to
the recent determination of Garmany et al. (1982) in the domain of
massive stars. The second mass function is derived from Serrano
and Peimbert (1981) and Lequeux (1979). The slope is x = 0.6 for
0.007 $M_\odot \leq M \leq 1.8$ $M_\odot$ and x = 2 above. The resulting yields and
characterizing parameters are reported in Table 1 for a few cases
of interest. A more detailed description of these calculations
is reported elsewhere (Chiosi and Matteucci, 1983).

3. Formulation of the problem
Following Twarog and Wheeler (1982), we analyze the data of Clegg
et al. (1981) in the context of a simple model of chemical evolu-

Table 1

Φ(M)	Pop I			Pop III		
	S	S	SPL	S	S	SPL
ζ	0.50	0.25	0.47	0.21	0.44	0.47
M_L	0.16	0.02	0.007	0.01	0.10	0.01
α *	1.5	1.5	1.5	0.	0.	0.
Y (He)	4.43(-2)	1.67(-2)	1.17(-2)	1.40(-2)	4.13(-2)	2.27(-2)
Y (C)	9.91(-3)	3.74(-3)	1.93(-3)	2.14(-3)	6.30(-3)	2.15(-3)
Y (O)	3.91(-2)	1.48(-2)	4.47(-3)	1.27(-3)	3.74(-2)	9.43(-3)
Y(Si-Fe)	4.75(-3)	1.80(-3)	6.59(-4)	1.06(-3)	3.12(-3)	1.20(-3)
Y (Z)	0.0695	0.0260	0.0098	0.0193	0.0567	0.0158
O/Si-Fe	8.2	8.2	6.8	12.0	12.0	7.8
C/Si-Fe	2.1	2.1	2.9	2.0	2.0	2.0

S: Salpeter   SPL: Serrano - Peimbert - Lequeux
* α of Renzini and Voli (1981)

tion of the galactic disk. Prior to this, a few important aspects
of the problem must be clarified. A potential source of ambiguity
in the comparison of theoretical results with the observational data
is the absolute abundances of various elements within the sun, as
the observational determinations are customarily referred to these
latter. We adopt the abundances of C, O and Fe for the sun given
by Cameron (1973) and assume the solar mass fraction of hydrogen
to be 0.73. Thus the abundances relative to H for the three elements
are $|C/H|_\odot = -2.22$, $|O/H|_\odot = -1.83$ and $|Fe/H|_\odot = -2.57$ in the
usual notation. Furthermore, as the solar abundances used by
Clegg et al. (1981) were not exactly the same as above, we have
normalized their data to our solar values, for the sake of internal
consistency. These revised data have been used to derive the following least squares fits

$$|C/H| = -0.122 + 0.795 |Fe/H| \qquad (3)$$
$$|O/H| = 0.046 + 0.450 |Fe/H| \qquad (4)$$

for $|Fe/H| \geq -1$. The $|O/Fe|$ versus $|Fe/H|$ relationship is accordin-
gly changed into

$$|O/Fe| = 0.046 - 0.65 |Fe/H| \qquad\qquad (5)$$

which for $|Fe/H| = -1$ gives $|O/Fe| = 0.60$. The reason for this
redefinition of the observational material is threefold: i) internal consistency;   ii) smoothing of the observational scatter;
iii) better derivation of the initial boundary conditions of the
model. As for the latest point, it will turn out that part of the
disagreement between theoretical results and observational data in
Twarog and Wheeler (1982) was due to inadequate initial values.
Further constraint to the chemical model is imposed by the $|Fe/H|$
versus age relation of Twarog (1980). This relation has been adapted to the assumption that the disk is 13×10^9 yr old and that the
age of the sun is 4.5×10^9 yr. The well known set of differential
equations is used to follow the chemical evolution of the galactic
disk.   These are

$$\dot{M} = f \qquad\qquad (6)$$
$$\dot{M}_g = -(1-R)\,\Psi + f \qquad\qquad (7)$$
$$\dot{X}_i = (Y_i(1-R)\,\Psi + (X_{if} - X_i)\,f\,)\,M_g^{-1} \qquad (8)$$

where f is the infall rate, Ψ is the star formation rate, R is the
return fraction of gas per stellar generation, X_i and X_{if} stand for
the abundance by mass of species i in the gas and infalling material
respectively, Y_i is the yield of species i, M_g is the current gas
mass and finally M is the current total mass. The following complementary relation is added to equations (6) to (8)

$$Y_i = A_i X_i + B_i \qquad\qquad (9)$$

which represents the major difference of the present formulation
with respect to standard ones. With this choice for Y_i, equations
(6) to (8) allow for analytical solutions when f and Ψ are assumed
constant with time. In the following we will discuss the problem
in two paradigm cases: rates of star formation and mass accretion
constant with time, and time dependent as described below.

 i) f and Ψ constant

If the rates of star formation and mass accretion are assumed
constant and proportional to each other ($f = \lambda \Psi$), the following
solution holds for the abundance of species i:

$$X_i = X_{if} \left(\frac{M_g}{M_{go}} \right)^{\kappa_i} + \left[1 - \left(\frac{M_g}{M_{go}} \right)^{\kappa_i} \right] \left[\frac{\beta \, X_{if}}{\beta - A_i} + \frac{B_i}{\beta - A_i} \right] \qquad (10)$$

where $\beta = \dfrac{\lambda}{1 - R}$, $\kappa_i = \dfrac{A_i - \beta}{\beta - 1}$ and

$$M_g = 1 - (1 - R)(1 - \beta) \, \Psi \, t \qquad (11)$$

Once Ψ , f, λ , M_{go}, X_{if} and $Y_i(X_i)$'s are assigned, the total
mass M, gas mass M_g and abundances X_i can be derived as functions
of time. First, to isolate the effect of updated yields, we calcu
late models with the same input parameters as in Twarog and Wheeler
(1982) and constant yields. Second, the effect of chemical yields
varying during the disk evolution is tested always adopting the
same input parameters as above. More precisely, the rate of star
formation is assumed 5 $M_\odot/pc^2/10^9$yr (Miller and Scalo, 1979). The
infall rate is 2 $M_\odot/pc^2/10^9$yr and in consequence $\lambda = 0.4$. M_{go} is
22 $M_\odot/pc^2$. With the above parameters, the present total mass is
48 $M_\odot/pc^2$, whereas the present gas mass is taken to be 6 $M_\odot/pc^2$.
The initial H abundance is 0.77, whereas the current abundance of
hydrogen is given by X = 0.77 - 0.005 t where t is in billion of
years. The return fraction R is derived from equation (11) and
with the adopted parameters it amounts to 0.35. To determine relations (9) we proceed as follows. We start imposing that at the
present time the yields and chemical abundances are those typical
of the young stars. For these latter, we assume $X_C = 4.5$ (-3),
$X_O = 1.1$ (-2) and $X_{Fe} = 2.0$ (-3) (Pagel and Edmunds, 1981). Conversely, while we assume that at the beginning of disk evolution, the
abundances are those given by relations (3) and (4) for $|Fe/H| = -1$,
the correspondent yields are considered as free parameters of the

problem. When a particular choice is made for these latter, the coefficients A_i and B_i can be fixed and the correspondent model can be calculated. The above procedure is repeated by varying the yields of the earliest epochs till when a guess model is derived reasonably matching all major observational constraints. This guess model provides $Y_i(X_i)$ relationships which are repeatedly adjusted till when the calculated abundances are in satisfactory accord with the least squares fits of $|C/H|$ and $|O/H|$ versus $|Fe/H|$. Table 2 summarizes the characterizing parameters for a few models of interest.

Table 2

Case	(M)	M_L	M_u	α	ζ	Y(C)	Y(O)	Y(Si-Fe)
Constant Yields								
A	S	0.02	100	1.5	0.25	3.74(-3)	1.48(-2)	1.80(-3)
B	SPL	0.007	100	1.5	0.47	1.93(-3)	4.47(-3)	6.59(-4)
Variable Yields (*)								
1	S	0.1	200	0.	0.44	6.30(-3)	3.74(-2)	3.12(-3)
	S	0.02	100	1.5	0.25	3.74(-3)	1.48(-2)	1.80(-3)
2	SPL	0.1	200	0.	0.56	2.74(-3)	1.20(-2)	1.53(-3)
	SPL	0.007	100	1.5	0.47	1.93(-3)	4.47(-3)	6.59(-4)

(*) The top row refers to the early epochs, whereas the bottom row refers to the present time.

Figure 1 shows the Fe versus age relation for models A and B, whereas Figure 2 presents the C to Fe and O to Fe histories for the same models. While the theoretical expectations for Fe and O are similar to those of Twarog and Wheeler (1982), we disagree in what C is concerned. In fact, their theoretical C abundances are too low when compared to the revised data. The opposite holds, when the original data of Clegg et al. (1981) are compared to the results of Twarog and Wheeler (1982). Since our models practically

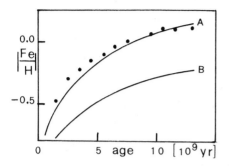

Fig. 1. Fe vs Age. The dots are the data of Twarog (1980)

coincides with those of Twarog and Wheeler (1982) at the Fe rich end of Fig. 2, they point out the effect of different initial abundances. Coming back to our models, the disagreement between theory and observations is alleviated if the case α=1 and/or α=0 of Renzini and Voli (1981) are used to derive the yield of C. In fact the production of C gets smaller with increasing α (efficient hot bottom burning). The main conclusions of this preliminary analysis are the following. In the context of standard nucleosynthesis, a Salpeter like mass function with ζ=0.25 ought to be favoured to account the time history of Fe. This choice is substanciated for young stars by the results of Garmany et al. (1982). Second, the enrichment histories cannot be reproduced with constant yields of C, O and Fe. This is equivalent to say that either the mass function or the stellar nucleosynthesis make up or both have varied over the history of the disk. Third, the chemical composition of the initial gas out of which the disk formed may represent a source of

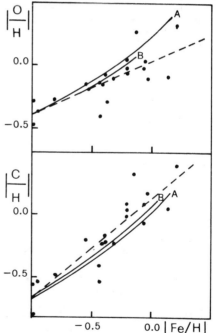

Fig. 2. O vs Fe (top) and C vs Fe relationships. The dots are the data of Clegg et al. (1981). The dashed lines show the linear fits of the data

uncertainty. The processes that led to O enrichment may have also
contributed to build up C in appreciable amounts. The assumption
of a C abundance as low as that of Fe may be incorrect. We discuss
now models calculated with the same input parameters, but with
variable yields (relation (9)) as presented in Table 2. The results
are schematically shown in Fig. 3 and 4 which diplay the $|Fe/H|$
versus age and $|O/H|$ versus $|Fe/H|$ and $|C/H|$ versus $|Fe/H|$ rela-
tions in the order. It is soon evident that the difficulty of O
overproduction is gratly alleviated, whereas the C deficiency still
remains. This latter can be partially removed by using other

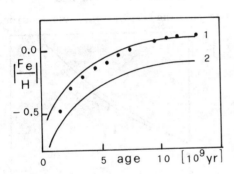

possible values for the yield
of C. Looking at models shown
in Fig. 3 and 4, model 1 of
Table 2 appears to match the
three major constraints.
Nevertheless, this model cannot
be safely used to describe the
chemical history of the galactic
disk as it rests on a very
simplified picture of galactic
evolution. In particular, it

Fig . 3. Fe vs Age. The dots
are the data of Twarog (1980)

assumes constant f and Ψ, and
it makes use of particular
choices for several of the in-

put parameters (M_{go}, present M_g and M, and return fraction R).

ii) time dependent f and Ψ
Following Chiosi (1980), the rate of mass accretion is let vary
with time as given by
$$f = (M - M_o)\, \tau^{-1}\, (1 - \exp(-t_g/\tau)^{-1}\, \exp(-t/\tau) \qquad (12)$$
where τ is the time scale of mass accretion, t_g is the disk age,
M and M_o are the present and initial mass respectively. This
formulation is particularly suited as it can be related to the

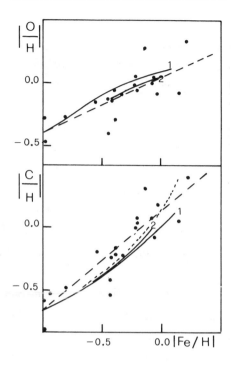

Fig. 4. O vs Fe (top) and C vs
Fe (bottom) relationships for
models with variable yields. The
dotted line is the case $\alpha = 0$.
The dots are the data of Clegg
et al. (1981). The dashed lines
show the linear fits of the data

dynamical nature of the process
of disk formation. The rate of
star formation and its variation
over the galaxy life is very
uncertain. Among the various
suggestions existing in the lite
rature, we assume the rate of
star formation to be proportional to the mass of gas

$$\Psi = \eta \, M_g \qquad (3)$$

where η is a free parameter.
Given M_0 and M, τ and η follow
from imposing that the model reproduces the present day values
of M_g/M, f and Ψ, and moreover
the ratio of the mean past to
present star formation rate.
We assume $M = 100 \, M_\odot/pc^2$ (Oort,
1960; Innanen, 1973), $M_g = 10$
$M_\odot/pc^2$ (Pagel and Patchett,
1975). $\Psi/<\Psi> \leq 2.5$ (Tinsley
1976; Mayor and Martinett,
1977). Given the above conditions
for f, Ψ, $\Psi/<\Psi>$ and M_g/M, the values for η and τ are 0.4
and 4×10^9 yr respectively. The results turn out to be almost independent from M_0, even though $M_0 < 10 \, M_\odot/pc^2$ is likely to be preferred. Model 1 of Table 2 is recalculated under the new assumptions.
The $|Fe/H|$ versus age relation is shown in Fig. 5. C versus Fe and
O versus Fe relations of this model are not displayed as they run
identical to the correspondent ones of Fig. 4. In particular this
model predicts a global metal yield $Y_z = 0.007$ and present metal
content $Z = 0.023$. Furthermore, $|O/Fe| = -0.56 \, |Fe/H| + 0.07$.

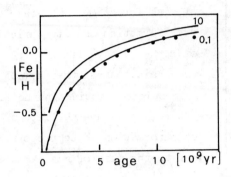

Fig. 5. Fe vs Age relation for
model 1 with variable f and Ψ.
Two values of M_o are indicated

As already pointed out, the
observational data seem to
require a more efficient C
production. To this purpose
we recompute the model repla-
cing the yield of C for the
present stellar populations
of Table 2 with $Y_C = 4.74(-3)$
which is derived with the
same mass function (slope x
and fraction ζ), but using
the data of Renzini and Voli
(1981) for $\alpha = 0$. As expected, C is more abundant and

it varies with respect to Fe as shown by the dotted line in Fig.4.
The main result of this analysis is that the yields of C, O and
Fe may have varied during the disk evolution as depicted in Fig.6.
We consider the model discussed in this section as fairly representing the true disk evolution as it obeys all major observational constraints. in fact, besides matching the chemical properties,
this model has $\Psi/ < \Psi >$ ratio equal to 1.5, which is below the
limit imposed by current estimates.

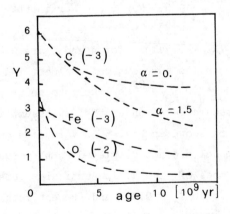

Fig. 6. Yields of C, O and Fe
as a function of age for model
1 with variable f and Ψ. Two
values of α are indicated

Due to the iterative procedure used to fix the run of chemical
yields with abundances and /or age, the final values shown in
Fig. 6 have somehow lost the memory of their input values. Furthermore, the yields predicted by our method even for the young stars
for which the best information is available, are significantly
different from those reported in Table 1. Therefore, it seems
worth examining whether they are still compatible with the underlying assumptions. Table 3 summarizes the input and output yields
for the present stellar generations.

Table 3 (*)

Case	Y(C)	Y(O)	Y(Fe)	C/Fe	O/Fe
$\alpha = 1.5$	3.74(-3)	1.48(-2)	1.80(-3)	2.1	8.2
	2.30(-3)	3.60(-3)	1.10(-3)	2.1	3.3
$\alpha = 0.$	4.75(-3)	1.48(-2)	1.80(-3)	2.6	8.2
	3.80(-3)	3.60(-3)	1.10(-3)	3.4	3.3

(*) top row input; bottom row output

5. Discussion of the yields

In the context of the nucleosynthesis prescription we have been
using, O and Fe are ejected by massive stars only ($M > 10\ M_\odot$),
whereas low and intermediate mass stars and massive stars concur
in about equal amounts to determine the yield of C. According to
our expectation (Table 3) the production of O by massive stars
should be about a factor of 4, whereas that of C and Fe a factor
of 1.6 lower than predicted by standard yields. Prior to any
other consideration, it should be pointed out that the mass function for massive stars is likely to be very similar to the one we
have adopted (Garmany et al. 1982). The fraction ζ is more uncertain. Therefore it is worth exploring whether different values for
ζ would give different yields and yield ratios. It is easy to
understand that while the yields change, their ratios do not , as
shown by the data reported in Table 1. As already suggested by

Twarog and Wheeler (1982), to lower the yields and to change their ratios, we can alter the mass function so that certain stars do not explode but perhaps collapse into black holes. We perform this exercise by systematically lowering the upper cutoff mass in the Salpeter mass function with $\zeta = 0.25$. This is shown in Fig. 7 as a function of M_u^*, the mass limit for exploding stars.

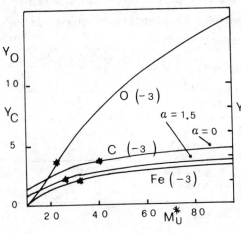

The asterisks visualize the cutoff mass needed to reproduce the desired yields. We are however left with the uncomfortably low masses that would be required. This case refers to massive stars evolved in occurrence of a moderate rate of mass loss by stellar wind. The inclusion of substancial mass loss would shift the cutoff mass to higher values as the total yields are lower. Even in this case the difficulty remains of different cutoff

Fig. 7. Yields of C, O and Fe for x = 1.35 and ζ = 0.25 as a function of M_u^*.

masses for different elements. On the other hand, the kind of nucleosynthesis in massive stars we have been using is based on the assumption that the collapse of the iron core is eventually capable of generating a strong outgoing shock that will eject and somehow reprocess all the material outside a mass shell of about 1.4-1.5 $M_\odot$. This assumption has been often questioned. A exhaustive review of the topic is by Woosley and Weaver (1982). It suffices here to mention that our view concerning the nucleosynthesis of heavy elements would be radically altered if stars more massive than say about 25 to 30 $M_\odot$ do not explode by the core bounce mechanism. To follow up on this idea and to explore the possibility that O and Fe are produced by stars in different ranges of mass, we present yields calculated under the following assumptions:

i) Massive stars do not eject Fe in appreciable amounts, whereas they are the major source of O and at less extent of C.

ii) Stars in the range 8 $M_\odot$ to 10 $M_\odot$ end up as suggested by Weaver et al. (1980) for a 9 $M_\odot$ star ejecting about 1.34 $M_\odot$ of Fe and 0.3 $M_\odot$ of He and leaving no remnant behind.

iii) Stars of lower mass are still suppose to contribute as indicated by Renzini and Voli (1981).

The resulting yields are presented in Table 4 for Salpeter's mass function with two values of ζ, three cases of mass loss by stellar wind in massive stars, and two values of α for low to intermediate mass stars.

Table 4

ζ	0.25	0.25	0.25	0.50	0.50	0.50
$< \dot{M} >$	0	low	high	0	low	high
Y(C) α =0	3.7(-3)	2.4(-3)	2.1(-3)	9.8(-3)	6.8(-3)	5.6(-3)
Y(C) α =1.5	3.3(-3)	2.1(-3)	1.5(-3)	8.7(-3)	5.6(-3)	4.3(-3)
Y(O)	1.2(-2)	5.1(-3)	3.9(-3)	3.1(-2)	1.5(-2)	1.0(-2)
Y(Si-Fe)	1.0(-3)	1.0(-3)	1.0(-3)	2.7(-3)	2.7(-3)	2.7(-3)
C/Fe α=0	3.6	2.3	2.0	3.5	2.5	2.0
C/Fe α=1.5	3.2	2.1	1.5	3.1	2.0	1.6
O/Fe	11.3	4.9	3.7	11.3	5.3	3.7

The comparison of the data of Fig. 6 with the entries of Table 4 may suggest that the case with high $<\dot{M}>$ and α =0 predicts yields that fairly agree with our expectation. Nevertheless, owing to the many uncertainties, we consider our analysis as very preliminary and deserving further investigation.

5. Conclusions

In this paper we have used the data of Clegg et al. (1981) to place constraints to the history of C, O and Fe over the lifetime of the disk. Our main conclusions and differences with respect of Twarog and Wheeler (1982) are that the yields of these elements may have

varied with time, thus accounting for Fe production without overproducing O and/or C even under current stellar mass functions.
However, the resulting yields suited for the observational requirement cannot be easily fitted into the classical nucleosynthesis
scenarios seeing massive stars as the dominant contributors to O
and Fe at the same time. Furthermore, the high $|O/Fe|$ ratio for
the most metal poor stars seems to require the kind of chemical
make up we have suggested for the earliest epochs. This implies
a true O overproduction like the one in our Pop III yields and an
initial mass function skewed in favour of massive stars. Notewhorty
our best model has the fraction ζ of massive stars decreasing
from the initial value of about 0.5 to the present value of 0.25.
Finally, the existence of low and intermediate mass stars in the
very first stellar generation cannot be excluded on the base of the
present analysis. On the contrary, the existence of such stars was
a distinct ingredient in the calculation of suitable chemical yields.
We conclude saying that the chemical history of the galactic disk
can perhaps be used as a probe of ealier enrichment.

This work has been financially supported by the Italian National
Council of Research C.N.R.

References

Arnett, W.D., 1978, Astrophys. J. 219, 1008

Bessel, M.S., Norris, J., 1982, Astrophys. J., 263,L29

Bond, J.R., Carr, B.J., Arnett, W.D., 1982, preprint

Cameron, A.G.W., 1973, Space Sci. Rev. 15, 121

Carr, B.J., Bond, J.R., Arnett, W.D., 1982,preprint

Chiosi, C., 1980, Astron. Astrophys. 83, 206

Chiosi, C., Caimmi, R., 1979, Astron. Astrophys. 80, 234

Chiosi, C., Matteucci, F., 1983, in preparation

Clegg, R.E.S., Lambert, D.L., Tomkin, J., 1981, Astrophys. J. 250, 262

Dearborn, D.S., Blake, J.B., 1979, Astrophys. J. 231, 193

El Eid, M.F., Fricke, K.J., Ober, W.W., 1982, preprint

Garmany, C.D., Conti, P.S., Chiosi, C., Astrophys. J., 263, 777

Innanen, K.A., 1973, Astrophys. Space Sci. 22, 393

Lequeux, J., 1979, Astron. Astrophys. 80, 35

Maeder, A., 1981, Astron. Astrophys. 101, 385

Mayor, M., Martinett, L., 1977, Astron. Astrophys. 55, 221

Miller, G.E., Scalo, J.M., 1979, Astrophys J. Suppl. 41, 3

Ober, W.W., El Eid, M.F., Fricke, K.J., 1982, preprint

Oort, J.H., 1960, Bull. Astron. Inst. Neth. 15, 45

Pagel, B.E.J., Edmunds, M.G., 1981, Ann. Rev. Astron. Astrophys. 19, 77

Pagel, B.E.J., Patchett, B.E., 1975, Monthly Notices Roy. Astron. Soc. 172, 13

Renzini, A., Voli, M., 1981, Astron. Astrophys. 94, 175

Salpeter, E.E., 1955, Astrophys. J., 121, 161

Serrano, A., Peimbert, M., 1981, Rev. Mex. Astron. Astrophys. 5, 109

Sneden, C., Lambert, D.L., Whitaker, R.W., 1979, Astrophys. J. 234, 964

Talbot, R.J.Jr., Arnett, W.D., 1973, Astrophys. J. 186, 51

Tinsley, B.M., 1976, Astrophys. J., 208, 797

Twarog, B.A., 1980, Astrophys. J., 242, 242

Twarog, B.A., Wheeler, JC., 1982, Astrophys. J. 261, 638

Weaver, T.A., Axelrod, T.S., Woosley, S.E., 1980, Proc. Texas Workshop on Type I Supernovae, ed. J.C. Wheeler, Univ. Texas Press, p. 113

Woosley, S.E., Weaver, T.A., 1982, in Supernovae: A survey of current research, ed. M.J. Rees and R.J. Stoneham, D. Reidel P.C., Holland, p. 77

"LINERS" AND ABUNDANCES IN GALACTIC NUCLEI

B.E.J. Pagel

Royal Greenwich Observatory, Herstmonceux, Sussex, U.K.

SUMMARY

[NII] λλ 6548, 6584 increase in strength relative to Hα in HII regions as the overall abundance of heavy elements is increased, but in some galactic nuclei λ6584 is still stronger than Hα, which never happens in conventional HII regions ionised by hot stars. These nuclei are either of the narrow-line or intermediate Seyfert type or of the more common variety called "LINERS" by Heckman or sometimes "Seyfert 3". Earlier observations and calculations of relative emission line strengths in LINERS suggested that they form a distinct group from Seyferts with a distinct excitation mechanism (shock heating as opposed to photo-ionisation by a power-law continuum) but recently several authors have shown that photo-ionisation may account for both. More recent observations confirm that LINERS are continuous with Seyferts and that the enhancement of [NII]/Hα is largely an excitation effect unless these nuclei differ in composition from nearby HII regions.

1. INTRODUCTION

In spiral galaxies with HII regions, the line intensity ratio [NII]/Hα decreases as one goes outwards from the central regions while [OIII]/Hβ increases. Both effects are explained following Searle[1] as consequences of a radial abundance gradient which is chiefly an outward decreasing ratio of heavy elements (typified by oxygen) to hydrogen. Variations in N/O also occur, but these are smaller than the variations on O/H.[2] For HII regions, especially giant ones, line ratios such as [OIII]/Hβ,

437

J. Audouze and J. Tran Thanh Van (eds.),
Formation and Evolution of Galaxies and Large Structures in the Universe, 437–444.
© 1984 by D. Reidel Publishing Company.

[OIII]/[NII] and ([OII] + [OIII])/Hβ can be calibrated against
the oxygen abundance.[3)4)5)6)]

Emission-line spectra in galactic nuclei themselves can
belong to one of three types.[7)] Late-type galaxies like M33,
M101 and probably M83 have low-excitation, oxygen-rich HII
regions in, or very close to, their nuclei, which simply continue
the trend shown by HII regions in the arms or reverse it to a
slight extent. M83 is classified as a "hot spot" galaxy[8)] with
separate bright knots in the nucleus which consist of supergiant
HII regions embedded in clouds of dust. Another hot-spot
galaxy, the barred spiral NGC 1365, has a true nucleus with a
Seyfert galaxy spectrum, surrounded by HII regions[9)] ; we do
not know how common such effects may be.

The Seyfert galaxies constitute the second type of nuclear
spectrum, where the nucleus is bright and, in the narrow-line
component of the spectrum, [OIII] vastly outshines Hβ, in
contrast to the nuclear HII regions where owing to the high
abundances [OIII] is always weak and often invisible. At the
same time [NII] $\lambda 6584$ in narrow-lined Seyferts is comparable in
strength to Hα, which is two or three times as strong as [NII]
in even low-excitation HII regions. These effects are
attributable to ionisation by a hard radiation spectrum
approximating a power law.[10)]

The third type of spectrum is that of the objects which
Heckman[7)] has called "Liners", i.e. low ionisation emission-line
regions, which are found in many Sb and Sc nuclei such as M81
and M51 as well as in elliptical galaxies like M87 and NGC1052,
often in association with a compact radio source. In these
cases [NII] is usually as strong as Hα or stronger[11)12)] and the
Hβ emission is so weak that it is often only detectable after
subtraction of a similar galaxy spectrum without emission lines.
[OIII] $\lambda 5007$ is stronger than Hβ, but by a smaller factor than
in a narrow-lined Seyfert. Following Koski and Osterbrock[13)]
and others, Heckman supposed that Liners result from shock
heating, as did Baldwin, Phillips and Terlevich[14)] (BPT) who
developed a classification scheme for emission-line spectra of
extragalactic objects in which Liners occupy a distinct region
of parameter space. Stauffer[15)] has found a few objects with
spectra intermediate between Seyferts and Liners which he
suggests could have mixed excitation mechanisms.

On the other hand, Péquignot[16)] and Ferland and Netzer[17)] have
suggested that Liners can be explained by photo-ionisation with
the same power law $F_\nu \sim \nu^{-1.5}$ as Seyferts but with lower
ionisation parameter, thereby making it appropriate to call
them Seyfert 3. (The ionisation parameter is the ratio of photon
density or flux at the Lyman limit to electron density.)

What I wish to do in this short talk is to present some recent AAT observations which bear on two questions:-

(i) Is shock excitation or power-law photoionisation the more plausible excitation mechanism for Liners? and

(ii) Is nitrogen overabundant in some or all of them?

Other recent observations, in particular of M81,[18] M51[19] and NGC 1365[9] will also be brought into the discussion.

2. OBSERVATIONS

The following emission-line galactic nuclei were observed at the Anglo-Australian Telescope using the Boksenberg Image Photon Counting System in October 1981 by M.G. Edmunds and M.M. Phillips or in August 1982 by B.E.J. Pagel:-

NGC	1087 (Sc)	Nuclear HII region
	1097 (SBb)	Liner
	1433 (SBa)	Liner
	1598	Liner
	1808 (SAB 0/a)	Hot spot HII with complex profiles
	6744 (Sbc)	Liner
	7552 (SBab)	Hot spot HII; X-ray source
	7590 (Sbc)	Liner

The observations were made with a 50 arc sec long slit which included not only the actual nuclei but also normal galaxy spectra including in some cases normal HII regions a few hundred parsecs away (1" $\equiv$ 100pc at 20 Mpc). Thus, in addition to comparing nuclear spectra with photo ionisation and shock models, we can also make some deductions about ambient element abundances from the spectra of the nearby HII regions. Observational line intensity ratios have been de-reddened assuming the conventional Case B Balmer decrement.

3. DISCUSSION

In this paper we shall consider mainly the following three diagrams: [OIII]/Hβ against [OII]/[OIII], [NII]/Hα against [OII]/[OIII] and [NII]/Hα against [OIII]/Hβ. The first two diagrams have been used both by BPT[14] and by Ferland and Netzer[17]; the third, which has also been considered by BPT, is easier to determine observationally, being independent of reddening and calibration uncertainties.

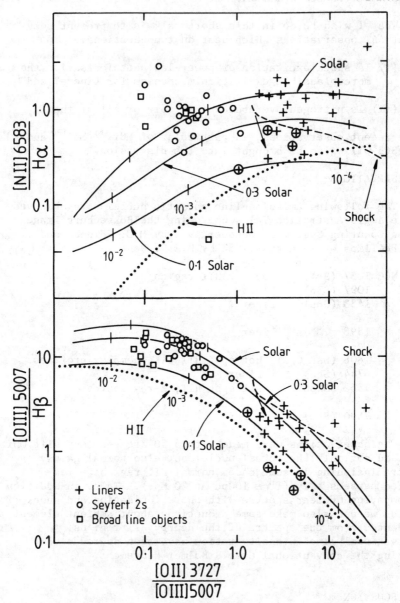

Fig 1.   Observed line ratios in active galactic nuclei after
Ferland and Netzer[17] together with predicted relationships for
power-law photoionisation models with $F_\nu \sim \nu^{-1.5}$ and different
metallicities (Ferland and Netzer: solid lines), shock models
(Shull and McKee[20]): broken lines; the arrow indicates the effect
of going down from solar metallicity to a factor of 3 lower) and
normal HII regions (Baldwin et al[14]: dotted lines).

Fig. 1 is a copy of Ferland and Netzer's Figure 2 whereon
I have superimposed the loci for HII regions from BPT and for
shock models from Shull and McKee. Most of the points that lie
low in the two panels represent objects that I believe to be HII
regions, because [OIII], [OII] and [NII] are not very strong and
[OI] $\lambda6300$/H$\alpha \leqslant 0.05$ (cf. BPT, Fig 4; Ferland and Netzer,Fig.3);
these points have been circled in the diagram. Bearing in mind
that other points may be affected by radiation from adjoining
HII regions, it is evident that either shock or power-law
photoionisation models fit the Liners reasonably well.
Furthermore, with photoionisation, most points are consistent
with something quite close to solar abundances, while with the
shock models an overabundance of nitrogen by a factor of 3 or so
would be indicated for a typical object. If this were indeed
the case, then one might expect it to show up also in spectra of
galactic nuclei that have HII regions instead of Liners.

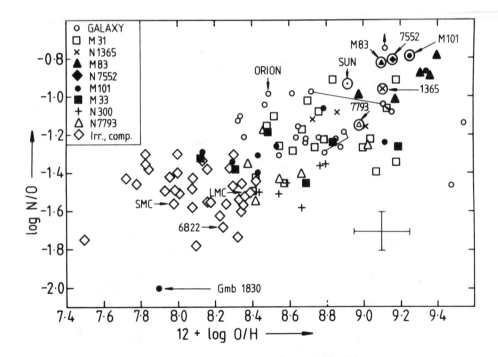

Fig. 2. Plot of N/O against O/H in galactic and extragalactic
         HII regions (cf Pagel and Edmunds[2]).

Fig. 2 is an updated version of a diagram given by Pagel and
Edmunds[2] in which galactic nuclear HII regions are indicated
by circled symbols with their names. Apparently N/O never
exceeds solar by more than 0.2 dex and (O/H, N/H) never exceed
solar by more than (0.3, 0.4) dex. The evidence is not exactly
compelling, but it does give us a certain predilection in
favour of photoionisation.

Another feature of Ferland and Netzer's figure, as of those
of BPT, is the apparent existence of a gap between Seyferts and
Liners. Stauffer[15] found some intermediate objects and such,
indeed, would be expected if photoionisation is dominant
throughout. A real gap, on the other hand, located (as it
apparently is) just where the shock locus undergoes a sharp
change of slope, would strongly suggest that the mechanism is
changing over at this point.

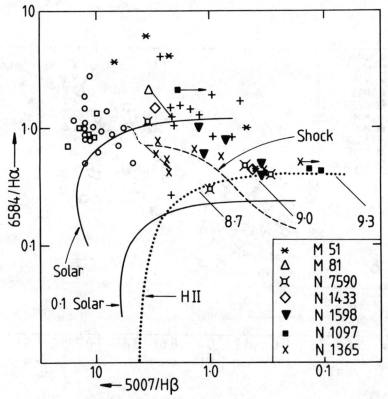

Fig. 3. [NII], [OIII] diagram for M51, M81, our objects and those
of Ferland and Netzer (after removal of suspected HII regions
among their objects). Solid, broken and dotted lines represent
power-law photoionisation, shocks and HII regions respectively
as in Fig. 1. For the HII regions, loci of constant oxygen abundances are marked by diagonal lines with (12+log (O/H)) indicated.

With these considerations in mind, we can now look and see
what the new data from the AAT and other sources do to the
appearance of the various BPT diagrams. We shall consider here
just one diagram, that of [NII]/Hα against [OIII]/Hβ (Fig 3).
On the diagram we plot the line ratios both in the nucleus
proper, where the relative strength of forbidden lines is a
maximum, and in outlying parts of the slit where the spectrum is
modified and eventually dominated by surrounding HII regions.
Our data, together with those of Peimbert and Torres-Peimbert[18]
on M81, evidently fill the gap between classical Seyferts and
Liners and fit the predicted curve for photoionisation with
solar abundances quite well; naturally variations by up to a
factor of 2 between different galaxies, as shown in Fig. 2, are
not excluded.

Within this factor, then, it seems reasonable to accept
that most Seyferts and Liners are galactic nuclei of normal
composition photo-ionised by a power-law continuum. However,
a few objects have even stronger [NII]/Hα than is accounted for
by this formula: most notable among these is M51[19], which
according to the diagram is a Seyfert at the very centre (7,3.5),
but comes closer to the Liner region from a radius of $5\frac{1}{2}$ arc sec
(∿ 250pc) outwards, which as Rose and Searle point out is just
as one might expect if the ionisation parameter decreases
outwards. Beyond $5\frac{1}{2}$" we get (as in other cases) a mixing line
with increasing contributions from conventional HII regions
until at a distance of 15" or about 700pc we have a pure HII
region with solar abundances (using the calibration of [OIII]/
[NII] in HII regions given by Pagel et al[4]).

Our own data show something similar happening in NGC1097,
1433, and 7590, with some indication that 1097 has the largest
overall abundances among the nuclear HII regions in the diagram,
consistent with the rather high point occupied by its Liner
nucleus. In other words, the nuclear abundances are in
agreement with the nearby HII region abundances in most cases.
From this I am tempted to conclude that, apart from overall heavy-
element or metallicity enhancements by up to a factor of 2 or so,
the great strength of [NII] lines in Liners is chiefly an
excitation effect due to hard-spectrum photoionisation rather
than an abundance effect. This could perhaps even apply to M51;
however, this extreme case appears to deserve further study.

I am grateful to the PATT for assigning time on the AAT
for this project and to the Director and staff of the Anglo-
Australian Observatory for their cooperation.

REFERENCES

1   Searle, L. 1971, Astrophys. J. 168, 327

2   Pagel B.E.J., and Edmunds, M.G. 1981, Ann Rev Astr.
     Astrophys. 19, 77

3   Alloin, D., Collin-Souffrin, S., Joly, M., and Vigroux,
     L. 1979, Astr. Astrophys., 78, 200

4   Pagel, B.E.J., Edmunds, M.G., and Smith, G. 1980. Mon.
     Not. R.Astr. Soc., 193, 219

5   Stasinska, G., Alloin D., Collin-Suffrin, S., and Joly
     M. 1981, Astr. Astrophys. 93, 362

6   McCall, M.L. 1982. Thesis, University of Texas, Austin

7   Heckman, T.M. 1980, Astr. Astrophys., 87, 142, 152

8   Sersic, J.L., and Pastoriza, M. 1965, Pub Astr. Soc
     Pacific, 77, 287

9   Edmunds, M.G., and Pagel, B.E.J. 1982, Mon. Not. R. Astr.
     Soc. 198, 1089

10   Koski, A.T. 1978, Astrophys J., 203, L49

11   Burbidge, E.M. and Burbidge, G.R. 1962, Astrophys. J.,
     135, 694

12   Burbidge, E.M. and Burbidge G.R., 1965, Astrophys. J.,
     142, 634

13   Koski, A.T., and Osterbrock, D.E. 1976, Astrophys. J.,
     203, L49

14   Baldwin, J.A., Phillips, M.M. and Terlevich, R. 1981,
     Pub Astr. Soc Pacific, 93, 5

15   Stauffer, J.R. 1982, Astrophys. J., 262, 66

16   Péquignot, D, 1983, unpublished

17   Ferland, G., and Netzer, H. 1983, Astrophys. J., 264 105

18   Peimbert, M., and Torres-Peimbert, S. 1981, Astrophys. J.,
     245, 845

19   Rose, J.A., and Searle, L. 1982 Astrophys. J., 253, 556

20   Shull M., and McKee C., 1979, Astrophys. J., 227, 131

AUTHOR INDEX

SUBJECT INDEX